Community Builders Handbook Series

INDUSTRIAL DEVELOPMENT HANDBOOK

Sponsored by the
Executive Group
of the
Industrial Council
of
ULI–the Urban Land Institute
1975

ULI–the Urban Land Institute, Washington, D.C.

About ULI–the Urban Land Institute

ULI–the Urban Land Institute is an independent, nonprofit research and educational organization incorporated on December 14, 1936 to improve the quality and standards of land use and development.

The Institute is committed to conduct practical research in the various fields of real estate knowledge; identify and interpret land use trends in relation to the changing economic, social, and civic needs of the people; and disseminate pertinent information leading to orderly and more efficient use and development of land.

ULI receives its financial support from membership dues, sale of publications, and contributions for research and panel services.

ULI Staff for Industrial Development Handbook

Managing Editor	Frank H. Spink, Jr.
Associate Editor	Nathaniel M. Griffin
Manuscript Editors	Diana W. Stavins Susan Regenbogen Rosinoff
Editorial Assistants	Jennifer Garbis Fransen Leslie N. Goldman
Production Manager	Robert L. Helms
Production Assistant	Sarah V. Lantz
Art Director	Carolyn E. Noe
Art Assistant	Michael S. Murtaugh
Graphic Consultant	Ken Fuller Scott-Fuller Communications, Inc.

First book in a new series of publications based on the
philosophy of the Community Builders Handbook
First Edition and First Printing 1947
Revised Printing 1948
Second or J. C. Nichols Memorial Edition and Second Revised Printing 1950
Third or The Members Edition and Third Revised Printing 1954
Fourth Revised Printing 1956
Fourth or The Executive Edition and Fifth Revised Printing 1960
Sixth Printing 1965
Fifty or Anniversary Edition and Seventh Revised Printing 1968

Other books in the series:
Shopping Center Development Handbook, 1977
Residential Development Handbook 1978
Downtown Development Handbook 1980
Recreation Development Handbook 1981
Office Development Handbook 1982
ULI Catalog Number-I02

1200 18th Street, N.W.
Washington, D.C. 20036

Second Printing 1977
Third Printing 1978
Fourth Printing 1982

International Standard Book Number 0-87420-562-X
Library of Congress Card Catalog Number LC 75-37218

Printed in the United States of America

Authors

Principal Authors

DONALD C. LOCHMOELLER, vice president and director, PGA Engineers, Inc., St. Louis, Missouri

DOROTHY A. MUNCY, consulting city planner, Arlington, Virginia

OAKLEIGH J. THORNE, president, American Real Estate Advisors, Inc., Washington, D.C.

MARK A. VIETS, vice president and director, Peckham-Guyton, Inc., Kansas City, Missouri

Contributing Authors

David F. Delaney, vice president, Finance, Cabot, Cabot and Forbes, Boston, Massachusetts

Nathaniel M. Griffin, associate, Technical Publications Division, ULI–the Urban Land Institute

Frank H. Spink, Jr., director, Technical Publications Division, ULI–the Urban Land Institute

with

The Industrial Council
of
ULI–the Urban Land Institute

Acknowledgments

The four principal authors have varied backgrounds and each contributed special technical expertise to the Handbook. The initial drafts for the Introduction, Background and Characteristics, and Planning sections were the primary responsibility of Dorothy A. Muncy; Engineering, Donald C. Lochmoeller; Development Strategy, Oakleigh J. Thorne; and Future Trends, Mark A. Viets. In addition, each of the principal authors had the opportunity to review and provide input into the work of the others.

In several instances, material by one author was located in another section for proper organization. All authors contributed to Future Trends. Within the Background and Characteristics section, Nathaniel M. Griffin had primary responsibility for the case studies and incorporation of new material. David F. Delaney contributed the Investment Decision and Financial Projections subsections of Development Strategy. As managing editor, Frank H. Spink, Jr. was responsible for overall coordination of authors, incorporation of the Council's reviews, and development of the final manuscript based upon Council consideration and input.

In addition to the principal and contributing authors, special thanks must be given to Robert E. Boley, executive vice president of the Society of Industrial Realtors and former executive director, ULI–the Urban Land Institute, Washington, D.C.; Donald N. Hohl, registered engineer, San Francisco, California; Sidney Robbins, industrial engineering consultant, Chicago, Illinois; Paul P. Shepherd, vice president, Cabot, Cabot and Forbes, San Francisco, California; and Leslie A. Smith, civil engineer, Seabrook, Maryland, for taking time to review various portions of the manuscript.

The Industrial Council of the Urban Land Institute provided continuing review as the manuscript proceeded from the first through final draft. A special acknowledgment is made to Marshall Bennett, who devoted many hours to a page-by-page review of the final manuscript and galleys.

About ULI Councils

Within ULI–the Urban Land Institute there are six councils: Industrial Council, Commercial and Office Development Council, New Communities Council, Recreational Development Council, Residential Council, and Urban Redevelopment Council. Each Council is composed of 35 members drawn from the ULI Sustaining membership. Council appointment is based on knowledge, experience, and a willingness to share. Developers, consultants, public officials, and academicians are included on each of the Councils to provide a broad perspective and interaction among various disciplines.

Executive Group of the Industrial Council—1975-1976

Chairman:

Marshall Bennett
Partner
Bennett and Kahnweiler
Chicago, Illinois

Vice Chairmen:

Lowell F. Eastman
Owner
Eastman & Associates, Inc.
Los Angeles, California

M. Sherman Eubanks
President
Crocker Land Company
San Francisco, California

R. John Griefen
Senior Vice President
Cabot, Cabot and Forbes Company
Boston, Massachusetts

Raymond W. Allan — Director, Real Estate, American Can Company, Greenwich, Connecticut

Fred H. Bielefeld — Director, Civic Affairs, Bethlehem Steel Corporation, Bethlehem, Pennsylvania

Gene F. Cermak — President, Liberty Properties, Greenville, South Carolina

William W. Dillard — Real Estate Broker and Developer, William Dillard & Associates, Dallas, Texas

Ben Dyer, P.E. — President, Ben Dyer Associates, Inc., Seabrook, Maryland

Andrew R. Evans — President, First Federal Savings and Loan Association of Pittsburgh, Pittsburgh, Pennsylvania

John S. French — Vice President and General Manager, Business Centers Division, Kaiser Aetna, Oakland, California

Harry G. Foden — Consultant, Arthur D. Little, Inc., Cambridge, Massachusetts

Carlos Fonts — Executive Vice President, Brookhollow Corporation, Dallas, Texas

Harry F. Higgins, AIP — Director of Planning, Arizona Department of Economic Planning and Development, Phoenix, Arizona

Cyril C. Herrmann — Vice President, Arthur D. Little, Inc., San Francisco, California

Julius R. Jensen — Industrial Development Consultant, Economic Development Corporation, San Diego, California

Donald E. Johnson	Real Estate Development Specialist E. I. DuPont de Nemours Company Wilmington, Delaware
Irving Korb	President Korb Company, Inc. Real Estate Consultants San Francisco, California
Laurence H. Lang	Chairman of the Board Cragin, Lang, Free & Smyth, Inc. Life Council Member Cleveland, Ohio
George S. Laporte	President Newport Associates, Inc. Hartford, Connecticut
O. G. Linde	President Southern Pacific Land Company Southern Pacific Development Company San Francisco, California
John E. Lynch	Director of Program Operations Office of Economic Adjustment Office of the Secretary of Defense Washington, D.C.
Edward J. McCaffery	Vice President General Electric Realty Corporation Schenectady, New York
Edmond J. McRickard	Senior Vice President Brooks Harvey & Company Inc. New York, New York
William F. Morton	Life Council Member Burlingame, California
John O'Donnell	General Partner John O'Donnell & Partners Newport Beach, California
Otto W. Pongrace	Director–Industrial Development Penn Central Transportation Company Philadelphia, Pennsylvania
William K. Robinson	Vice President and Secretary George Henkle Associates, Inc. Lebanon, Ohio
Howard G. Roepke	Professor Department of Geography University of Illinois
Willard G. Rouse, III	Director Rouse & Associates Bala Cynwyd, Pennsylvania
B. Raymond Sayer	Vice President and Personnel Manager The Austin Company Life Council Member Cleveland, Ohio
James R. Scott	Vice President Real Estate & Industrial Development Santa Fe Railway Chicago, Illinois
Raymond Sydansk	General Manager Real Estate Department Armstrong Cork Company Lancaster, Pennsylvania
Alexis P. Victors	Vice President–Industrial Development The Western Pacific Railroad San Francisco, California
Robert Walker, Jr.	Vice President Cushman & Wakefield, Inc. San Francisco, California
Thomas C. Wolff, Jr.	Senior Vice President The Irvine Company Newport Beach, California
James A. Wuenker	Director of Area Development The Cincinnati Gas & Electric Company Cincinnati, Ohio

Contents

List of Illustrations

A Brief History of the Community Builders Handbook

Publication of the Industrial Development Handbook—Community Builders Handbook Series, marks the beginning of a new period in the history of the Community Builders Handbook. The Community Builders Handbook was first published in 1947. The objective of the Handbook was to share the experience and knowledge of developers toward the improvement of land use and development practices. It was developed under the sponsorship of the Community Builders Council, now Residential Council, which had been formed in 1944. The first edition was 205 pages in length and sparsely illustrated. It represented a major accomplishment since, for the first time, a book was made available which described the development of residential communities and shopping centers.

The second edition, the J. C. Nichols Memorial Edition, was published in 1950. It incorporated a modest revision and updating of the original text. In 1954, the third, or Member's, edition provided a significant expansion of coverage to 315 pages. The fourth, or Executive, edition of 1960 continued this expansion in response to the increasing complexity of development practices. By this time, the Handbook had grown to 476 pages but continued to focus, as had earlier editions, on residential and shopping center development. The fifth, or Anniversary, edition was published in 1968. It had grown in size to 526 pages and had been broadened in terms of coverage. In addition to sections on residential and shopping center development, new material discussing a variety of special types of land development was added. Included was a section on industrial development drawing on the experience of ULI's Industrial Council, which had been formed in 1951. The Industrial Council had sponsored other ULI publications but was now represented for the first time in the Community Builders Handbook.

The Community Builders Handbook achieved wide recognition as a major reference source and textbook on land use and development practice based on the practical experience and accumulated knowledge of leading practitioners in the field. In 1965, as work on the 1968 edition was beginning, ULI was experiencing major growth in membership and a significant expansion of areas of interest. The development industry was reaching maturity and new directions for the Institute's future were being examined. By 1970, a decision was made to publish future editions of the Community Builders Handbook in separate volumes in order to provide the expanded and more comprehensive coverage not possible in a single text.

In 1972, the original three Councils of ULI, the Community Builders Council, the Central City Council, and the Industrial Council, were reorganized into six councils which recognized the new diversity in development activities and in ULI's membership. This gave further impetus to the multi-volume, Community Builders Handbook Series. The Industrial Development Handbook is the first volume of the new series. Additional volumes, each focusing upon a specific land use type, will be forthcoming at regular intervals in the future.

Frank H. Spink, Jr.
Managing Editor
Community Builders Handbook Series

Foreword

The precepts which led to preparation of the original Community Builders Handbook as a practical reference manual for land development are those which have been followed in the development of the Industrial Development Handbook.

The Handbook is largely devoted to builder/developer procedures within the framework of the public regulatory and approval process. The primary objective has been to describe the evolution of industrial developments to current practice; describing the planning, engineering, financial, and marketing methodology of the development process; setting forth recommended practices and standards of excellence, and identifying typical and special project examples.

ULI has traditionally been a leader in recommending new and better development practices. This leadership role requires foresight and an understanding of past, present, and future trends. This Handbook will assist the developer, planner, public official, financial officer, and all others in making development decisions today which will be responsive to community needs in the present and future.

The development of this volume has been a major commitment of the Industrial Council for almost three years. The initial funding for authorship of the Handbook was provided by a grant from the Urban Land Research Foundation in early 1973. Ultimately, four principal authors and three contributing authors participated in the development of the completed manuscript. The contributions of various authors are set forth for the reader's benefit under Acknowledgments.

The first draft manuscript was completed in June 1974 and submitted for review to all members of the Industrial Council. The results of this initial review were incorporated into a second draft which was then submitted to committees of the Council. The entire Council discussed "future trends" as a major topic at their Fall Executive Group Meeting in 1974. Council input led to a major revision of this section. In late 1974 after the committee's review, the editors and authors met with the Council chairman and Lowell Eastman, council vice chairman for publications. This meeting led to additions and modifications. Because of the importance of the Future Trends section, it was once again discussed in final draft form at the Spring Executive Group Meeting in 1975.

As a consequence of this review process, the Industrial Development Handbook has become a cooperative effort of the entire Industrial Council. As a group they have concurred on its content. Their individual knowledge and experience is incorporated throughout the book and specific viewpoints have been attributed to Council members.

We of the Industrial Council of ULI–the Urban Land Institute believe this book will be a major contribution to the understanding of the place of industrial development in the community development process.

Marshall Bennett
Chairman
Industrial Council 1975

1. Introduction

Objective

The principal purpose of this Handbook is to help communities and developers achieve better planned work centers for the production and distribution of goods. As such, this Handbook is addressed to four groups of readers, all of whom are decision makers who shape the physical form and economic future of the community:

- development community—private developers and their consulting planners, engineers, architects, and finance-management advisors engaged in the development of industrial parks and employment centers;
- governmental bodies—legislators, public officials, and planning agencies who set local, state, and national land use policy;
- industrial management—those who make decisions on the location and design of research, manufacturing, and distribution facilities; and,
- the public—citizens' groups who are concerned about the total environment of their community—social and economic, as well as biologic and atmospheric.

In the postwar period, the industrial park has come to typify the grouped siting of industrial activity within urban areas. The term *industrial park* implies careful planning and brings to mind extensive low-rise buildings which are attractive in appearance, located in a landscaped setting of wide lawns, and interconnected by means of broad boulevards. Many parks can be found which are well-planned and reflect a sophisticated awareness of the potential inherent in this form of development. Many others, however, reflect scant effort to benefit from the mistakes of others, and exhibit serious and elementary defects. This is at least partially due to the rapid changes in basic concepts and regulatory controls taking place in the field.

Industrial parks have not evolved in a steady progression. As new uses have been accommodated, some uses commonly included in earlier parks have been excluded. For example, heavy industry, once part of early planned industrial districts, is seldom provided for in the newest industrial parks. At the same time, however, new facilities such as offices, service areas, shopping centers, and recreational facilities, are being added to enhance the convenience of the planned industrial neighborhood for tenants and visitors. Faced with a growing scarcity of land, materials, fuel, funds and increasingly more complex land use, environmental, and pollution controls, industrial park developers must anticipate additional changes in the form, function, and location of planned employment centers. Evaluation of successful examples of operating industrial parks, analysis of the functions to be performed within the industrial environment, an understanding of park design, engineering, and development strategies, and an awareness of changing human needs are all essential steps in understanding the current status and future prospects of this development form.

Three primary sources have been utilized in preparation of the *Industrial Development Handbook*:

- experience of leading industrial park developers and their professional associates, many of whom are active members of the Urban Land Institute's Industrial Council, and many who have been building and managing industrial parks for many years;
- basic source material, including recognized technical publications on industrial parks, published by ULI-the Urban Land Institute, the Society of Industrial Realtors, and the National Association of Industrial Parks; and,
- authors' experience and judgment gained through many years of active work in the field; and, material contributed by the ULI technical staff.

1-1 The Houston Ship Channel—an industrial area stretching 25 miles between the city of Houston and Galveston Bay which was created by dredging Buffalo Bayou. The ship channel is comprised of scores of chemical plants and port-oriented manufacturing and warehousing companies. Houston's rapid growth is derived, in large measure, from the success of this extensive industrial area.

Definitions

At the outset, the differences among *industrial areas, planned employment centers,* and *industrial parks* need to be delineated. Many misconceptions have arisen concerning their respective functions and characteristics.

Industrial Areas

In older, heavily urbanized areas, large concentrations of freestanding factories, warehouses, and supply yards are frequently found intermixed with commercial and service establishments. Such areas, which vary in size according to the size of the communities in which they are located, usually were developed in the pre-highway era and located adjacent to rail lines. These *industrial areas* are characterized by multilevel masonry construction and high levels of site coverage. Physical and structural obsolescence are continuing problems in these areas. In the central city, efficient operation is hampered by crowded streets and insufficient areas for parking and truck loading. Sometimes these industrial areas are adversely affected by a renewal process which has not been oriented toward creation of viable employment centers.

Newer industrial areas present a far different picture. In the suburbs, industrial areas typically consist of a combination of industrial plants, research and development laboratories, industrial parks and office buildings, found among vacant parcels of land. Such areas have been zoned for permissible uses, which has prevented the more extreme intermixture of deleterious uses found in the inner city areas but has not acted as a positive force to insure high standards of development. An increasing share of total industrial employment is found in suburban industrial areas. Whether old, or of recent vintage, such existing aggregations of employment are typically not developed under a detailed master plan.

Industrial Parks

Industrial parks and their antecedents have had many names: *manufacturing district, planned industrial district, organized industrial district, industrial estate,* and, since about 1950, *industrial park.* In the early 1960s, the phrases *research park* and *research and development park* were added to indicate development oriented primarily toward research laboratories and similar technological activities. More recently, the phrases *business park,* and *office park* have been introduced in part to indicate a variety of industrial and related service activities, but also to de-emphasize the industrial character of many of the occupants. Outside the United States, the terms *industrial estate* and *trading estate* have spread from Great Britain, the country of origin, to other industrial countries.

Whatever the name, the presumption is that the *industrial park* is a project which has been planned and developed as an optimal environment for industrial occupants.

Various national organizations concerned with urban planning and industrial land development have prepared definitions of an industrial park. A chronological sequence of definitions of industrial parks and their predecessors follows:

Urban Land Institute—Industrial Council's 1952 Definition

An *industrial district,* in the usual sense, is any area dominated by industrial activity. But when such an area incorporates a deliberate allocation to restricted use coupled with proprietary control, its designation takes on special significance. This distinction is best referred to by the term *planned or organized industrial district.*

A planned industrial district may be further described as a suitably located tract of land subdivided and promoted for industrial use by a sponsoring managerial organization. In this sense industrial district connotes a restricted use of improved land over which there is a proprietor who devotes himself to the area's planning and development. The term may thus refer to either the tract or to the operating company.[1]

National Industrial Zoning Committee's Definition for an Industrial Park

In 1952 the National Industrial Zoning Committee adopted a definition for industrial parks which was amended in 1965, and endorsed in 1966 by all member organizations. The National Industrial Zoning Committee was a nonprofit group composed of delegate committees from eight nationwide organizations whose professional members were active in the fields of urban planning or industrial development. These organizations include:

American Society of Planning Officials	ASPO
American Industrial Development Council	AIDC
American Institute of Planners	AIP
American Railway Development Association	ARDA

[1] J. Ross McKeever and Milburn L. Forth, *Planned Industrial Districts,* Technical Bulletin no. 19 (Washington, D.C.: Urban Land Institute, 1952), p. 4.

1-2 Oakland Ridge Industrial Center—one of four industrial parks under development in the new community of Columbia, Maryland. These parks house 110 firms employing 5,000 persons and are characterized by groupings of single buildings located in heavily-landscaped suburban settings. Shown in this view is Hittman Associates, Inc., a research and development firm in the nucleonics field.

American Society of Civil Engineers	ASCE
Industrial Development Research Council	IDRC
Society of Industrial Realtors	SIR
Urban Land Institute	ULI

The principal product of NIZC was a series of pamphlets written by its volunteer professional members, for distribution to the general public, on such subjects as industrial zoning, industrial performance standards, industrial off-street parking and loading, and industrial parks.

The revised definition for an industrial park issued by NIZC reads:

> An industrial park is a tract of land, the control and administration of which are vested in a single body, suitable for industrial use because of location, topography, proper zoning, availability of utilities, and accessibility to transportation.
>
> The uses permitted are regulated by protective minimum restrictions including size of site, parking and loading regulations, and building setback lines from front, side, and rear yards.
>
> The front yards, and side yards adjacent to streets, are to be landscaped in conformance to planning standards set for the park.
>
> All requirements are to be compatible with the community and surrounding land uses in accordance with a comprehensive plan to enable a group of industries to operate efficiently within it.[2]

U.S. Department of Commerce 1954 Definition

> An *organized* or *planned industrial district* is a tract of land which is subdivided and developed according to a comprehensive plan for the use of a community of industries, with streets, rail lead tracks, and utilities installed before sites are sold to prospective occupants.
>
> The comprehensive plan must provide for adequate control of the area and buildings through restrictions and zoning with a view to protecting the investments of both developers of the district and industries occupying the improved sites.[3]

Dartmouth Conference on Industrial Parks

In 1958 the Dartmouth College Conference on Industrial Parks produced two definitions of industrial parks. The compromise definition agreed upon

---

[2] *Planning for Industrial Parks* (Columbus, Ohio: National Industrial Zoning Committee, 1967), p. 2.

[3] Theodore K. Pasma, *Organized Industrial Districts: A Tool for Community Development* (Washington, D.C.: U.S. Department of Commerce, Area Development Division, 1954), p. 1.

by the participants added some essential details to earlier definitions, but did not enlarge the concept.

> An *organized* or *planned industrial district* is a tract of land which is subdivided and developed according to a comprehensive plan for the use of a community of industries. The plan includes detailed provision for streets, rail lead tracks and utilities, either installed before the sites are sold or otherwise assured to prospective occupants.
>
> The comprehensive plan must insure adequate control of the area and buildings through zoning, private restrictions incorporated as legal requirements in deeds of sale or leases, and the provision of continuing management, all with a view to protecting the investments of both developers of the district and industries occupying the improved sites. The management handles negotiations with local governmental authorities on behalf of the tenants, and it may erect buildings prior to sale or lease for speculative purposes.[4]

A second definition, however, written in the conference report, synthesized diverse opinions expressed at the conference, and more accurately identified all the characteristics of the modern well-planned industrial park.

> An *industrial park* is a planned or organized industrial district with a comprehensive plan which is designed to insure compatibility between the industrial operations therein and the existing activities and character of the community in which the park is located. The plan must provide for streets designed to facilitate truck and other traffic, proper setbacks, lot size minimums, land use ratio minimums, architectural provisions, landscaping requirements, and specific use requirements, all for the purpose of promoting the degrees of openness and park-like character which are appropriate to harmonious integration into the neighborhood.
>
> The industrial park must be of sufficient size and must be suitably zoned to protect the areas surrounding it from being devoted to lower uses. The management is charged with the continuing responsibility of preserving compatibility between the park and the community as well as protecting the investments of the developer and the tenants.[5]

This expanded definition emphasizes the importance of physical land use planning details to achieve efficiency and compatibility, and stresses the responsibility of industrial park management to operate the park by standards which protect the neighborhood and community. According to this definition, the elements essential to a well-planned industrial park are:

- comprehensive plan,
- compatibility among industrial operations in park,
- compatibility between industrial park and existing activities and character of community,

1-3 The Irvine Industrial Complex—an aerial view of the largest industrial district now under development on the West Coast.

- design controls:
  - streets designed for trucks and other traffic,
  - building setbacks,
  - lot size minimum,
  - land area/building coverage ratio minimum,
  - architectural control provisions,
  - landscaping provisions,
  - specific use requirements,
- openness and park-like character,
- harmonious integration into neighborhood,
- zoning to protect surrounding area,
- continuing responsibility of park management to:
  - preserve compatibility between park and community,
  - protect investment of developer and tenants.

---

[4] William Lee Baldwin, *A Report on the Dartmouth College Conference on Industrial Parks* (Cambridge, Mass.: Arthur D. Little, 1958), p. 27.

[5] Many of the ideas in this second definition were included in ULI's contribution to the Dartmouth conference. See, Robert E. Boley, "Effects of Industrial Parks on the Community," *Urban Land*, vol. 17, no. 10 (Washington, D.C., 1958), pp. 3-6.

1-4 Site plan of the Irvine Industrial Complex—a 6,300-acre industrial district now under development at three locations on the Irvine Ranch property in Orange County, California. The central section of the District has unparalleled transportation advantages, including the Newport and San Diego Freeways, the Orange County Airport, and the mainline of the Atchison, Topeka and Santa Fe Railway.

National Association of Industrial Parks' Industrial Park Criteria and Standards

The National Association of Industrial Parks accepts the NIZC definition of an industrial park. The NAIP has also adopted the following code of standards to which its members are asked to adhere.

NATIONAL ASSOCIATION OF INDUSTRIAL PARKS
Industrial Park Criteria and Standards[6]

CRITERIA

An industrial park is the assembly of land, under one continuing control, to provide facilities for business and industry consistent with a master plan and restrictions, resulting in the creation of a physical environment achieving the following objectives:

consistency with community goals,
efficient business and industrial operations,
human scale and values,
compatibility with natural environments,
achieving and sustaining highest land values.

STANDARDS

PHYSICAL	OPERATIONAL
Paved streets	One management
Adequate utility systems for multiple industrial users	Master plan
Setback or percentage coverage requirements	Design review process
Landscaping	Enforcement process
Off-street parking	Community relations
Architectural control through approval of: appearance, siting, building materials, signs	Restrictions
	Legal representation
	Effective supervision of environmental conditions

Planned Employment Centers

The *planned employment center* is fundamentally different from *mixed-use industrial areas* as found in both city and suburb. Its design, occupancy, and operation are carefully planned to accommodate problems of materials availability and handling. Environmental protection and enhancement are emphasized. Also, maximum application of technological innovations is made. The planned employment center is responsive to changes in social values and is much more than an expansion of the industrial park although an industrial park or parks may constitute one or several elements of the planned employment center. The planned employment center is a multi-use district designed to provide for employment needs of urban areas. It should be self-sufficient to the degree that basic employee requirements are provided within its boundaries and carefully designed to provide an optimal environment for a wide range of commercial and industrial needs. It should be fully integrated with the larger community, of which it is a part.

Not surprisingly, some of the best planned employment centers are found in large-scale PUDs and new communities, where employment needs have been given serious consideration from the outset. To be successful as a designed multiple land use area as well as a business venture, planned employment centers should exhibit most of the following qualities:

- comprehensive planning both beyond the periphery as well as within;
- joint action and investment both by public agencies and the private sector;
- resource oriented toward market, transportation, or materials;
- developed at a prime location;
- serve as a major land use element in the community and in regional land use plans;
- a focal point for multi-modal transportation facilities for goods and people;
- provide proximity opportunity for industries linked by process, product, or service;
- designed to insure optimum efficiency for occupant industries;
- provide on-site essential services both public and private for industries, employees, and clients to minimize individual vehicular movements;
- provide pleasant, convenient, and satisfying work conditions for employees, including the opportunity for a short journey to work;
- walk-to-work residential areas;
- engineered and operated to insure protection and enhancement of the physical environment;
- function as a center, sharing facilities, minimizing redundancy;
- have a durable as well as attractive appearance;
- continue to provide profitable return to the community in taxes, wages, and local sales; and,
- provide a reasonable return to the entrepreneur and the investors.

[6] *Proceedings of the National Association of Industrial Parks*, Fourth Annual Seminar and Convention (Arlington, Virginia, 1971).

Summary

The Handbook has been divided into six sections: Introduction, Background and Characteristics, Planning, Engineering and Design, Development Strategy, and Future Trends. Following is a summary of the contents of each section.

Section 1—Introduction

The Introduction explains the objectives of the Handbook, identifies the primary sources utilized, and defines the three major types of industrial concentrations: industrial area, industrial park, and planned employment center.

Section 2—Background and Characteristics

Background and Characteristics traces the history of planned industrial developments, evaluates the evolution of industrial land use controls, and describes the basic characteristics of the industrial environment. This section includes nine case studies of recent industrial developments, each of which illustrates a subcategory of planned industrial development.

Section 3—Planning

Planning covers the steps involved in initiating an industrial development project. Items discussed include site selection, major locational factors, and location guidelines considered important by leading industrial developers. The importance of community and regional land use plans to the developer in selecting his site is explained. The steps in the preliminary planning analysis and the zoning process are described. Design guidelines for planned employment centers are provided.

Section 4—Engineering and Design

Engineering and Design provides the developer and his staff with insight into the engineering site design considerations which are a vital part of industrial land development. The primary intent of this section is to provide the developer with a checklist, rather than specific engineering design methods, which can be applied to a parcel of land under consideration for development. Design guidelines are discussed in detail.

Section 5—Development Strategy

Development Strategy outlines the procedure to be followed in obtaining the necessary financing for an industrial development. Also discussed are strategies for marketing and managing the project and operational techniques to be applied from the outset, through the absorption period, and concluding with sellout and transfer of operational responsibilities to the tenants themselves.

Section 6—Future Trends

Future Trends deals with the continuing evolution of industrial development practices. The subject is considered in light of present trends and probable future impacts on industrial development. Among these trends are accommodating higher energy costs; innovations in design which hold forth the promise of more efficient and attractive industrial environments; technological changes in transportation, communication, and materials handling; improved industrial marketing techniques; and similar factors.

2.
Background and Characteristics

Origins of the Modern Industrial Park

Today's modern industrial park is the evolutionary product of almost 80 years of development experience in the United States and abroad. During these eight decades, private real estate developers, transportation companies, and public agencies have successfully demonstrated that a well-located, properly serviced tract of land, carefully designed for the use of many individual industries can:

- benefit the public interest through efficient land management;
- insure compatible operation of productive activities essential to an urban society;
- provide a marketable product needed by small, medium, and even large industrial firms;
- assist the community by attracting new employment opportunities;
- expand the fiscal capacity of local and state government by adding new investment to the property tax base;
- contribute to community appearance;
- respect the natural environment; and,
- provide a reasonable return on investment to the entrepreneurial developer while accomplishing these other objectives.

The purpose of this historical summary is threefold: first, to remind us that during the period when many persons were decrying the effects of the industrial revolution upon the urban landscape, some industrial leaders, private land developers and public planners were working to bring order and form to new industrial areas so that these areas would be efficient, pleasant, and compatible places to work; second, to note that many admirable features of modern industrial parks were first introduced early in this century; and third, to identify features of early planned industrial areas which have continuing validity and should remain as important elements in present and future industrial development.

Trafford Park Estates, Ltd.—1896

The first planned industrial estate was begun in 1896 in Manchester, England when a private company, Trafford Park Estates, Ltd., purchased a 1,200-acre country estate on the Manchester Ship Channel adjoining the docks. The Ship Channel, opened two years earlier, provided access for ocean vessels to reach Manchester which was 54 miles inland and as such the port closest to the main concentration of British industry. Although the world's first planned industrial development was served by navigable deep water, few subsequent parks have been water transportation oriented.

This industrial district, served by over 35 miles of railroad, was dominated by heavy manufacturing. Trafford Park Estates remained the world's largest planned industrial estate until the 1950s when larger ones were developed in the United States and Canada.

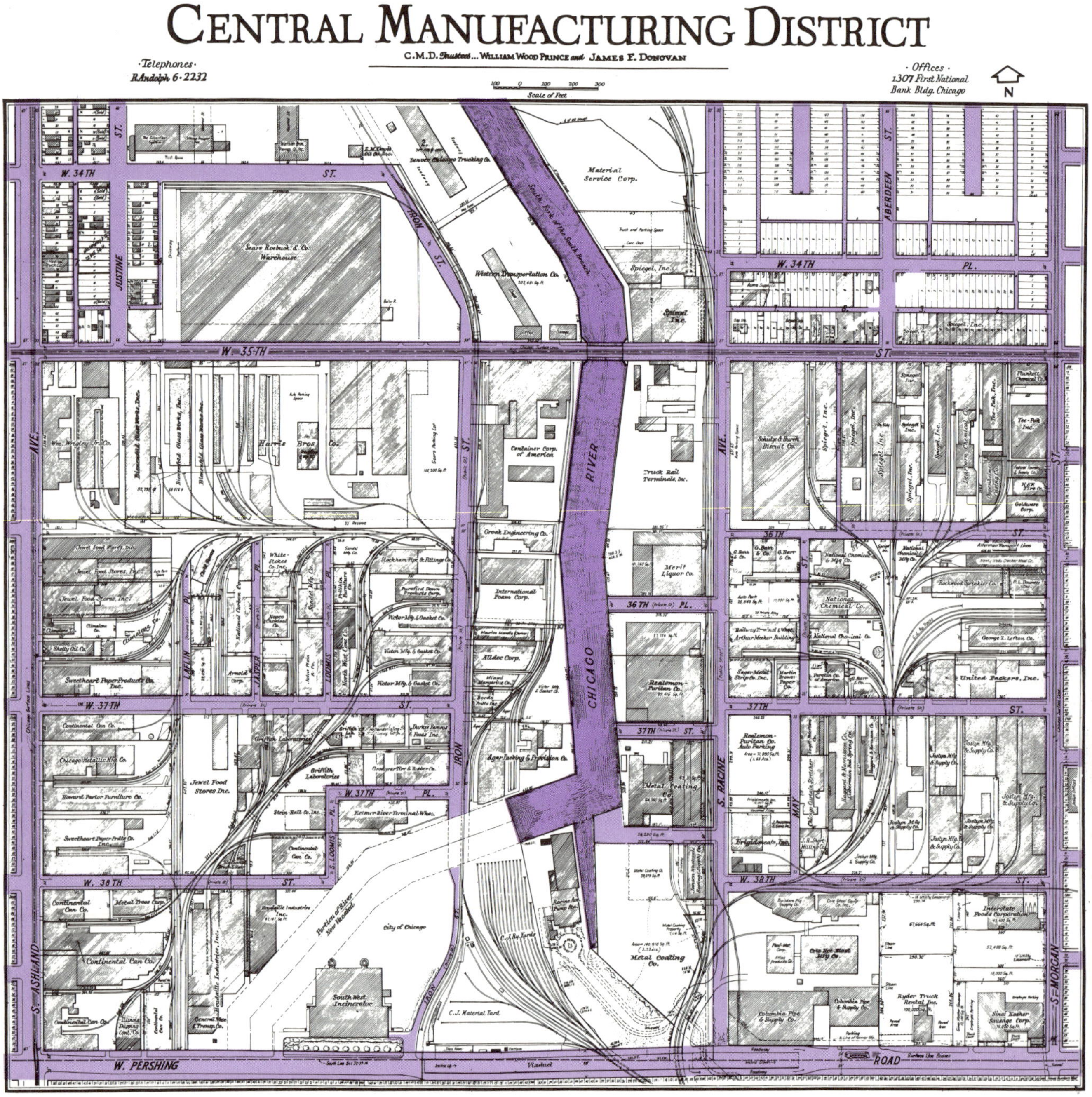

2-1 Central Manufacturing District—the original East District, developed in 1902.

2-2 A view of the Pershing Road Industrial District, perhaps the most noteworthy predecessor of the modern industrial park.

Central Manufacturing Districts (CMD)—Chicago

In the United States, pioneer planned industrial districts began in Chicago shortly after the turn of the century.

Original East District—1902

Founded by Frederick Prince, the Central Manufacturing—Original East District was undertaken by representatives of the Union Stock Yard and Transit Company to attract additional freight for the company's belt rail line which served the Yard.[1]

The 260-acre tract was located less than three and one-half miles southwest of downtown Chicago. Buildings had a uniform height of four stories, designed to facilitate the gravity flow production practice then prevalent. Private rail sidings served each building, and the streets followed a gridiron pattern. The CMD provided electric power and steam for its industries from a generating plant located on Stock Yard land nearby.[2] From the beginning, landscaping, planting strips, and ornamental street lighting were an integral part of the design. Most occupants were engaged in manufacturing, but warehouses were included also. Although a navigable fork of the Chicago River flowed south and southwest through the middle of the original Central Manufacturing District, no mention is made in various histories of whether barge transport was used in this pioneer American industrial district.

Pershing Road—Second Central Manufacturing District

By 1910, the management of the Central Manufacturing District had acquired 98 acres for its second, and probably best known industrial district, the Pershing Road development. Diagonally opposite the Original East District, Pershing Road District opened in 1916. Although considerably smaller than the 260-acre Original East District, the Pershing Road District had several unique features:[3]

- a central power unit to provide steam and electric current for light, heat, and power to serve the entire tract;
- a central sprinkler system to protect all structures in the tract and to avoid unsightly tanks;
- a centrally located freight station for convenience of LCL [less than carload] freight originated, or consigned to industries, in the district;
- standardized structural design based on the minimum height of six stories and basement, complete with plumbing, light, heat, and sprinkler systems, and high-speed elevators;

---

[1] J. Ross McKeever and Milburn L. Forth, *Planned Industrial Districts,* Technical Bulletin no. 19 (Washington, D.C.: Urban Land Institute, 1952), pp. 18-22.

[2] Robert E. Boley, *Industrial Districts: Principles in Practice,* Technical Bulletin no. 44 (Washington, D.C.: Urban Land Institute, 1962), pp. 28-34.

[3] Ibid., p. 32.

- underground traffic system, with concrete tunnels which supplemented surface facilities by connecting all buildings with each other and with the central freight station, and which provided fast, safe transport of goods and materials; and,
- underground utility system of concrete tunnels which conveyed utilities from the central power unit to all plants, to avoid unsightly surface accessories.

This pioneer industrial district had notable locational characteristics which were forerunners of trends developed several decades later. The classification yard of the Chicago Junction Railroad at the rear of the district provided primary access; however,

- CMD-Pershing Road had a mile of frontage on a principal east-west highway. All major buildings were oriented to this thoroughfare, thus presenting a handsome facade which preceded the freeway frontage trend by over 40 years.
- McKinley Park was another location asset. It provided a green park similar to those included in corporate headquarters estates built in the 1960s. This public park, with its large trees, lake and extensive recreation fields, faced the front of the Central Manufacturing District on the opposite side of Pershing Road, providing a spacious front yard in scale with six-story and higher structures in the District. This permanent public protective greenbelt extended for almost two-thirds of CMD's mile-long frontage.

Clearing Industrial District (CID) Original Bedford Park—Chicago

The other pioneer industrial district development in the Chicago area was the Clearing Industrial District founded by Ogden Mills and the Isham family and subdivided about the same time as the initial Central Manufacturing District.[4] In contrast to the CMD, which was backed by belt rail line interests, the CID was organized by private real estate developers. The 530-acre CID, twice the size of CMD's first district, was opened in 1909. Its location, adjacent to the classification yards of the Belt Railway of Chicago, was important to its early success. Chicago Municipal Airport, located immediately north of CID, began as a mailport in the 1920s and served as Chicago's major commercial airport during the 1930s and 1940s. The continued growth of CID was further enhanced by its proximity to this emerging major form of transportation. South Clearing District, begun in 1948, added 1,300 acres to the original district.

The Clearing Industrial District's major contribution was the superblock concept. The property, with one-quarter mile of depth and bounded by West 65th Street at the front and the Chicago Junction Railroad at the rear, was divided for its full three-mile length into one-quarter mile square blocks, each containing 40 acres. No internal road paralleling either the rail line or West 65th Street was built.

[4] Ibid., pp. 35-40.

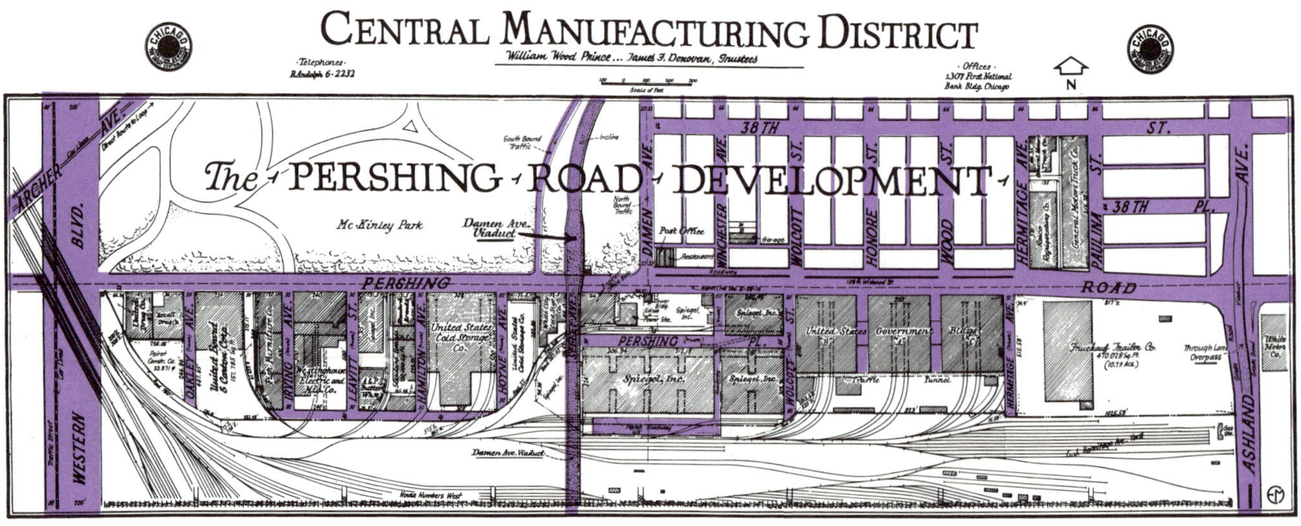

2-3 Site plan of the Pershing Road development. Note the extensive frontage on McKinley Park and avoidance of street/rail crossings.

2-4 Slough Estates—a pioneering English industrial district.

Each 40-acre superblock permitted diagonal entry of the main rail lead, a design concept with significant advantages, including:

- elimination of grade rail crossing of internal streets;
- improved personnel safety;
- minimization of internal vehicular traffic delays;
- absorption by site purchaser of only one-half of rail lead curvature; and,
- a variety of site sizes and depths in each block.

While this layout resulted in some irregular areas, these areas could be used for outside storage.

Most occupants were engaged in manufacturing, but warehouses also were built.

Slough Estates, England—1920

The Trading Estate in Slough Berkshire, England, was possibly the first planned industrial district built upon the former site of a military service depot.[5] In 1920, the 602-acre site was purchased as a part of a transaction involving the acquisition of surplus military vehicles. Subsequently, when the refurbishment of the military vehicles was completed, the buildings were leased to industry and thus, the industrial estate was started.

During the late 1920s, new factory construction was accompanied by the building of houses for key workers on part of the estate land. Later, houses were built in other parts of Slough.[6] The 1974 population of Slough was approximately 101,800.

Slough Estates, a profit-making organization, derives its income from property rentals and a wide range of utilities and services it provides to industrial tenants, including:

- electricity provided from a power station on the estate;

[5] In the United States, military bases have been converted to industrial and other uses. For those assisted by the United States government's Economic Adjustment Program during the period 1967-1973, see Appendix B.

[6] William Bredo, *Industrial Estates, Tool for Industrialization* (Menlo Park, California: Stanford Research Institute, 1960), p. 182.

- steam for processing and heating also provided by an on-site power plant;
- water supply from wells drilled on property; a large reservoir has been constructed;
- government-operated training center on site, functioning since 1927, to service a number of occupational trades represented in industries on the estate;
- trouble-shooting technical service provided by estate to assist industrial managers in solving special problems on request;
- administration building with five banks, post office, customs and excise office, conference room, board room, and club room; and,
- large social center, welfare, and sports facilities including an 80-acre playing field.

Railroad Sponsored Districts—1920-40

The success of the pre-World War I, rail-served districts in Chicago encouraged the railroads to develop several rail-oriented districts in the 1920s and 1930s. Among the largest and most representative of the districts developed during that period is the Los Angeles Central Manufacturing District.

The developer of the Central Manufacturing Districts in Chicago established the original Los Angeles Central Manufacturing District in the Vernon District in 1922. The original 280-acre parcel, part of an early cattle ranch known as Rancho San Antonio, was acquired by the Atchison, Topeka and Santa Fe Railroad in 1928, and in 1930 the Bandini Area was added. Both districts were connected to transcontinental rail service by the Los Angeles Junction Railway, a Santa Fe affiliate.

The gridiron street layout of the original Vernon District—with rail service to every site whether very small or large—was to become a pattern for many future industrial parks. Although five-foot sidewalks were provided in the original Vernon District, these were not built in the expansion. This omission has been made in most postwar industrial parks.

Street rights-of-way ranged from 40 to 100 feet, with paved area width of 40 to 55 feet. Rail was brought in on 17, 20, and 34-foot rights of way, with 50 feet for a lead plus spurs. Although good-sized for the time, the blocks proved to be too short, resulting in too many grade-crossings. The block size also resulted in some rail-served lots being too shallow. Lot depths range from 100 feet to a maximum of 400 feet. Sites were from one-quarter acre to 10 acres in size. Building sizes were between 8,000 to 125,000 square feet (a large warehouse for that time), with the majority in the 20,000 to 35,000-square-foot range.

The Los Angeles CMD continued this pattern, expanded to 3,600 acres, and by 1960 was surrounded by dense urban development.

During the 1920s, the ornamental street lighting and diagonal lead tracks pioneered in the early Chicago districts were discarded. Also, landscaping and planting strips were reduced. Except for diagonal lead tracks, such features were not added again to the industrial environment until the postwar era.

2-5 Central Manufacturing District, Los Angeles, established in the 1920s.

The Continuing Evolution

Evolution continued during the 1930s, but following World War II, the pace of change quickened and the modern industrial park as an outgrowth of earlier industrial districts, emerged as the major new trend in industrial development. Without attempting to provide a complete list or to identify their contributions, the following projects, from the 1950s, are representative of those in the forefront of the evolutionary process.[7]

PROJECT	DEVELOPMENT COMMENCED
New England Industrial Center—Needham, Massachusetts	1949
Stanford Industrial Park—Palo Alto, California	1951
Fair Lawn Industrial Park—Fair Lawn, New Jersey	1952

[7] Boley, *Industrial Districts: Principles in Practice*, contains detailed discussion of all these projects. In addition, most of these projects are referenced elsewhere in this text as examples or case studies. See Index.

Bohannon Industrial Park— Menlo Park, California	1954
Tampa Industrial Park— Tampa, Florida	1956
Centex Industrial Park— Elk Grove Village, Illinois	1957
Crocker Industrial Park— Brisbane, California	1959

The continuation of the evolutionary process in the 1960s and into the 1970s is discussed throughout the book. This evolution was both encouraged and constrained by the parallel evolution of land use controls.

Industrial Land Use Controls

Increasingly, the location, development, and operation of industrial parks and employment centers has become subject to a proliferation of restrictions from state and national legislation and local administrative decisions. Air and water quality, availability of public transportation and electric power, have become major concerns, along with increasing limitations on public water supply and sewage treatment capacity. Land use controls, once strictly local, now are being applied by regional, state, and federal agencies. Imposition of this wide array of controls has been critical in upgrading overall industrial environment standards, but also has led to increased developer time and cost required to obtain the necessary approvals. These costs are passed on to future industrial occupants in the form of higher sales, or rental, prices.

From a wider perspective, these trends reflect the evolution of industrial land use controls over an extended period. A review of past practice and policy helps to place the current situation in proper perspective. What have been the objectives and standards of industrial land use controls in the past? Have these controls done the job? What changes have occurred? Are present regulations and practices relevant for the decade ahead with its probable shortages of space, energy, and materials? Local land use controls fall into two categories: immediate code standards as exemplified by zoning ordinances and subdivision regulations; and, long-range philosophical judgments of the comprehensive plan which show where the community will permit future industry to go, approximately how much land will be set aside for future industry, and which type of industries will be accepted.

Early Planning and Zoning Practices

Until the postwar era, much land zoned for industry was located adjacent to railroad lines in river bottoms, in deteriorated areas, or in locations having little or no potential for residential development because of topographic constraints, inaccessibility, and other factors. Seldom was industry allowed to use front sites on a prominent location. Frequently, industrial land was leftover land, unusable for any other purpose.

The land on either side of the railroad tracks usually was zoned for industry, regardless of terrain. On thousands of local zoning maps across the country, industrial zoning tended to start 200 feet back from the main highway, so that a 200-foot-wide commercial strip could front on the road. The reasoning was that the commercial zone provided a buffer between industry and the community. The fallacy of employing a "buffer land use" instead of setting specific development standards for an industrial zone, continues to obscure real planning requirements intended to achieve compatibility.

Unprotected Industrial Use Districts

The negative community attitude toward the future location for industry also affected the use provisions of the zoning ordinance. From the first comprehensive zoning ordinance adopted in New York City in 1916, until the middle of this century, zoning in most communities was based upon a philosophy of rank order of land uses. The single house on a large lot was identified as the highest and best use; therefore, only this use was permitted in the most restricted zone. Following in the social value ranking were row houses, apartments, commercial uses, and at the bottom, industry. Each lower district, however, permitted all uses below it on the ranking scale. As a result of this cumulative concept of zoning, with higher uses permitted by right in the lowest use zone, large tracts of land zoned for industry could not be protected from intrusions by housing or roadside business uses.

Among the first to call attention to the need for protection of vacant land zoned for industry were two urban planners, Seward H. Mott and Max S. Wehrly, both former executive directors of the Urban Land Institute.

> There has been, and still is, the tendency . . . to use the industrial district as a residue or catchall.

The theory that residential uses can be allowed to occupy potential industrial areas until such time as needed, and that industry will then come in and acquire the land [and houses or small commercial uses] for industrial use, has proved to be largely erroneous.[8]

Mott and Wehrly also noted, "industrial zoning has been more often both inadequate (in amount) and improperly located."

Exclusive Industrial Zoning

Mott and Wehrly's pleas for prohibition of residences in industrial zones, and for better community planning for industrial land, were continued by seven professional organizations which formed a working group known as the National Industrial Zoning Committee. In 1951, the NIZC drafted twelve "Principles of Industrial Zoning," which were endorsed by these professional organizations:

American Industrial Development Council
American Institute of Planners
American Railway Development Association
American Society of Civil Engineers
American Society of Planning Officials
Society of Industrial Realtors
ULI-the Urban Land Institute

The twelve "Principles of Industrial Zoning" are:

Principle 1. Most communities require a certain amount of industrial development to produce a sound economy.

Principle 2. Zoning controls are basic tools in the reservation of space for industry; guidance of industrial location into a desirable pattern; and, provision of related facilities and areas needed for convenient and balanced economy.

Principle 3. Industrial use is a legitimate land use possessing integrity comparable to other classes of land use established under zoning and is entitled to protection against encroachment.

Principle 4. Through proper zoning, industrial and residential areas can be good neighbors.

Principle 5. Industry will continue to grow and most industries will require larger areas in the future.

Principle 6. There is need for a reclassification of industry based on modern manufacturing processes and the prevailing policy of plant construction in order to determine the desirability for inclusion in a given area.

Principle 7. Industrial potentialities of lands bearing a favorable relationship to transportation should be recognized in the zoning process.

Principle 8. Industrial zoning and highway planning should go hand in hand.

Principle 9. Special consideration should be given to the street layout in industrial areas.

Principle 10. Zoning ordinances should be permissive rather than prohibitive.

Principle 11. A good zoning ordinance should be sufficiently definite to convey to a landowner a clear concept of what he can do with his land.

Principle 12. Industrial zoning can be most effective when considered on a metropolitan basis.

Although first published in 1951, these principles are still relevant to today's problems in planning land for industry.

By the early 1960s, many communities had revised their zoning ordinances to prohibit residences in industrial areas. Likewise, most or all commercial uses were usually prohibited in the most exclusive industrial zones, in order to protect the industrial land reserve from invasion by strip commercial uses along the frontage on major roads and to prevent shopping centers (a new strong competing use for highway sites) from usurping sites specifically reserved and zoned for industry.[9]

In recommending that industrial land be protected for future industrial development by exclusive zoning, Mott and Wehrly also explained that

> . . . the exclusion of residential uses might also apply to certain types of commercial development. Baseball parks, skating rinks, and similar enterprises needing large sites will often seek locations in industrial districts . . . (with a) diversion of land designated for industrial use to other purposes.[10]

[8] Seward H. Mott and Max S. Wehrly, *The Prohibition of Residential Developments in Industrial Districts*, Technical Bulletin no. 10 (Washington, D.C.: Urban Land Institute, 1948), p. 2.

[9] Unfortunately this practice excluded those commercial uses needed to serve industrial uses and, thus, constrained the evolution toward planned employment centers.

[10] Mott and Wehrly, *Prohibition of Residential Developments*, p. 7.

Mott and Wehrly, however, did not propose to exclude business facilities which serve industry and employees, and which are often necessary adjuncts to industrial development.

Recent Innovations in Industrial Land Use Controls

Performance Standards

At the end of the 1950s a new permissive element was introduced into industrial zoning—performance standards. In 1951, the National Industrial Zoning Committee had recommended in Principles 6 and 10 that industry not be arbitrarily divided into light and heavy categories, and that the effect of an industry on its environment is more important as a means of classification than any material or product category listed in the zoning code. Under performance standards zoning, the customary list of excluded industries would be replaced by specific measurable standards. Any industry which could meet these standards would be permitted, and others which could not would be automatically excluded.

In 1954, a paper entitled *Performance Standards in Industrial Zoning,* authored by the late Dennis O'Harrow, former executive director of the American Society of Planning Officials, received considerable attention in both urban planning and industrial development fields.[11] This paper encouraged suburban counties to amend their zoning codes, providing for industrial performance standards within the industrial sections of the ordinance. O'Harrow recommended that the zoning ordinance specify performance standards for eleven types of potential industrial nuisances: noise, smoke, odor, dust and dirt, noxious gases, glare and heat, fire hazards, industrial wastes (solid and liquid), transportation and traffic, aesthetics, and psychological.

The specific numerical standards enumerated in these updated zoning ordinances contrasted sharply with the general phrasing about nuisance uses found in most zoning codes and covenants for industrial parks. Such regulations usually read:

> No noxious or offensive activities shall be conducted on any site, nor shall anything be done thereon which may be or become an annoyance or nuisance by reason of unsightliness or excessive emission of fumes, odors, glare, vibration, gases, radiation, dust, liquid waste, smoke, or noise.

These performance standards applied to each individual new building, or to any expansion of an industrial facility. The review and approval procedure usually required an applicant for a building permit, or for a certificate of occupancy, to submit a plan of development. The plan would include a description of the proposed operation, including all machinery, processes, and products, and, specifications for the mechanisms and techniques to be used in restricting the emission of dangerous and objectionable elements. This would apply to laboratories, offices in industrial areas, and warehouses, as well as to manufacturing establishments.

The adoption of performance standards by the city of Chicago, in 1957 and 1958, and New York City in 1960, added credence to this complex but more objective system of evaluating industries. Both Chicago and New York applied performance standards to all their industrial zones; however, the levels of required performance varied somewhat between the most restricted and the heavy manufacturing zones.

Suburban counties, however, often adopted performance standards only for industrial park zones. Montgomery County, Maryland, a residential suburb in metropolitan Washington, D.C., included performance standards when it adopted an I-3 Zone-Industrial Park, in 1959.[12]

The conditions which led to widespread acceptance of the performance standards concept in zoning are:

- Citizens were becoming active in environmental movements, and often assumed an anti-industrial growth posture at rezoning hearings.
- City officials, faced with a large out-migration of industries, wanted to retain this vital part of the tax base. Much vacant land was zoned for industry to encourage expansion of existing plants, and hopefully to provide for new firms. Performance standards were adopted as a means of developing compatible economic growth within the city's limited space.
- Suburban officials, overwhelmed with costs of servicing the rapid growth of new homes, sought a protective type of industrial zoning to assure compatibility of sorely needed new industries with largely residential suburban communities.

---

[11] Dennis O'Harrow, *Performance Standards in Industrial Zoning* (Chicago: American Society of Planning Officials, 1954).

[12] See Appendix C for industrial ordinance provisions and performance standards.

- Public officials, utilities, and others responsible for economic development, hoped to educate the citizenry that industrial growth meant more than photogenic research laboratories, and that a wide range of industries must be accommodated on choice sites. Performance standards could assure good industrial neighbors.
- Planners looked to performance standards as a planning tool, albeit a complex and possibly cumbersome one, to provide better locations for industrial parks and individual industrial plants. The buffer would be provided on the industrial property so that homes and most industry could co-exist compatibly.

The Industrial Park Zone

About the same time that performance standards were being first adopted in some areas, a new industrial zone was receiving much wider acceptance into zoning ordinances of suburban counties and communities. Many planners and zoning administrators felt that the space and aesthetic standards of the Industrial Park Zone would be more easily administered than the engineering requirements of performance standards. Furthermore, the development plan review provided by law would assure good-neighbor, well-planned industrial parks.

The Industrial Park Zone differs from performance standards in four respects:

- A development plan for the entire site—showing subdivision, and often placement of structures—is required frequently in the Industrial Park Zone;
- Specific performance levels for air, water, vibration, etc., are usually omitted or included by reference to state agency standards; or in the case of electromagnetic radiation, to federal standards;
- Aesthetic standards in the Industrial Park Zone included landscaping, open space, building orientation, etc.; and,
- There are limitations on amount, type, and timing of commercial facilities in industrial parks.

The name of the new zone varied, sometimes identified as MR—Manufacturing Restricted, MLR—Manufacturing Light Restricted, or IR—Industrial Restricted. Occasionally, industrial occupancy in the new zone was severely limited, as indicated in the name RD—Research and Development Zone, or SP—Science Park. Whatever the name, the purpose was the same: to provide for the planned subdivision of an area of land, usually a minimum of 25 to 100 acres, for a number of industrial sites, to be developed under space standards considerably more restricted than the usual industrial districts permitted by the zoning ordinance. To achieve a park-like environment, low structural density (building coverage and floor area ratios) and large landscaped areas were required.

Space standards and other land use controls in the new zone, often modeled after examples of the site planning for either a single manufacturing plant, research and development laboratories, or corporate headquarters, were frequently excessive in terms of open space requirements.

The text of an Industrial Park Zone ordinance includes the following requirements to be met by an applicant developer:

- a comprehensive development plan for the entire tract—which requires approval;
- site development plan approval for each lot;
- lists of permitted and prohibited uses;
- minimum tract size and minimum lot area;
- minimum ratio of ground coverage to lot size, and sometimes minimum floor space;
- required yards (setbacks) and landscaped areas;
- highway access for park, and street frontage for lots;
- building height limits and structural exceptions;
- space requirements between buildings on same lot;
- sign controls, and limitations on outdoor lighting;
- off-street parking and loading standards, including location, design, and number of these facilities; and,
- restrictions on outdoor storage.

The following two restrictions, frequently contained in Industrial Park Zones, present problems for the industrial developer:

- requirement of a very detailed comprehensive development plan before rezoning and essential permits are granted to begin park construction; and,
- open space requirements including yards, structural density, floor area ratio, and occasionally percent of site to be landscaped.

With regard to the first difficulty, a developer seldom has both the financing and a major occupant for the park in hand before he begins the preliminary investment in options, planning, engineering, market analysis, and the application for rezoning. Thus, an industrial park developer is not prepared to fix rigid lot lines at this stage, as is customary in a residential subdivision. Some sites for future large plants, or for large single-tenant buildings, must be custom shaped and cannot be achieved by combining small preplanned lots on a standard gridiron site plan.

More flexibility, therefore, needs to be introduced into industrial park zoning requirements. Changes in the development plans which do not violate basic standards should not require further public hearings once zoning is granted. These changes can be reviewed administratively. The developer, however, should understand the qualities and characteristics of his land so thoroughly that he is able to present alternative development plans any one of which would result in an industrial park equal to the basic objectives and spirit of the community's zoning ordinance.

The second difficulty relates to provision and utilization of open spaces. Rather than having density and open space requirements which focus upon the amount of land which cannot be developed, built upon, or paved, the emphasis should be placed upon the utility of the open space. Most industrial park zoning standards require large yards, sometimes 200 to 300 feet wide at the periphery of an industrial park if it is contiguous to a residential area. Front yard setbacks of 100 feet are not uncommon. The result of these overly restrictive zoning requirements is the division of open space into small separate parcels. The exchange of some of this excess yard area for one or more large open space areas could provide recreation area for all occupants of the park to enjoy. If the industrial park site has a unique topographic feature or enjoys an outstanding scenic view, this open space can be shared with the public which would result in a more aesthetically pleasing community asset than deeper or wider individual yards.

An example of this approach is Cabot, Cabot and Forbes' South San Francisco Industrial Park, which provides two types of shared open space. Adapting the site layout to a hilly terrain, three graded earthen banks extend through the park, planted to screen back yards and rail spurs from view of adjacent industries.

This is probably the first industrial park to preserve land at the high point of the site and to enable the public to visit and enjoy magnificent vistas of San Francisco Bay. The hilltop is also the location of a 92-foot high steel sculpture which has become a landmark on the Bay.

Another land-saving opportunity which will require a space credit adjustment in the Industrial Park Zone text is the shift of off-street parking for employees and visitors from parking lots to roof parking, under-building parking, parking garages, or shared parking.

The low structural density requirement often set in industrial park zones—25 percent to 35 percent—coverage, in large part is an allowance for the land area required for employee parking. It must be kept in mind, however, that roof and covered parking is a function of land cost and only very expensive land can justify the added cost for this kind of parking. The zoning ordinance requirements for off-street parking are very important protections provided to the company which decides to locate in an industrial park. An important extra benefit for moving to a planned industrial park is to get away from the traffic congestion caused by on-street parking in the older industrial areas.

It takes more land area to park an employee's car than it does to provide and service his or her work space in the plant (including the proportionate share of floor space of services within the plant, such as office area, clinic, and cafeteria). When land can be saved by providing parking under or over an office or industrial building, or in a multistory garage, the zoning ordinance should credit the industrial park developer or the industrial occupant with the area of saved land, permitting some of it to be used for plant expansion, and some for a true open space use such as active or passive recreation.

Planned Multiple-Use Zones: PUD, Town Sector, and New Town

The industrial park *developer* is seeking flexibility in land use controls, particularly for:

- increase in site coverage;
- mix of land uses, such as commercial and office buildings with manufacturing and warehouse facilities in the same park; and,
- some elasticity in development and subdivision planning, without pre-commitment to exact acreage to be rezoned for industrial or commercial, and to delay precise complete subdivision layout until a major client lease or sale is settled.

This flexibility is being achieved in some of the new multiple-use zoning districts which are usually applied to tracts of land ranging from 400 to several thousand acres.

The *planner* is seeking flexibility in land use controls to accommodate compatible mixtures of uses within the industrial environment. The purpose is to:

- save distance and time (which also means saving land);
- achieve useful open space rather than small, little used cosmetic greenery;
- design for harmonious and useful relationships among different land uses, rather than neutral unrelated compatibility; and,
- reduce the journey to work for some people.

These goals have more chance for realization in the fluid multiple-use zoning.

Not all the conveniences sought by the industrial occupant can be provided within an industrial park permitting only industrial occupants. There are social as well as economic justifications to permit a range of commercial facilities to serve the industries and employees of the industrial parks. Industrial occupants have other concerns about land uses in the industrial park neighborhood. Enlightened industrial management has long been concerned about the lengthening journey to work for both executives and employees. Production managers in new suburban facilities have worried about the long daily work trip of experienced workers whose homes remain near the former in-town plant. Office managers and research directors are disturbed that clerks and junior professionals may have difficulty finding apartments they can afford near the office or laboratory. Industry needs a variety of housing (in type, size, and price) near the industrial plant for employees who have a wide range in salaries, age, and life styles.

The Planned Unit Development (PUD) Zone

Residential developers in many suburban areas are utilizing the PUD concept which permits—under an approved development plan—a mixture of land uses with a range of density and open space requirements. The application of this zoning concept to industrial development offers a unique opportunity to the developer to prepare a comprehensive plan for light industry including a variety of housing types and commercial activities. The whole is designed to be favorably related to each of the parts.

However, PUD has been used primarily for mixtures of townhouses, garden and highrise apartments, and neighborhood centers. Most PUD zones do not list light industries as permitted uses. For example, Howard County, Maryland has a PUD-type zone named Planned Community (PC) District (minimum area requirement—400 acres) which specifically excludes all industrial uses, including industries permitted in the MR (Manufacturing Restricted) District, which is the most restricted industrial zone. Yet, the MR District was established with tough development standards which would permit light industries to locate next to residential zones.[13]

2-6 Reston's 1,300-acre Center for Government, Business, and Industry is being developed under Fairfax County, Virginia's PUD zoning provisions.

If PUD ordinances are liberalized to include industrial uses, then residences, places of work, stores and services, could be integrated to allow a measure of neighborhood self-sufficiency. These industrial uses should be served by a street network designed to prevent intrusion of industrial traffic into residential and commercial areas. An industrial park next to a commercial center next to a residential development—but not planned together—does not accomplish this.

Town Sector Zone

Some zoning ordinances provide a PUD-type district with a larger minimum area requirement. In such a district, industry is permitted within a comprehensively planned multi-use development. However, the tendency often is to limit severely the percent of land within the sector which may be occupied by industrial-type activities.

In the Town Sector Zone of Montgomery County, Maryland (minimum required area—1,500 acres),

[13] *Zoning Regulations of Howard County, Maryland,* fifth edition (1971), section 17A PC District, and section 14 MR District.

not more than 5 percent of the total area of the planned Town Sector may be devoted to industrial purposes or other major employment facilities. Thus, in a Town Sector which has 1,500 acres, or more than 2 square miles, only 75 acres may be used for industry or other major employment facilities. In this same zone, twice as much land may be developed for commercial purposes—up to 10 percent of the total area of the Town Sector.[14] Whether this ratio is reasonable for Montgomery County must be based on the community's needs. In most balanced communities this would appear too low.

New Town District

Howard County, Maryland has recognized the importance of planned industrial areas as a significant part of its New Town District (see Figure Howard County, Maryland). Each new Town District (minimum required area—2,500 acres) is required to provide a minimum of 10 percent of the total New Town area for industrial uses. The maximum area for industry is 20 percent of the New Town area. In contrast, commercial uses have a required minimum of 2 percent and a maximum of 10 percent.

Example—NEW TOWN DISTRICT ZONE

Howard County, Maryland
Required Minimum and Maximum Percentage of Total Area in New Town District for Permitted Uses

PERMITTED USE	REQUIRED TOTAL AREA IN NEW TOWN DISTRICT MINIMUM %	MAXIMUM %
Open Space Uses	20	NA
Single Family—Low Density	15	NA
Single Family—Medium Density	25	NA
Apartments	NA	10
Commercial (B-1, B-2, T-1, and SC Uses)	2	10
Industrial Uses (MR and M-1 Uses)	10	20
Other uses presently permitted in any zoning district other than those permitted only in M-2 or T-2 Districts	NA	15

* NA means Not Applicable

Each New Town District must also provide adequate public transportation facilities and public water and sewer systems in areas shown on the final development plan. As used in Section 17.018 of Zoning Regulations, the term *open space uses* is defined as being those uses which do not involve any extensive coverage of land with buildings—for example, all lands devoted to raising of crops, agricultural uses, parks, playing fields, golf courses, and any other outdoor recreational uses (whether any such uses are publicly or privately owned or operated for profit), as well as all lands covered by lakes, rivers, or streams, and all lands devoted to public or community uses. Open land designated for residential uses shall be considered qualified as *open space use* only if it is held for the common use of the public or persons residing in the particular locality within the community, and if it is larger than two acres in size. For the purpose of meeting the 20 percent requirement imposed above:

- the term *open space uses* shall not include parking lots, vehicular rights-of-way, amusement parks, golf driving ranges, or drive-in movies; and,
- at least 50 percent of the land area so designated shall be available to the public at large.

Protective Covenants

Covenants provide the industrial developer with private land use controls in the form of a legal instrument to protect future occupants of his development, its investors, and the community interest. Covenants are legally enforceable standards adopted by the developer and applicable to a specified property. They supplement municipal regulations. Covenants can apply to virtually every aspect of industrial development including site coverage, architectural design, parking requirements, signs, and landscaping. Covenants can be filed either with the plat or with each deed. Through the use of covenants, the developer can upgrade the community's industrial development standards by setting a successful example. It should be noted that attorneys representing some industrial firms prefer such standards to be part of the public law, and publicly enforced. Citizens' groups also tend to prefer that industrial land standards be upgraded through amendment of zoning codes, with enforcement provided under public law.

There are several reasons why covenants represent a valuable industrial land development control tool and are frequently preferred to zoning regulations. Addition to or modification of the adopted text of the official zoning ordinance usually requires a considerable amount of time for prepara-

[14] Montgomery County, Maryland Zoning Ordinance. Section 111-25 Town Sector Zone.

tion, discussion, revision, public hearings, more revision, and final adoption of the text by the local legislative body. A developer may not be able to delay his investment for a choice location until: (1) a revision to the zoning has been adopted; and, (2) his re-zoning application for a zoning *map* amendment goes through the usual progression of conferences and public hearings. An industrial client, wishing to build as soon as possible in the industrial park, can seldom afford such a long delay.

Another reason for the use of covenants is that the zoning text, which sets standards for an industrial park zone, must be suitable for general application in the city or county. For this reason, other developers or land owners may object to the general application of certain standards in the zoning *text* revisions. Furthermore, such general standards in a zoning text seldom protect topographic features which might be unique to a particular site. Thus an industrial park developer, wishing to provide limitations on occupancy or protection of unique aspects of his site, should have detailed protective covenants prepared especially suited to his design concept and to his site.

What should be included in the covenants? Any space standard which the developer thinks is not properly covered in the zoning code may be included: such as setbacks, side yards, off-street parking, off-street truck loading, percent of site covered by structures, and height of buildings. Covenants can decrease intensity of land use, but cannot increase site coverage beyond that provided in the zoning ordinance. Thus, provision of rooftop parking or parking structures in exchange for increased density cannot be accomplished through covenants.

Protection of natural or man-made features such as groves of trees, rock outcroppings, streams, lakes, or historic monuments, can be provided in covenants. Aesthetic matters, such as design approval of buildings or building materials, and provision for landscaping and maintenance, are usually covered in covenants. Of special interest is a set of design standards for signs prepared for Bennett and Kahnweiler for Inverness, their new multi-purpose industrial park in Arapahoe County, Colorado.[15] Inclusion of performance standards covering air, noise, water, glare, radiation, etc., will require the advice of specialists in each field, and usually are not necessary because of existing local, state, and federal regulations.

The developer is cautioned, however, to avoid adopting other covenants until they have been studied for suitability for his particular site. The timing of covenant adoption and land area to be covered also requires special consideration. First, no covenants should go into effect until the rezoning has been granted. The time period for renewal of covenants (traditionally 20 years) should probably be less in view of rapid social, economic, and technological changes. With provision to modify or continue, ten years seems reasonable but might be considered too short. If the site is very large, and is to be subdivided into special development areas with a wide range of land uses (offices/commercial, laboratories and headquarters buildings, light industry, warehouses and truck terminals, heavy industry, and residential), then separate covenants for each area should be considered.

Technical standards and design aspects of covenants are formulated by the planner, developer, and engineering consultant. The final draft should not be prepared until all studies related to the rezoning of the site have been completed. An attorney should assist in the drafting of the covenants and is responsible for preparing the final document in legal form. The attorney should ensure a strong enforcement section in the covenants to protect future occupants of the industrial park, land owners adjacent to the park, and the management of the park. The final set of covenants should reflect and be responsive to reasonable concerns of adjacent land owners, occupants, and citizens' organizations.

Industrial Park Characteristics

Two aspects of industrial parks are often misunderstood: the *land uses* suitable for inclusion in an industrial park, and the optimum *size* of industrial parks.

Use Patterns in Industrial Parks

Manufacturing in Industrial Parks

Manufacturing plants have operated successfully in planned industrial areas for several decades. The range of manufacturing activities in industrial parks is wide: of the 20 major industrial groups covering manufacturing in the *Standard Industrial*

[15] See Appendices D, E, F, and G for sign standards and protective covenants.

Classification Manual,[16] 18 major groups are represented by manufacturing plants in industrial parks:

MAJOR GROUPS:	MANUFACTURING ACTIVITY
No. 20	Food and kindred products
22	Textile mill products
23	Apparel and other finished products made from fabrics and similar materials
24	Lumber and wood products, except furniture
25	Furniture and fixtures
26	Paper and allied products
27	Printing, publishing, and allied industries
28	Chemicals and allied products
30	Rubber and miscellaneous plastic products
31	Leather and leather products
32	Stone, clay, glass, and concrete products
33	Primary metal industries
34	Fabricated metal products, except machinery and transportation equipment
35	Machinery except electrical
36	Electrical and electronic machinery, equipment, and supplies
37	Transportation equipment and parts
38	Measuring, analyzing, and controlling instruments; photographic, medical, and optical goods; watches and clocks
39	Miscellaneous manufacturing industries

Only two SIC major industrial groups in manufacturing have little or no representation in industrial parks:

21	Tobacco manufactures
29	Petroleum refining and related industries

Specific manufacturing plants often found in industrial parks include: biscuit and cracker plants, breweries, extracts and sauces, meat processing, candy, soft drink bottling plants, freezing and canning of specialty foods; pharmaceutical, film processing, color printing, lithography, plastic extrusion, glass manufacture, paper conversion-containers, envelopes, boxes; toys, aluminum casting, luggage, furniture; cutting and forming of steel, precision tools, storage batteries; clothing, bed linens, mattresses, custom draperies; heating and air conditioning equipment, custom built kitchens, electric lighting, switchgear assembly, electronic equipment; small metal parts, jewelry manufacture, office furniture, office equipment, and this list is not complete.

Typically, however, manufacturing activities accommodated within the framework of the industrial park either have efficiently eliminated any deleterious features or were free of them from the beginning.

Industrial parks in the Chicago metropolitan area report that manufacturing is the dominant industrial activity accounting for 55 percent of the firms in the region's industrial parks.[17] Warehouse and distribution uses ranked second—representing 27 percent of the firms.

In contrast to the region-wide dominance of manufacturing in industrial parks, warehousing and distribution was the major use in industrial parks located:

- in the city of Chicago—50 percent of the firms;
- in the western area of suburban Cook County, adjacent to Chicago—64 percent of the firms;
- within a five-mile radius of O'Hare Airport—56 percent of the firms.

While there was no detailed analysis of this geographic concentration of distribution activity, a few conclusions seem warranted:

- Concentration of distribution facilities in Chicago's industrial parks would be expected in a region large in area and population, partly because of Chicago's role as a hub of rail transportation and also because of the highway radii to service both central and suburban markets.
- Dominance of distribution firms in industrial parks within five miles of O'Hare International Airport would be expected because of increasing importance of air freight, both in and outbound.

The significance of the 64 percent dominance of

16 *Standard Industrial Classification Manual,* Executive Office of the President, Office of Management and Budget (Washington, D.C.: U.S. Government Printing Office, 1972). This manual defines and classifies industries, and is used by government, industry, and trade associations in gathering statistics.

17 Ronald R. Pollina, *Industrial Park Development: Locational Factors in Chicago Metropolitan Area* (Chicago: Continental-Illinois National Bank and Trust Company, 1973).

distribution firms in industrial parks located in western suburban Cook County could be attributed to:

- the industrial park developers in this section of the region who offered fast service with ready land and buildings to meet the needs of clients—for sale or lease;
- the Interstate highway system which shifted the geographic center of distribution from the inner area of Chicago to the western section of Cook County for distributors serving multistate inter-regional markets as well as multistate "Chicagoland".

Montbello Industrial Park

Industrial parks dominated by manufacturing include Bayport, Texas; Irvine Industrial Complex, California; and Montbello, Colorado.

Montbello is a 510-acre Denver project opened to development by Bennett and Kahnweiler in 1966. It is part of a 3,000-acre tract annexed to the city which has been comprehensively planned for a mix of residential, commercial, and industrial uses. Montbello Industrial Park has one mile of frontage on Interstate 70. It is served by the Union Pacific Railroad, and is adjacent to Stapleton International Airport. Of 450 acres available for development after subtraction of 60 acres for streets, rail, and utilities, approximately 90 percent has now been improved. Over 100 companies with over 5 million square feet of factory, office, and warehouse space, employ 7,000 persons at Montbello.

While the park has many distribution facilities, 36 percent of the occupants are engaged in manufacturing. The largest tenant is the Samsonite Company which manufactures luggage. Samsonite's one million square feet of space occupies a 100-acre prestige site adjacent to I-70. Other products produced at Montbello include precision metal components for business equipment and scientific apparatus, polishing wax, paperboard containers, power regulators, plastic fabrication, bakery bread, custom kitchen cabinets, envelopes, and folding metal furniture.

Adjacent to the industrial area, and separated by a major collector street, are 2,500 acres being developed as a middle-income residential community. Garden apartments and single-family homes in a wide range of prices are provided. Many employees in the industrial park reside in this area. Also adjacent are a 160-acre office park including a 150-room inn with restaurant, and a community shopping center.

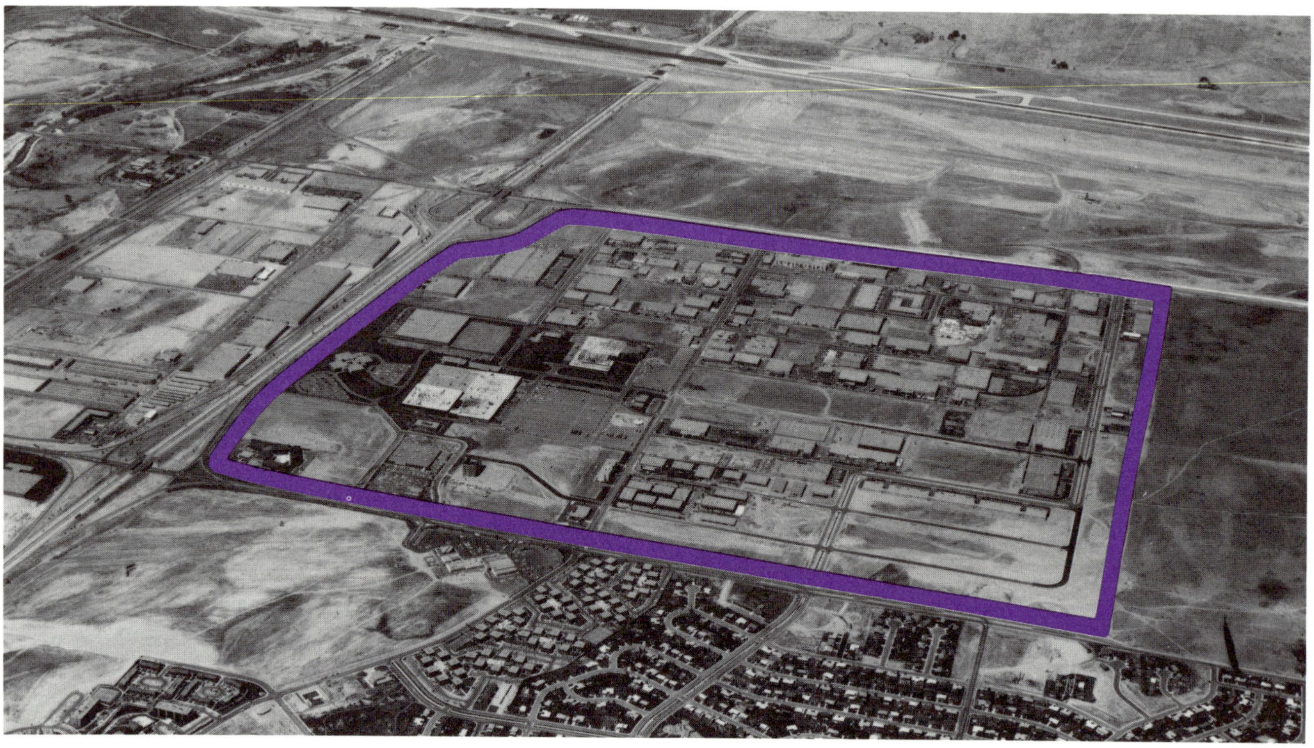

2-7 Montbello Industrial Park is predominately manufacturing in character. The largest single occupant, with frontage on Interstate 70, is the Samsonite Corporation.

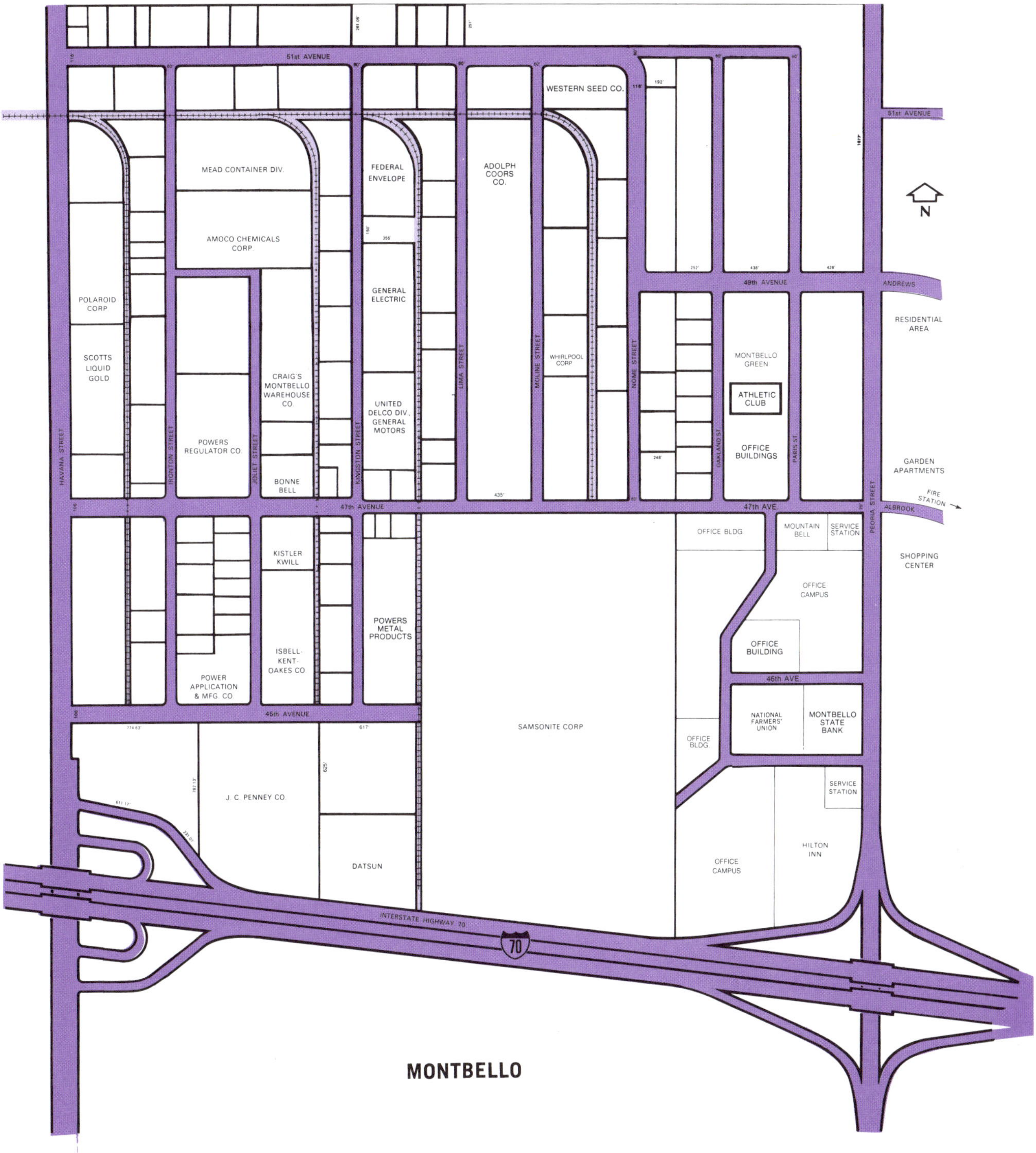

2-8 Montbello Industrial Park site plan.

2-9 A view of the Samsonite facility at Montbello from the freeway.

Warehousing in Industrial Parks

Most industrial parks are dominated by warehousing and distribution activities rather than production facilities. The reason is because these uses can be more readily accommodated within the organizational framework of a planned industrial park. These activities do not typically generate nuisances such as noise, odor, and smoke which are characteristic of certain industrial processes. Proper planning makes possible the accommodation of vehicular traffic which is usually the biggest problem generated by a warehouse and distribution park. In addition, these activities constitute an easily identifiable group of uses because most major material corporations require regional distribution facilities.

Warehousing and distribution functions are characterized by relatively low ratios of employment relative to building coverage. Property taxes generated, therefore, tend to be less than for more intensively utilized industrial property. For these reasons, there is resistance today in some communities to building more warehouses. Locations seeking to expand their employment base recognize that manufacturing pays higher wages and employs more workers. Other communities feel that commercial and office areas have more aesthetic appeal and have less truck traffic. What is needed is a wider understanding that industrial parks, properly located and specifically planned for a mix of uses, can have a pleasant appearance with minimum traffic and other ill effects upon the community.

In order to deal with this problem, it is important to recognize that both warehousing and manufacturing enterprises need balanced industrial parks for all types of industrial activities. The shortage and high cost of land for development, citizen opposition to industrial rezoning, difficulties of obtaining water and sewer extension, the increase in environmental controls, and the shortages in materials and fuels, are all impediments to the "go it alone" industrial site selection.

Los Angeles Industrial Center

The Los Angeles Industrial Center is typical of a planned industrial park which is oriented primarily toward warehousing and distribution. LAIC is 549 acres in size. Of the 391 acres which had been developed through 1974, 210 have been occupied by distributors in warehousing facilities. Light manufacturing accounts for another 80 acres. Ready access to both local and regional markets is a primary consideration for warehouse and distribution tenants. LAIC—located halfway between Los Angeles International Airport and the Port of Los Angeles—was a particular attraction for uses of this kind. The project is bisected by a freeway and has good rail access which permits ready movement of goods. The street system and lot configuration were designed to accommodate the movements of semi-tractor trailer trucks. Minimum site dimensions are 250 feet of depth by 200 feet of width, and individual parcels range from one to 20 acres in size.[18]

2-10 Truck-served warehouse building.

Offices and Other Uses

Offices, commercial functions, and certain public and semi-public uses are emerging as important new elements of the industrial environment. This trend is based upon an awareness that industry generates a demand for goods and services beyond what can be provided by the individual industrial concern.

Public and semi-public facilities now being located within industrial parks include fire stations,

[18] See case study example pp. 31-35.

post offices, police stations, and educational and technical schools. Proximity to the area they serve means that they can provide a higher and more efficient level of service. The city of Compton fire station and police station in LAIC serves the entire park and adjacent municipal areas.

Commercial functions now located within industrial parks provide a number of services beneficial to both employees and visitors. Among the most common are hotels and motels, restaurants, and services such as barber shops, drugstores, and beauty parlors. The nature and extent of these commercial activities will depend on the size of the project and the number of persons employed.

The presence of office buildings and office park areas within industrial parks illustrates the increasingly fine line to be drawn between industrial and commercial activities. The implementation of performance standards and industrial park zoning has had the effect of excluding most industrial uses that would be incompatible with offices. The rationale for combining these functions on a single site is based upon the office services which can be made available to the industrial neighbors. What remains is a mixture of warehouses, distribution facilities, and light manufacturing concerns which are wholly compatible with offices. From small offices built in conjunction with manufacturing operations, this trend has expanded to the point that many industrial parks now contain large office developments. These expanded industrial parks trend toward the *employment center* concept.

Hunt Valley Business Community

Hunt Valley Business Community is an excellent example of a project which was initiated as a conventional industrial park with primarily manufacturing tenants but subsequently evolved into an employment center combining office, manufacturing, and commercial activities within its 250-acre site.

Formerly called the Greater Baltimore Industrial Park, the project is located on Interstate 83 in Baltimore County, Maryland—15 miles north of Baltimore City and 16 miles south of the Pennsylvania state line. Begun in 1962 as a planned development for manufacturing and distribution facilities, recent years have seen the addition of five office towers, the Hunt Valley Inn, and an array of services to accommodate park occupants.

The Hunt Valley Inn, opened in the spring of 1971, provides a broader base of service to persons who are part of this business community. During the early industrial stage, when internal sites in the industrial park were developed with factories and distribution facilities, McCormick Company set aside a 17-acre site at the front entrance to the park for the Inn. Already a successful hospitality and convention center, the Inn differs in many respects from the typical chain motel or commercial hotel often found in larger industrial parks. The Inn contains 182 guest rooms, six suites, and five times more public space than bedroom space. Extensive conference facilities include: ten private meeting/dining rooms; a large room seating 540 for meetings or 400 for banquets, and capable of being subdivided for smaller groups; and a 10,000-square-foot ballroom, which accommodates 1,000 at a banquet or 1,250 for meetings, and also can be used as two separate rooms. These meeting space facilities are used by resident industries to hold employee training conferences, staff and stockholders meetings, and customer or dealer meetings to display and explain new products or services. This public space is only a short walking distance from offices and plants.

The swimming pool and tennis courts at the Inn may be used without cost by any employee in the business park on a first-come-first-served basis, although guests at the Inn have priority.

Other public and commercial services for the park industries are provided in the underground sky-lighted and carpeted walkways which link the Inn with five adjacent tower office buildings totaling 525,000 square feet of Gross Building Area (GBA). These office buildings house a full range of office tenants in a distinctly urban setting which is in no way affected by the proximity of industry. Facilities within the office complex include a cafeteria, snack bars, and classrooms used by the University of Maryland-Baltimore County to serve the needs of the industrial and office employees in Hunt Valley Business Community; and, a stationery store, and fast reproduction and printing service. Full banking service is provided in Executive Plaza I.

Probably the most important service is the full-scale post office to expedite the daily flow of mail generated by offices and industries in the park. The C & P Telephone Company's data processing center, a park tenant, mails over 1,500,000 home and business phone bills for the entire state of Maryland every month. Although Baltimore County has no incorporated cities or towns—Hunt Valley, Mary-

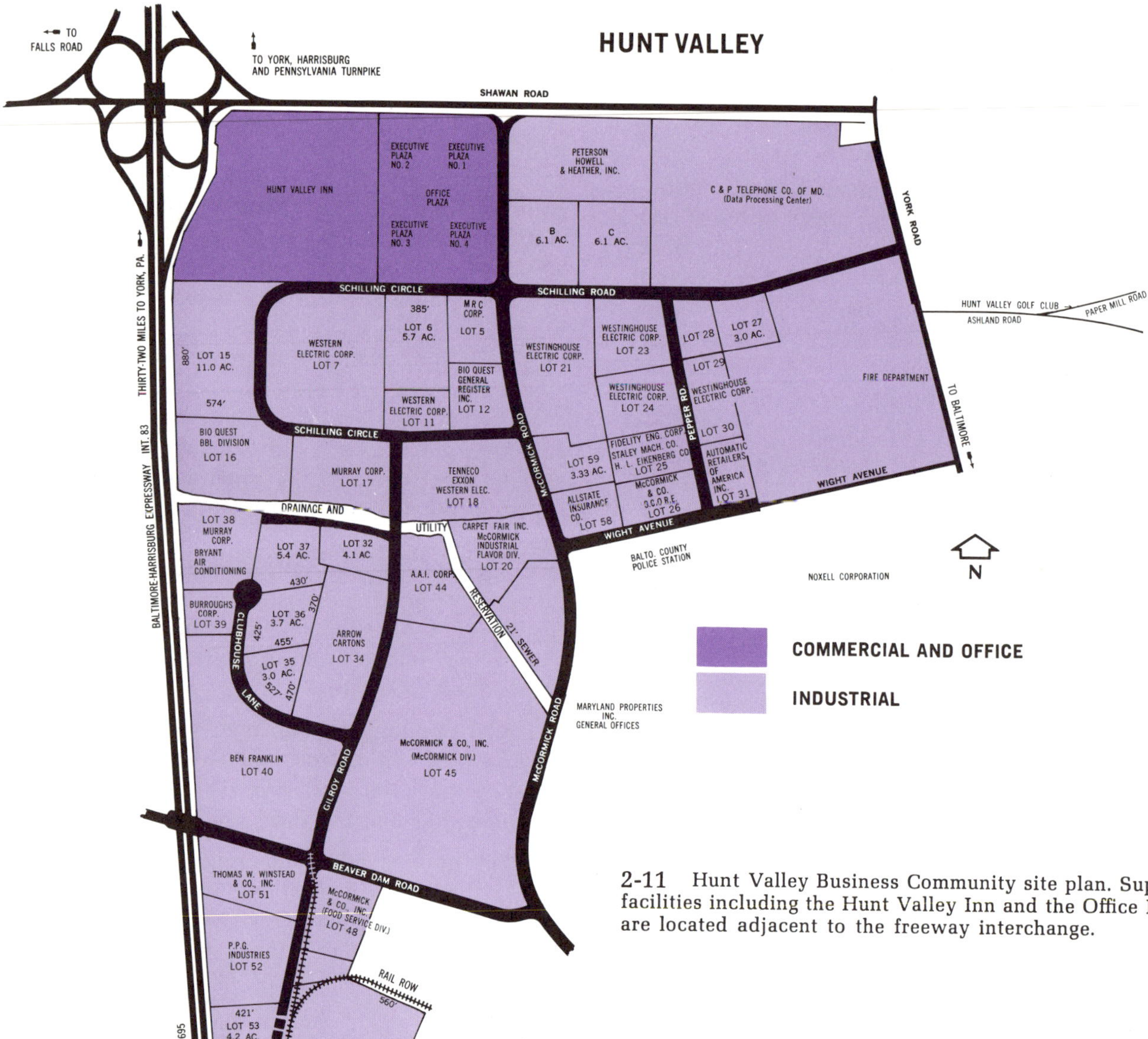

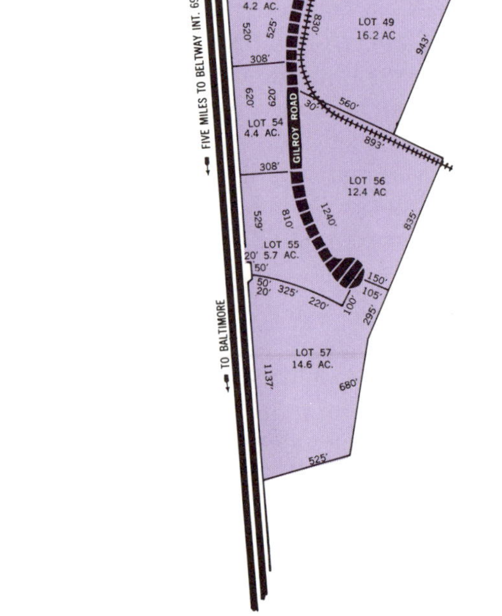

2-11 Hunt Valley Business Community site plan. Support facilities including the Hunt Valley Inn and the Office Plaza are located adjacent to the freeway interchange.

2-12 The Hunt Valley Inn.

land 21031—is the legal mailing address for all occupants of this business community.

The Central Medical Center specializing in industrial medicine is situated on the ground floor of Executive Plaza I. It is comprised of a group of twelve physicians: internists, surgeons, a radiologist, an orthopedic surgeon, a cardiologist, and a physical therapist. The center gives pre-employment physicals and executive physicals.

As should be apparent, the policy emphasis in these facilities is upon services and conveniences to facilitate the daily operation of industries in the park. If a needed service or activity is identified, it can represent a significant business opportunity for the appropriate entrepreneur. Future occupants are selected with care. As yet, there is no shopping center although this possibility is now being considered. Should a shopping center be added, it will be oriented toward serving the business community, and therefore quite different from the typical group of apparel, gift, and food shops.

Size of Industrial Parks

The most comprehensive survey of industrial parks to date, *Industrial Districts Restudied,* contains a comprehensive survey of 272 planned industrial districts. This 1961 report includes information of project by type of sponsor.[19] The acreage of planned industrial districts ranged from a 261-acre average for railroad-sponsored districts, to a 478-acre average for districts sponsored jointly by government-private and/or local community groups. The adjusted average was 319 acres. It should be noted that extremely large and atypical districts were omitted in computing district size average, to avoid distortion.

In 1972, the National Association of Industrial Parks made a short survey of their membership, which included an inquiry about the number of

[19] Robert E. Boley, *Industrial Districts Restudied,* Technical Bulletin no. 41 (Washington, D.C.: Urban Land Institute, 1961).

2-13

INDUSTRIAL DISTRICTS BY SPONSOR AND SIZE CATEGORIES †

Size of District	5-49	50-99	100-199	200-499	500-999	1,000-2,499	2,500-4,999	5,000-9,999	10,000-25,000	Total	Average District (Adjusted)
	NUMBER OF DISTRICTS BY SIZE—in acres										
Private real estate development	16	17	25	12	9	6	2	1	0	88	332*
Railroads	2	20	24	27	5	3	0	1	0	82	261*
Private local community groups	10	9	10	13	2	3	1	0	0	48	294
Government sponsored groups	5	6	4	6	3	3	1	1	1	30	398*
Local government	(1)	(4)	(2)	(5)	(2)	(3)	(0)	(0)	(0)	(17)	(397)
Federal, state, and redevelopment authority, and territorial governments	(4)	(2)	(2)	(0)	(0)	(0)	(0)	(0)	(0)	(8)	(59)
Port authority	(0)	(0)	(0)	(2)	(0)	(0)	(1)	(1)	(1)	(5)	(1,309)*
Joint government, private local community groups	1	4	3	2	4	2	0	0	0	16	478
Other (university utility company, etc.)	2	0	1	2	2	0	1	0	0	8	300*
Totals											
Number of Districts	36	54	67	62	25	17	5	3	1	272	319*
Percent of Total	13.2	20.6	24.6	22.8	9.2	6.3	1.8	1.1	0.4	100%	
Cumulative Percent of Total by Size	13.2	33.8	58.4	81.2	90.4	96.7	98.5	99.6	100		

† This table is a combination of Tables 8, 9, and 10—with the addition of the cumulative percent of districts by size category—from Boley's *Industrial Districts Restudied,* Technical Bulletin No. 41, Urban Land Institute, 1961.

* To avoid distortion, extremely large and atypical districts were omitted in computing district size averages.

parks per member, and the acreage in industrial parks per member. Based on data received in 117 replies, which was approximately 50 percent of the NAIP 1972 membership, the findings were:

- 369 industrial parks, averaging 3.15 parks per member; and,
- 115,273 acres in 369 industrial parks, averaging 312 acres per industrial park.

A comparison of the 1960-1961 ULI study, and the short 1972 questionnaire of NAIP, indicates the average size of industrial parks remained almost the same over the 12-year period: 319 acres (ULI-1960) compared to 312 acres (NAIP-1972).

R. John Griefen, vice president of Cabot, Cabot and Forbes, has compared industrial land acquisitions during the decade of the 1950s on Route 128, the first circumferential outside Boston, with recent industrial land acquisitions along I-495, the outer circumferential route:

> From past experience, CC&F knows that early acquisitions on a new highway tend to be large, but the comparison of Route 128 and I-495 indicated development on I-495 would need to be substantially larger than the average park on Route 128, inasmuch as the acquisition of large sites seemed to be sustained over a period of time, and is supported by companies that want to have sufficient expansion land this time, unlike short-sighted commitments along Route 128, ten to fifteen years ago.[20]

Cabot, Cabot and Forbes' first industrial park, New England Industrial Center, was 140 acres in size and was acquired in 1949, with Route 128 still in the planning stage. More recent CC&F industrial parks are much larger. Two eastern Massachusetts projects—Interstate 290 Industrial Center and Interstate 95 Industrial Center at I-495—are 320 and 750 acres respectively.

Very large projects are not uncommon. Sizeable projects now under development include three by Bennett and Kahnweiler—Centex Industrial Park, Elk Grove Village, Illinois (2,250 acres); Atlanta Gateway Park (3,000 acres); and Governor's Gateway Park, Park Forest South, Illinois (1,000 acres). Evans-Pitcairn's 1,400-acre Iron Mountain Industrial Park in Pennsylvania's Lehigh Valley is five times the size of their well-established East Gate and Mid-Atlantic Parks in New Jersey. On the extreme end of the spectrum of size in industrial parks are the Irvine Company's 6,300-acre Irvine Industrial Complex in Orange County, California and the Friendswood Development Company's (Exxon) 10,500-acre Bayport project on Galveston Bay near Houston.

Although these citations of individual projects are not necessarily a true representation of the trend toward larger industrial projects, they clearly show that projects in excess of 500 acres are commonplace.

Despite the apparent shift toward large-scale industrial developments, it should be kept in mind that very large industrial parks usually are warranted only at unique locations, such as:

- adjacent to major public investment in transport facilities, including:
  confluence of Interstate highways,
  air carrier airports,
  major shipping channel—either deep draft or barge, or in proximity to deep water;
- recent extensions of rapid transit, which justify large employment areas in vicinity of terminals;
- linked industry uses in vicinity of:
  petrol/chemical plants,
  power plants,
  natural resource deposits,
  metallurgical centers; and,
- relatively level land, serviceable by rail, in a region of rugged or otherwise difficult terrain.

Usually, the public interest is best served by a wide regional distribution of industrial parks and industrial sites, rather than a few very large industrial parks, in order to:

- reduce the journey to work;
- avoid large concentrations of industrial traffic; and
- provide a variety of locations with different transport orientations:
  freeway-interstate system,
  rail,
  airport—air cargo, or general
  sea or river port—deep water or barge,
  rapid transit or other public transport services,
  pipeline,
- provide a choice of environments and topography for a wide range of industrial facility standards.

Perhaps more important than average size is range of size and implications of size. In the Boley study, 33 percent of the projects were less than

[20] R. John Griefen, "Creating an Industrial Park," *Lawyers Title News*, May/June (Richmond, Virginia: Lawyers Title Insurance Corp., 1972). Also reprinted in *SIR News* (Washington, D.C.: Society of Industrial Realtors, August 1972), p. 4.

100 acres. At this scale, it will be less likely that the development will include a wide range of non-industrial uses unless their economic support is from users outside the development. The larger project can more readily and more logically achieve the objectives of the planned employment center suggested earlier.

Ultimately, the size of development must be based on market demand which will allow reasonable absorption over a logical development period based on the needs of the community; the constraints of available sites; the costs of acquiring, holding and developing larger projects; and, the need to accommodate new regulations. This latter factor may be either an encouragement toward larger projects or a discouragement—an encouragement because larger scale will more readily allow resolution of environmental and other issues on site; and a discouragement because larger projects may cause communities to place greater burdens for resolution of problems on the developer. Smaller projects would require the community to solve the problems that can be dealt with only at greater increments of scale.

Industrial Development Case Studies

Many noteworthy and innovative industrial developments have been initiated in the United States during recent years. To identify or include them all is not the purpose of the case studies. The following case studies have been selected for two reasons: first, to provide a cross section of various types and sizes of planned industrial developments; and second, within each project development practices or concepts have been highlighted as innovations, examples of solutions to problems or evolving ideas or development, and financing or marketing techniques. In each case study, the statistics reflect conditions as of January 1975.

Los Angeles Industrial Center

(Conventional Industrial Park)

The Los Angeles Industrial Center (LAIC) is a 549-acre planned industrial park which includes warehousing and distribution facilities, light manufacturing, research and development, and limited commercial uses, including a motel, bank, office building, and police and fire stations.

Of the 390.8 acres of the site which have been developed through 1974, 209.8 acres are occupied by distribution facilities; 79.5 acres are devoted to light manufacturing; 95.5 acres have been set aside for streets and a freeway widening project; 3.0 acres are in office use; and 3.0 acres are occupied by a motel. An additional 14 parcels occupying 25.8 percent of the site remain available for development while 16.8 acres, representing 3.1 percent of the site, consist of unusable land.

The site was acquired by the developer in May 1968 and construction began approximately five months later. Almost 71.1 percent of the park is now complete and occupied by 58 companies housed in an aggregate of 4.45 million square feet of building space. Based upon the past rate of growth, LAIC should be virtually complete by the end of 1976.[21]

Location and Site Description

The Los Angeles Industrial Center is situated in Compton, California, an incorporated community

[21] *Project Reference File*, vol. 3, no. 5 (Washington, D.C.: Urban Land Institute, 1973).

2-14 A landscaped boulevard extends the length of the Los Angeles Industrial Center and provides the predominant visual focus of the development.

located between Los Angeles and Long Beach. Downtown Los Angeles is approximately 14 miles to the north, and downtown Long Beach is approximately eight miles to the south. The site is readily accessible via freeways to the Los Angeles International Airport, the Long Beach Airport, and the Port of Los Angeles. East-west access is provided by Artesia Boulevard while north-south access is provided by the Long Beach and Harbor freeways, which have exits onto Artesia two and three miles from the Center respectively.

The LAIC site, which was vacant prior to acquisition by Cabot, Cabot and Forbes, is roughly 1.5 miles long and .7 miles wide. Artesia Boulevard, which bisects the Center, has been rebuilt to freeway standards and has three interchanges providing access to LAIC. The Compton Creek Flood Control Channel, the Pacific Electric Railroad, and Alameda Street form the eastern boundary of the property. Adjacent to LAIC on the north is a low-to-moderate income residential neighborhood, while areas to the west and south are vacant or being used for industrial purposes.

Planning and Engineering Features

The LAIC was master planned from the outset in order to provide an optimal environment for industrial land uses. In addition to devising a site layout, major development considerations included provision of streets and utilities, landscaping, and extension of rail spurs to serve the site.

The LAIC has been laid out on a modified grid system in order to create the square and rectangular parcels best suited for warehouse and industrial development. Minimum site dimensions at the LAIC are 250 feet of depth and 200 feet of width. Individual parcels range from one to twenty acres in size.

There are 5.8 miles of streets in the Center, most of which were built by the developer. Two collector streets intersect near the center of LAIC. One of these, Walnut Street, is four lanes wide and has a series of landscaped median islands giving the impression of an uninterrupted median strip. The second collector, Wilmington Avenue, predated the development of LAIC and was a major design constraint. Local streets are a minimum of 66 feet wide and provide access to all parts of the Center. Most parcels are served by the 3.2 miles of railroad spurs which connect with the Pacific Electric Railroad and the mainline of the Southern Pacific Railroad.

Water is supplied to the Center by the Dominguez Water Company via a looped system of 10-inch and 12-inch mains interconnected to the Los Angeles Metropolitan District. Natural gas is supplied through high pressure mains from local oil fields. The developer installed a complete system of sanitary sewer mains which connect with the main trunk system of the Los Angeles County Sanitation District. The underground storm drainage system drains into the Compton Creek Flood Channel. Fire protection is provided by a 5,000-square-foot station of the Compton Fire Department located within the Center.

Marketing

A prospective occupant of the LAIC has the option of either purchasing a site and building his own facility, or utilizing the total unified services offered by Cabot, Cabot and Forbes for building design, engineering, and construction with financing available on a longterm lease or purchase basis.

A primary objective of the market strategy has been to overcome the generally negative image of the Compton community. When the property was purchased, it was an unincorporated section of Los Angeles County. The developer chose to annex to Compton because it was felt locational advantages and the stability of local government outweighed any disadvantages attached to locating in a predominantly non-white community. Tenants of the park have been encouraged to hire community residents and an estimated eight percent of the LAIC workforce lives in Compton.

In order to capitalize on the Center's ready access to the Port of Los Angeles, and to participate in the rapidly growing trade between the West Coast and the Far East, the developer has initiated an active program soliciting Japanese firms to locate in the Center. All brochure materials have been translated into Japanese and CC&F personnel have carried out solicitations in Japan.

The developer tries to maintain an inventory of between 100,000 to 200,000 square feet of unoccupied space at LAIC. Although the total fluctuates according to market trends, these buildings are constructed as shells and the interior finished to the clients' specifications.

PROJECT DATA

Los Angeles Industrial Center

LAND USE INFORMATION

Site Area 549 gross acres
475 net acres

Land Uses	Acres	Percent
Distribution	209.8	38.2
Light manufacturing	79.5	14.5
Office	3.0	.5
Motel	3.0	.5
Freeway taking	43.1	7.8
Streets, right-of-way, rail, slope, etc.	52.4	9.6
Total land used	390.8	71.1
Remaining usable land	141.4	25.8
Unusable land	16.8	3.1
Total	549.0	100.0

Total no. tenants 58
Average sq. ft. per tenant 76,000

ECONOMIC INFORMATION

Site Cost $14,750,000 [1]
Site Improvement Cost $ 4,200,000 [2]
Land Values:
1969: $1.50 per square foot[3]
1974: $2.25 per square foot[3]

Notes:
[1] Acquisition cost in 1968 does not include taxes, interest, and carrying charges which must be capitalized to arrive at effective land costs.
[2] Estimated total development costs include roads, utilities, planning, landscaping, etc.
[3] Typical land prices on fully improved prepared sites.

2-15

Development Controls and Zoning

Protective covenants have been used by the developer for two reasons: (1) to insure that firms locating within the Center improve their property in accordance with pre-established standards; and (2) to achieve overall compatibility and continuity of architectural design, lot layout, and landscaping. Building coverage is limited to a maximum of 50 percent of the site. Setbacks of 50 feet from major arterials, 40 feet from local streets, and 15 feet from common property lines are required. Two-thirds of the square footage within front setback areas must be devoted to lawns and landscaping and all other open spaces must be similarly landscaped. Parking and loading space, sufficient for the facilities, must be provided on sides of the building not fronting on the street. Architectural controls include a requirement that all buildings be finished with tilt up concrete and there is a prohibition on variations in roof lines visible from the street. Uniform sign standards have been established and objectionable uses are prohibited.

These controls are spelled out in restrictive covenants which are enforced by the developer to whom all plans must be submitted for review.

Financing Program

The acquisition and development of LAIC have been financed by a three-tier program involving three different types of lending institutions.

The acquisition was financed by a sale-leaseback of the raw land which was accomplished through a major national insurance company. The leaseback agreement requires that a certain percentage of the total land area be converted into sites with buildings which will revert to the insurance company at the expiration of the lease. The remaining land may be repurchased from the insurance company on a lot basis in order to be able to make land sales to users who prefer to own rather than lease.

Land development costs have been partially financed through an interim-term loan from a real estate investment trust.

Construction of the very active inventory-building program in LAIC has been financed by the Los Angeles commercial banks. They provide credit lines which are drawn against to construct the buildings and to hold the buildings during the lease-up period. After lease-up, these inventory buildings, as well as build-to-suit structures are financed by permanent mortgages by conventional lenders. These, in most cases, have been California savings and loan institutions.

Developer Observations

The developer, in evaluating his project, offers the following observations which might benefit others involved in the development of similar projects:

- The benefits of a strategic location proximate to freeways, harbors, airports, and rail lines have far outweighed the disadvantages of being situated in a community with a slightly higher tax rate and a variety of urban problems. The developer has been able to maintain good relations with the city of Compton.
- In a project of this size, provision of ancillary facilities such as hotels, restaurants, and five police installations becomes critically important.
- The developer has found that maintenance of close contact with clients is imperative, both to insure that initial requirements are satisfactorily fulfilled and to oversee changing needs for building alterations and expansions.
- The landscaped islands along LAIC's primary collector have been a unique and successful design feature.

2-16 Arcata Communications is characteristic of buildings found in the LAIC. Note the landscaping, berms, and concrete aggregate finish applied to the structure.

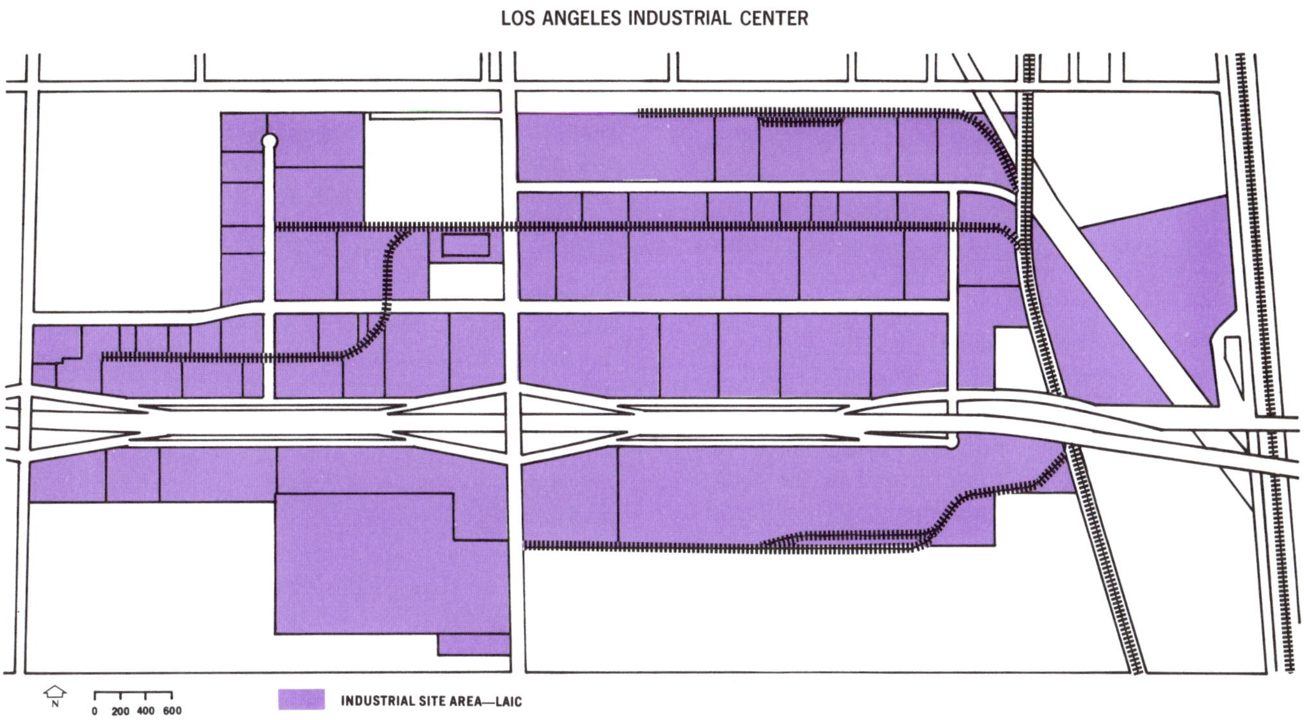

2-17 Los Angeles Industrial Center site plan.

New England Industrial Center
(Freeway-Oriented)

History and Development

The New England Industrial Center (NEIC) in Needham, Massachusetts was one of the first planned suburban industrial parks in New England. This 148-acre tract on Route 128 was developed by Cabot, Cabot and Forbes, a Boston industrial real estate firm which pioneered the concept of fully planned industrial developments.

In 1949, Route 128 was only in the planning stages by the Commonwealth of Massachusetts. The highway was intended originally as a limited access express route to link communities north and south of Boston, circumventing the congestion of the city. Recognizing the development potential of land along Route 128 as a location for expanding business and industry, CC&F assembled an aggregate site of over 140 acres fronting on the highway. For over two years, the firm conducted an intensive educational campaign within the town of Needham to secure the zoning necessary for the establishment of an industrial center.

The NEIC offered to industry the opportunity to acquire large land sites on which to construct new facilities designed for operation in attractive suburban surroundings. Location on a limited-access expressway permitted rapid product distribution to suburban metropolitan markets throughout the Greater Boston area as well as markets in New England, New York state, and points west.

Other special assets of the Center include: easy accessibility to labor resources of the surrounding residential communities; tax rates and assessment policies lower than those in the city of Boston; and highway frontage with excellent advertising value. The developer offered sites in the Center with all utilities installed. Spur track rail extensions were made available.

A prime factor in the remarkable growth of the NEIC was the fact that the developer provided an integrated staff for a comprehensive building program. The one organization offered sites ready for construction as well as such services as building design, engineering, and financing. This coordinated package plan program enabled a company to deal with just one firm for all phases of facilities planning, resulting in savings of both time and money.

The CC&F program incorporated three alternative approaches for prospective occupants:

- a company could acquire a site in the Center and handle its own building design and construction as long as it conformed to the architectural standards established by the developer;
- a company could purchase a site in the Center and have CC&F design and construct a building to the company's requirements and for the company's ownership upon completion; and,
- a company could lease a site and lease a building provided by CC&F to the company's requirements.

Location and General Plan

The NEIC is located at the Needham-Newton town line, only a two-mile drive from the city limits of Boston. The Center has 4,500 feet of frontage on the east side of Route 128, Boston's circumferential highway. Major arterials are situated to the north and south of NEIC, and the Charles River is located to the east. The Center has a direct cloverleaf access onto Route 128, and is only a four-minute drive north of the Massachusetts Turnpike. The substantial labor resources of Greater Boston are within easy commuting distance of the Center via Route 128.

As its initial approach, CC&F set up a master development plan which included establishment of grades and drainage, water and sewerage lines, railroad spurs, roads, electric power and telephone lines (at the rear of building sites), and street layouts. A paramount consideration in the plan was flexibility, an essential factor in a long-range undertaking. Previously, the site had been used as a sand and gravel pit and a substantial amount of site work was necessary to prepare it for reuse as an industrial park. This decision to make an en-

2-18 The Upjohn Company, one of 50 occupants of the New England Industrial Center. The 33,604-square foot building houses approximately 70 employees.

2-19 New England Industrial Center occupies a strategic location on Boston's circumferential freeway, Route 128.

vironmental rehabilitation of derelict land preceded by more than a decade the National Sand and Gravel Association program to educate their member firms about the redevelopment potentials of worked out quarries.

Status of Development

Total area, exclusive of streets, amounts to 148.7 acres of which 140.6 are occupied by a total of 50 different corporations and businesses employing approximately 2,600 persons. Construction began in 1952 and virtually all sites have been sold and built upon. Care was taken to minimize the number of rail crossings at street grades by running the spur lines parallel to the streets whenever possible. Individual parcels at NEIC range in size from 32,000 square feet to 23.7 acres. Total Gross Building Area (GBA) amounts to 1.4 million square feet and the Floor Area Ratio (FAR) is .24, a figure which provides an indication of the single-story building configuration characteristic of suburban industrial parks. Most sites, except those occupied by highway-oriented commercial activities, are rail served. Occupants of NEIC are predominantly composed of sales offices, and warehousing and distribution facilities of both regional and national firms. Over nine acres along Highland Avenue which are adjacent to the freeway exchange, however, have been developed for commercial uses, including a restaurant, bank, motel, and automobile dealership. This area draws support from the industrial park as well as from traffic along both arterial roads and Route 128. An 8.1-acre tract in the center of NEIC remains undeveloped.

Control of Development

In order to ensure stabilization of property values and prevent depreciation and economic loss to the long-term investments of its clients, definite protective provisions were established and incorporated into all leases and deeds. These represent a greater protection than that afforded by local zoning ordinances. Protective covenants are now a commonplace feature of industrial development; but they represent a noteworthy innovation at the time NEIC was established. The Center's protective covenants govern the following conditions:

Setbacks: Building lines are maintained 50 feet back from Route 128 and 40 feet back from all other streets.

Landscaping: A minimum of two-thirds of the square foot area within the setback areas of each building must be in lawns and landscaping.

Light and Air: Provisions are made on each lot to provide adequate light and air between all buildings. Building lines are maintained 15 feet from side lot lines.

Loading Docks: Insofar as possible, facilities for handling all shipping by rail or truck are on those sides of each building which do not front the street.

Construction Materials: To avoid the unsightly appearance and costly maintenance problems of inferior construction and to maintain architectural integration, the exterior walls of buildings are finished with brick.

Plan Approval: To maintain landscaping and architectural compatibility, a primary consideration for the protection of the occupants of a planned industrial development, plans pertaining to construction, landscaping, signs, and alterations are subject to prior approval by the developer.

Uniform Sign Standards: In order to preserve the aesthetic value of buildings and landscaping, minimum standards with respect to signs have been adopted. These pertain to size, type, and location of signs and do not interfere with the effective advertising value of properly designed signs. These standards provide the following restrictions on illuminated signs:

- no letter in a sign may exceed seven feet in height, or project more than nine feet above the top of the highest exterior wall of the building;
- no sign may extend more than 49 feet above the ground nor be more than 50 feet in length;
- no sign may be used which is of the traveling, animated, or flashing light types; and,
- no gaseous tube type sign may have an intensity greater than one foot-candle at four feet from the sign.

Objectionable Uses: No open storage is permitted. No use is allowed which is objectionable by reason of noise, odor, vibration, smoke, or the hazardous nature of the operation. Care is taken to select industrial neighbors whose operations are compatible.

Floor Area Ratio: Provision is made for a .50 FAR. This assures sufficient space for parking, loading, and unloading facilities without using adjacent streets. It also encourages the reservation of additional land for future plant expansions.

Future Adjustments: In order to permit minor variances where desirable and not detrimental to the Center, and to permit adjustments that may become necessary to maintain values over the years and meet changed conditions, exclusive enforcement rights are held by the trustees of the Center rather than by neighboring occupants. This provides for greater flexibility should the need for minor variances ever develop.

Developer Observations

- Cabot, Cabot and Forbes' decision to locate the Center in Needham was predicated on the advantages offered by Route 128. Since NEIC's establishment, the firm has developed a total of fifteen industrial centers along, or in close proximity to, Boston's circumferential beltway.
- This strategic location at a point midway between New England areas north and south of Boston provides easy trucking access to these regions. Also, the Center is ideally positioned for access to New York and the West because of its proximity to Route 9 and the Massachusetts Turnpike. Another advantage is the excellent source of employees for the plants within the Center, which is provided by residential growth in surrounding suburbs.
- The beltway offers distinct benefits both in regard to production distribution and employee convenience—the latter greatly aiding firms in the Center in the recruitment and retention of desirable personnel.

- Certainly, the economic impact of this limited-access expressway must be accorded its rightful significance in connection with the development of the New England Industrial Center; but it would be a mistake to attribute the success of this development solely to the presence of Route 128. Many other vital factors are involved. These include: sound planning and careful control of the Center through protective covenants legally incorporated into all leases and sales agreements to guarantee the preservation of attractive surroundings; provisions for utilities, rail and trucking facilities, parking, and expansion; and a package plan which offers a variety of alternative sale, lease, and service arrangements.
- One test of the success of the location, site planning, and management-maintenance practices of an industrial park is the continuity of the original occupancy. Comparison of the companies located in New England Industrial Center in 1960 with a current list of occupants shows that only four of the 37 firms in the 1960 list have subsequently vacated their premises. Two retained ownership and subleased, and two sold. In each case, other industries already located in the park acquired the vacant buildings for expansion of their activities.

2-20 NEW ENGLAND INDUSTRIAL CENTER

CLIENT	USE	NO. OF EMPLOYEES	SQ. FT.
The Harris-Fandel Company (rail)	Distribution	30	26,110
Howard Johnson's Restaurant	Restaurant	35	3,160
Firestone Tire & Rubber Company (rail)	Sales and distribution	45	121,500
The General Tire & Rubber Company (rail)	Warehousing	23	39,000
General Motors Corporation (rail)	Office	30	4,560
Decatur Hopkins Bigelow Dowse Company (rail)	Distribution and warehousing	150	116,000
Abbott Laboratories	Warehousing	18	22,900
A. R. Booth & Company	Sales	6	1,560
Berejik Motors, Inc.	Sales and service	20	12,500
American Can Company (rail)	Manufacturing	450	413,000
B. F. Goodrich Company	Sales	30	12,000
The Upjohn Company	Office	70	33,604
Union Carbide Corporation	Sales	90	21,400
GTE Sylvania (rail)	Electronics	1,300	357,500
The Singer Company	Office	24	32,620
A. Sandler Company	Warehousing and office	100	80,080
RCA Corporation	Sales	29	5,800
Norfolk County Trust Company	Bank	10	3,600
Needham Motor Inn	Hotel	35	39,416
Goodyear Tire & Rubber Company	Sales	30	4,545
Manpower, Inc.	Office	4	1,100
MacIntyre, Fay & Thayer	Office	60	8,827
Macdonald-Wadman Company	Distribution	20	10,300
Brunswick Corporation	Laboratory	10	13,087
Merck & Company	Warehousing and shipping	30	25,500
Kraft Foods (rail)	Distribution, sales, and office		60,000
TOTAL		2,649	1,469,669

2-21 New England Industrial Center, Needham, Massachusetts site plan. This park pioneered suburban freeway locations for industrial development. To the north of the NEIC across the Charles River is another CC&F development, the Newton Industrial Center.

Stanford Industrial Park

(Research Park)

Stanford Industrial Park, located approximately 32 miles south of San Francisco, is perhaps the most well-known research park in the United States. This 770-acre project is being developed as part of a program for increasing the revenues from the original 8,800 acres of land granted to the University at its founding in 1885 by Senator Leland Stanford, on condition that the land could never be alienated. During more than two decades of continuing development, almost 80 percent of total site area has been leased to over 80 tenants. At the end of 1974, approximately 16,000 persons were employed at the Stanford Industrial Park. A majority of tenants are in scientific, technical, and research-oriented industries. Many tenants have links with the University. The site of Stanford Industrial Park was used for little more than cattle pastures until after World War II when the attractive peninsula south of San Francisco became one of the fastest growing areas in the United States. The University property was being surrounded rapidly by new commercial, residential, and industrial developments. Nearby cities were edging their boundaries toward the western foothills. At the same time, inflation and the higher costs of education were straining the University's budget. The opportunity was identified for developing the income-producing potential of the land.

In response to these development pressures, a master plan was drawn up initially designating 210 acres of University property for industrial development. In 1953, a faculty advisory committee on land and building development was organized to make recommendations to the University president on all land and building proposals. Actual industrial construction began two years earlier when Varian Associates, an electronics firm, leased 10 acres in the southeast corner of the property. Subsequent expansions have brought the park to its current size of 770 acres.

Development Policies and Procedures

The Stanford University Board of Trustees laid down general policies to guide the University's land development program in 1954. Among other things, the policy stated "that the aim of the development shall be to produce in the ultimate a community of which the University Trustees and all those who have its welfare at heart can be proud and that will, by reason of the fact that it is a University project, serve in an important way as an educational example in the field of community development."

Zoning provisions for Stanford Industrial Park have been developed cooperatively between the city of Palo Alto and the University. The park falls within two different zones—the (LM) and (LM-5) Light Manufacturing districts, of which some of the most important provisions are:

	LM	LM-5
Minimum lot size	1 acre	5 acres
Maximum building coverage	40 percent	20 percent
Open green area	—	30 percent of total lot area
Off-street parking	1 space/250 sq. ft. gross building floor area	1 space/200 sq. ft. gross building floor area
Minimum front setback	50 feet	100 feet
Minimum rear and side yard setbacks	20 feet	40 feet

Use of the LM-5 zone, which provides for reduced industrial densities, was necessitated by the extension of development into steeper terrain south and west of Junipero Serra Boulevard. For parking space, the requirement of one car space for each 200 square feet of gross building area has proven more ample than needed in the warehousing operations of the publishing and printing house tenants. Surfaced streets in the development are 40 feet wide on 60-foot rights-of-way; sidewalks are five feet wide.

Operation and Management

The development of Stanford Industrial Park is managed by the Land Development Office, which functions under the University's Vice President for Business Affairs. To attract tenants, the University depends on the prestige of association with the University, the beautiful location, and the pleasant climate of the San Francisco peninsula. Since the Park's inception, the University has accepted members into the industrial community with great care and as development progressed, the standards for admission have become even more rigid. If a would-be tenant passes the preliminary and informal tests, he is invited to submit an over-all plan detailing the type, size, and location of buildings, along with plans for setbacks, roads, off-street parking, and green areas. The plans are considered by the Land Development Office, the University Planning Department, and then by the President's Faculty Advisory Committee on Land and Building Development before being submitted to the President and the Board of Trustees for review and approval. The University Committee on Land and Building Development composed of six faculty members, three students, and two staff members, represents an accommodation to the demand for greater student participation in the operation of the University.

Clients retain their own architects for building and site development plans. The final blueprints approved by the University are incorporated into the lease. As construction proceeds, the Land Development Office supervises the job to insure its completion according to the exacting standards specified.

Originally, industrial land was leased for 99 years, but when demand mounted, the length of leases was reduced to 51 years. The University recognizes that shorter leases will permit an earlier recovery of the land in the event of changing expansion needs of the University. The fact that the land reverts to the University on the termination of the lease gives the University an interest in the property throughout the life of the lease. Currently, the prepaid lease fees range from $30,000 to $175,000 per acre and are added to the University's endowment funds. The lessee pays for all taxes,

2-22 Stanford Industrial Park—a primary orientation toward research and development.

2-23 Park occupants provide bicycle racks for their employees.

assessments, utility extensions, and service. Utility systems—including gas, water, electricity, and sanitary sewer—are owned and operated by the city of Palo Alto.

The University must explicitly approve any additions or material modifications of the exterior design of buildings. Tenants must keep and maintain neatly and in good repair all buildings, grounds, sidewalks, roads, and landscaped areas and are required to prevent the development of nuisances. Lessees may not remove earth or trees, except by permission, or place any sign or advertisement on roofs, exterior walls, or grounds.

Observations

Stanford Industrial Park has become one of the best known research-oriented industrial developments in the United States. Nevertheless, like nearly all other such developments, Stanford does not cater exclusively to research firms, but in fact has a wide variety of tenants and operations.

Despite the restrictions imposed—land cannot be purchased by its occupants—and the lack of active solicitation on the part of the University, Stanford Industrial Park has attracted many desirable tenants. None of the tenants has required rail service, a not uncommon circumstance among operations which do not produce high-bulk, low-value products.

Possibly, the most unique feature of the Stanford Industrial Park is its sponsorship by an institution of higher learning. The University has done a commendable job in creating a community asset which not only benefits financially the city of Palo Alto, but through careful planning and rigid controls also fits harmononiously into the existing land use pattern.

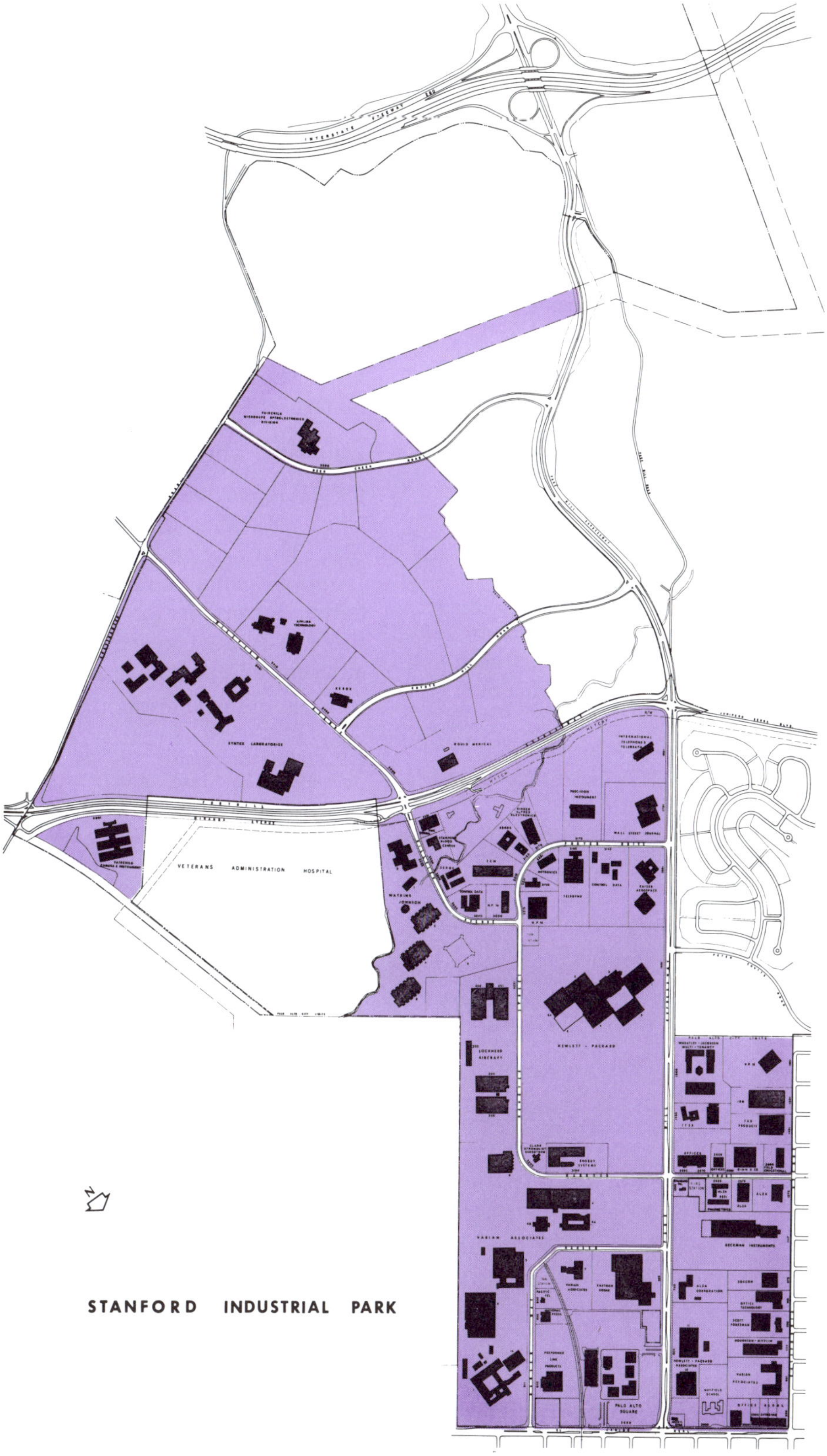

2-24 Stanford Industrial Park site plan. The adjacent Stanford University campus is situated to the north and west of the industrial park. (The SE corner of the page.)

Cross Westchester Executive Park

(Multi-Tenant Occupancy)

Many companies have space requirements which fluctuate rapidly or are too small to warrant construction of their own building. For such firms, the opportunity to occupy 2,500 to 5,000 square feet of space in a large multi-tenant building, which has been designed for maximum flexibility, can be very attractive. To date, however, most industrial parks have been composed primarily of free standing, single tenant buildings and little attempt has been made to fully accommodate the small firms. For the most part, they must find space in older industrial districts. Cross Westchester Executive Park, a 90-acre suburban light industrial-office development, combining single and multi-tenant buildings, constitutes a noteworthy example of how small industrial firms can be accommodated in a comprehensively planned campus-like setting.[22]

Started in 1967, CWEP now has 190 companies occupying 29 buildings on a 90-acre site. Over two-thirds of the 1.5 million square feet of gross building area (GBA) is in multiple occupancy. The 19 multi-tenant structures average between 20,000 and 90,000 square feet of GBA and contain between three and 20 tenants each. A 100-room motel is located on the site and there are nine single-tenant buildings.

The initial development concept envisioned a standard light industrial park configuration. This approach was modified in response to strong market demand for land zoned for light industry but offering quality office space in combination. Since the developer built the structures himself, there was no need to rely on the credit rating of prospective tenants to obtain financing as is the case with many industrial parks.

Multi-tenant space is completely self-contained with separate heating, air conditioning, utility meters, entrances, and loading facilities.

Location and Site Description

CWEP is located in suburban Westchester County, New York near the intersection of the New York State Thruway and Cross Westchester Expressway —approximately 40 minutes from downtown Manhattan. Several major corporate headquarters facilities are located nearby. The park is immediately adjacent to a 95,000-square-foot shopping center, a condominium apartment townhouse development,

[22] *Project Reference File*, vol. 2, no. 20 (Washington, D.C.: Urban Land Institute, 1972).

2-25 Westchester Plaza Campus—six single-story, multi-tenant buildings totaling 225,000 square feet at the Cross Westchester Executive Park.

PROJECT DATA

Cross Westchester Executive Park

LAND USE INFORMATION

Site Area 90 acres
Building Coverage 36.5%
Gross Building Area (GBA)

Office	490,000 sq. ft.
Warehousing	395,000 sq. ft.
Manufacturing	615,000 sq. ft.
Total	1.5 million sq. ft.

Buildings

Single occupancy	9
Multi-occupancy	19
Motel	1
Total	29

Parking

Warehouse	1 space per 1,000 sq. ft. of GBA
Manufacturing	2 spaces per 1,000 sq. ft. of GBA
Office	3 spaces per 1,000 sq. ft. of GBA

TENANT INFORMATION

Total number of tenants 190
Number of multi-tenant occupants 181
Average size 5,300 sq. ft. (average GBA)
Number of single tenants 9
Average size 56,000 sq. ft. (average GBA)
Largest tenant 100,000 sq. ft. GBA
Smallest tenant 800 sq. ft. GBA

ECONOMIC INFORMATION

Land costs $100,000 per developed acre
Site improvements $20,000 per acre

LEASE INFORMATION

Free-standing, built-to-suit warehouse space rates vary from $2.25 to $2.50 per square foot net lease.

Multi-tenant warehouse space rates vary from $3.60 per square foot with 10 percent office to $6.75 per square foot with more office space for a net usable space lease since no common areas are involved.

Full-service office space rates in a multi-story office building vary from $9.00 to $9.50 per square foot for a modified gross lease.

Building Development Costs
(costs based on most recently completed buildings)

Shell $20.00 per square foot
Office $34.00 per square foot

2-26

and a tract held by the New York City Water Authority. The site, generally level with some gently sloping terrain, was profitably utilized to provide separate entrances and concealed loading areas for single-story structures. Road widths are varied to restrict truck traffic to certain areas while providing adequate circulation for automobiles.

Planning and Design

There are unique design and site layout problems, presented by multi-tenant buildings, serving a multiplicity of users varying from low employee density warehouses to full service office tenants. Most difficult to accommodate is separate office-type entrances and screened loading areas while providing flexibility for unidentified future tenants.

Structural design provides for both interior and exterior flexibility through a system of interchangeable modular panels. These components can be shifted to provide new entrances, loading areas, fenestration, and internal partitions as tenant requirements charge. Most individual units are separated by means of steel stud partitions faced by layers of gypsum board. No footings are required masonry walls have been installed, but these too for these walls. Where sound is a problem, fixed can be removed if a tenant requires. Interior spaces are designed to permit consolidation of office, manufacturing, and warehousing activities under one roof.

Structures are sited to avoid loading areas and entrances of separate buildings facing one another. In addition, attractive landscaping, split-level buildings on sloping terrain, and interior loading docks effectively screen most loading areas. There are 40-foot setbacks along the public right of way. A 200-foot setback was required along the bordering residential areas. Most buildings have been designed by the same architect to create a unified appearance. Considerable design diversity, however, avoids the monotony of a single style. Facade building designs typical of most industrial park developments were precluded because of the multi-tenant approach, which exposes all sides of the buildings. A unified signage program has been implemented with the developer providing directional and directory signs and permitting, after approval, business logos and ground mounted identification signs of up to six square feet in size.

Building design and layout have changed as the

2-27 **A single tenant office and light manufacturing facility with 80,000 square feet.**

development process has proceeded. Continuous window walls have replaced brick and other solid panels on the front and sides of buildings. Straight-in-line loading areas were changed to interior loading courts in U-shaped buildings. Depths of more recent buildings vary from the 100-foot standard. One wing of the U-shape might have the 100-foot depth, while the other wing has a depth of 140 feet or more and the base of the U might be as shallow as 60 feet to accommodate small spaces and spaces requiring an expanse of exterior wall. Ceiling heights in the deep wing have been raised from the original 14 feet to 23-foot clear to permit the use of forklift equipment and the stacking of pallets and containerized shipments.

Marketing and Development

A key theme in marketing is diversity of size, function, and hours of operation. Available tenant space requirements range from 800 square feet to 100,000 square feet with the average multi-tenant building user occupying approximately 5,300 square feet. The variety of tenants provides mutual benefits to each and the 24-hour operation of some firms serves to reduce considerably parking requirements and peak-hour traffic congestion. The 24-hour use of the park also minimizes security requirements.

The marketing approach stresses the fact that the small user can obtain the same features of an industrial park which are, for the most part, available only to large space users in traditional parks. Separate entrance and loading facilities are a particular attraction to the small firm which cannot justify the development of a free standing site, but wishes to avoid the uneconomical bottlenecks of common facilities in standard multi-tenant buildings. The independence provided by separate utilities and the freedom to use the building at any hour are additional market attractions.

Developer Observations

The developer, in evaluating his project, offers the following observations which might benefit others involved in the development of similar projects:

- The success of the project indicates a large market for small space-multifunctional users which are largely unserved by standard industrial parks. The space requirements are substantially different from both established standards and rules-of-thumb and must be carefully analyzed and considered in early planning.

- It is beneficial to set up a development feedback system so that tenant needs, construction and maintenance problems, and other difficulties are brought to the attention of the architects and others on the developer's team and incorporated in later building designs.

- Initially, parking for executives was placed at the front entrances and employee parking was massed in a parking lot which took advantage of a 200-foot screening strip mandated by the planning board. This worked well for the first buildings which were adjacent to this parking area. With subsequent buildings, areas of congestion developed due to the unwillingness of employees to walk to parking, even when the distance was less than 400 feet. In later designs, parking was distributed far more evenly, minimizing the walking distances.

- Balancing tenants by their hours of operation and functional space requirements provides a more economical use of land because of reduced parking requirements. Security needs have been minimized. The staggered hours have eliminated almost totally peak-hour traffic problems.

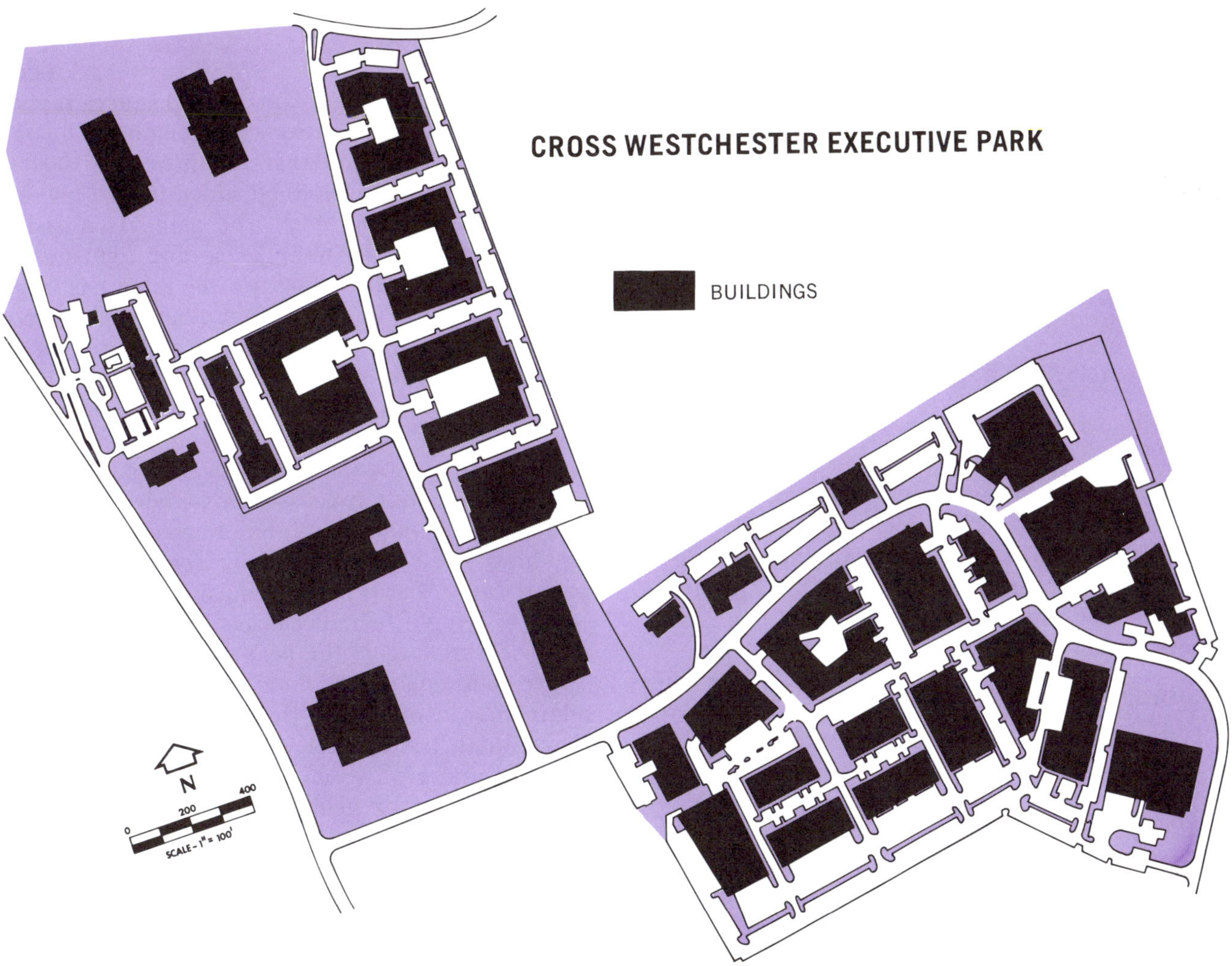

2-28 Cross Westchester Executive Park site plan.

Anchorage Business Park

(Industrial Development in the Smaller Community)

Since the mid-1960s, industrial developers have found a particularly favorable climate for industrial park development in smaller metropolitan areas (100,000 to 250,000 population). In major urban centers, management, land, and labor costs tend to be high, but because of the competitive situation, the developer is not always able to offset these higher costs with adequate rental income and therefore, economic value. Demand for industrial land also tends to be cyclical because it is tied to expansion and in-migration of new industry. In less highly developed locations, on the other hand, there is often a substantial, unmet demand for improved industrial space because much existing space is functionally obsolete. By providing a better industrial environment, the developer can attract many firms already located in the community; therefore, he does not have to rely so heavily upon in-migration and expansion. It is also possible to establish somewhat higher rates than

2-29 Industrial development typical of the Anchorage, Alaska area.

those prevailing for existing industrial space. Likewise, planning commissions and city councils tend to be sympathetic if it is apparent that the developer is attempting to satisfy a local need and has taken the municipality's requirements into consideration.

Anchorage Business Park

Anchorage Business Park, occupying an 18-acre site in the city of Anchorage, Alaska illustrates many opportunities inherent in developing within a smaller community. It also exhibits many features which are highly adaptable for use in smaller industrial parks. Despite the impetus to rapid growth provided by the North Slope oil strike, most industrial activity in Anchorage is located in somewhat obsolete facilities. In part, this has been due to the fact most industrial land is owned by the state and federal governments and leased to tenants for maximum five-year periods. Occupants, therefore, have little motivation to improve their property. A further rationale for well-planned suburban industrial/office park development has been provided by the rapid growth, lack of parking, congestion, and higher costs characteristic of Anchorage's downtown. The industrial developer, therefore, has an opportunity to provide space for professional firms, printers and sales offices, and similar organizations which do not have to be situated downtown.

Development Strategy, Design, and Engineering

At the outset, it was determined the development strategy for Anchorage Business Park would be to aim for both the office and industrial markets by developing a park consisting of one-story buildings adaptable to a combination of offices, light storage, and industrial use.

Development began in 1973. Two phases comprising seven buildings with 100,000 square feet of GBA (gross building area) were completed by the end of 1974. Five additional buildings with 100,000 square feet of GBA are to be added in 1975, which will complete the project. The plan visualized a 50/50 breakdown between office and light manufacturing space. In actuality, the ratio has been 75/25 in favor of office users. A 10,000-square-foot restaurant/cocktail lounge will be built in 1975 on one corner of the park. A two-acre site has been set aside for a banking/general commercial building.

Owing to the rigors of the Alaskan climate, the entire construction process must be telescoped into a four and one-half month building season extending from mid-May to October. The entire site is underlain by peat (tundra or muskeg) to a depth of three to twelve feet. Because of cost, it was decided to surcharge[23] the peat under parking areas rather than replace it with mechanically compacted nonfrost-susceptible fill. Under building slabs and foundations, however, it was necessary to remove the peat and replace it with nonfrost-susceptible fill. Two inches of rigid insulation was applied to all exteriors from the frost line at 42 inches below grade and extending to the roof line.

Tilt-up concrete panel construction was used. Both the front and rear of buildings are substantially the same except when truck doors have been provided in the back of building areas being used for light storage.

Developer Observations

The developer, in evaluating his project, offers the following observations which might benefit others involved in the development of similar projects:

- Developing in an area where there is no local back-up for most construction materials demands strong inventory control, larger inventory, and purchasing management. Almost all items used in construction are ordered, and in some cases fabricated, in Seattle and shipped by barge or hydrotrain from Seattle—a two-week proposition once the material is on the water. Given the short building season, careful planning and scheduling is essential.

[23] This is a method of forced settlement or compaction using soils or other granular material, piled on top of undisturbed ground, left for a period of time, and then removed.

- From previous experience in Los Angeles and Seattle, the owners determined they could pre-install 25 percent offices and market the facilities. However, it was soon apparent that the market was not deep enough to design for only a certain segment that would fit the developer's plans. The park is now expected to be 75 percent office space. Ninety percent of all offices are custom installed, with the developers experiencing order changes with almost every tenant.
- To develop in an area as far away as Anchorage, strong local representation is necessary—usually in a more structured manner than an employee-employer relationship. In this case, the developer chose a local partner, who shares in a substantial portion of both profits and losses. Further, local talent is used whenever possible. At Anchorage Business Park, the contractor, soils and civil engineer, and legal and accounting personnel are all Alaskan. The architect and structural engineer are from Seattle, but have a history of Alaskan work.
- Pioneering in a market creates some special problems. The product was the first business park in Anchorage; therefore, an educational program was necessary. Scale models of the development were made and are displayed on a rotating basis at the airport and in the lobbies of the two major hotels. The owners sponsor the morning stock market report and a world news radio broadcast. These, along with a sales brochure and personal calls on business, have provided the primary entry to the market.

PROJECT DATA

Anchorage Business Park

LAND USE INFORMATION

Site Area 18.0 acres

Gross Building Area (GBA)

Business Park	180,545 sq. ft.
Commercial	26,450 sq. ft.
Total	206,995 sq. ft.

Parking

Business Parking	576 cars = one space per 313 sq. ft. of GBA
Commercial	172 cars = one space per 154 sq. ft. of GBA
Total	748 cars

Land Use Plan	Acres	Percentage of Project
Boulevard	2.5	13.9
Building Area	4.8	26.6
Parking	7.5	41.7
Landscaping	3.2	17.8
Total	18.0	100.0

2-30

2-31 Anchorage Business Park provides a modern setting for light industrial and office uses. Free-standing buildings front on a central boulevard and have extensive areas of surface parking.

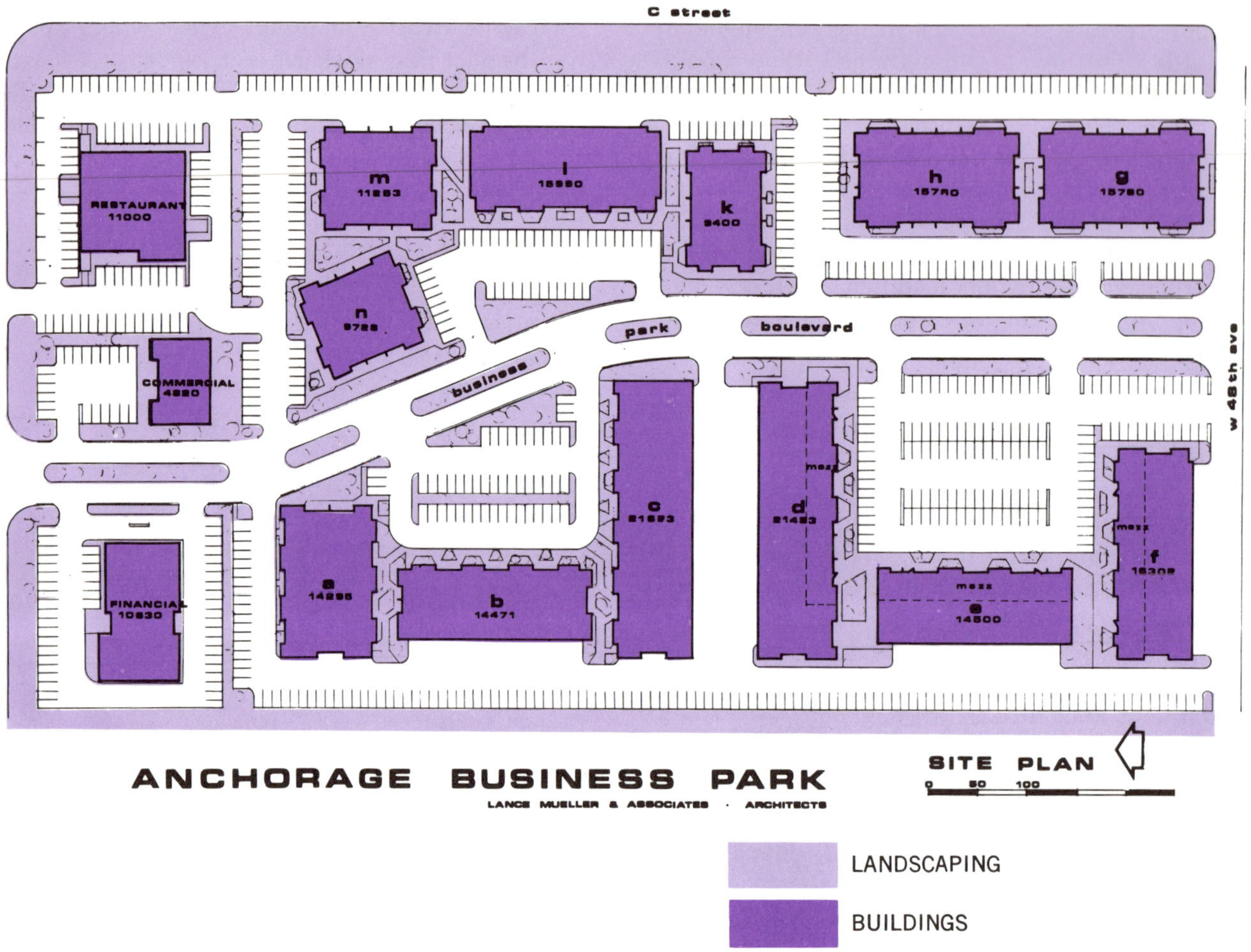

2-32 Anchorage Business Park site plan.

Koll/Irvine Center

(Industrial Condominium)

General Description

Koll/Irvine Center is a 51.4-acre condominium development offering to industrial land users the advantages of property ownership combined with the conveniences of association provided maintenance and administration. One of the first such projects in the United States, Koll/Irvine Center was inspired by the success of the condominium concept in residential land development over the past few years.

Koll/Irvine Center is comprised of 101 attached and detached units ranging from 4,000 square feet to 8,000 square feet in size. Purchasers acquire only their building and the land under it. The remainder of the site—70.5 percent of the total land area—includes landscaping, parking, and loading areas. It is owned and maintained by the community association.

The project was conceived during 1971. Implementation began in mid-1972. Actual construction was started in 1973, and three of four phases were complete by the end of 1974. The final phase, comprising 22 units, is scheduled for completion in August 1975.[24]

[24] *Project Reference File,* vol. 4, no. 19 (Washington, D.C.: Urban Land Institute, 1974).

The Site

Koll/Irvine Center is part of the Irvine Industrial Complex, the 6,300-acre industrial district now under development at three locations on the Irvine Ranch property in Orange County, California. The condominium is located within the largest of these industrial areas, a contiguous 4,000-acre property which has been under continuous development since 1967. More than 250 acres of industrial land are being absorbed within the Irvine Industrial Complex each year.

Koll/Irvine Center is located at the intersection of Red Hill Avenue and the San Diego Freeway. The Orange County Airport is adjacent to the property on the east. Other industrial uses are found to the south and west.

Site Planning and Building Design

The plan is designed to project an image of simplicity and order, enhanced by extensive landscaping. Condominium buildings have been sited in a series of U-shaped clusters except where aviation easements require placement of buildings in a linear configuration adjacent to the airport property line. Open space, which defines parking areas and enhances the setting of individual buildings, is distributed throughout the site. Berms have been used to screen parked vehicles from view. A central focus for the open space framework was created by grading and landscaping a mound at the intersection of the two major streets.

Buildings are of tilt-up concrete construction with slab floors. For the sake of variety, roof lines were varied, screens were used to conceal air-conditioning units, and several different surface textures were used on the walls.

Depending upon size, each building unit is provided with either one or two truck doors. The one-door configuration appears to be sufficient for most small condominium owners, although two doors are being used on some units in the final two phases.

Most buildings are separated by an eight-foot gap. When the project was being planned, it was felt that offering detached structures would be one way of fostering the pride of ownership which was a central element in the overall development strat-

2-33 Industrial condominium buildings, free-standing and attached, with units ranging from 4,000 to 8,000 square feet. Purchasers become members of an association which has overall responsibility for maintenance and enforcement of CC&Rs.

2-34 Buildings feature tilt-up concrete construction. Heavy emphasis is placed on compatible exterior appearance, uniform signage, and extensive landscaping.

egy. Experience now suggests that many buyers will accept attached units, and more of them probably will be included in future industrial condominium projects by the Irvine Company.

Condominium Association

Each unit owner is a member of the community association and has one vote. Dues in 1975 were $.03 per square foot per month. Actual operation of the association is carried out by a five-member board of directors—all of whom will be elected by the members as soon as 80 percent of the units have been sold. Duties and responsibilities of the association include maintenance of common areas and building exteriors, provision of water service and trash collection, establishment of a contingency fund, maintenance of insurance policies, and enforcement of covenants, conditions and restrictions (CC&Rs). The CC&Rs provide for the establishment of the association as well as a five-member architectural committee responsible for approval of exterior building modifications. In addition, the CC&Rs specify performance standards designed to maintain Koll/Irvine Center as a location for clean industry and to preserve its compatibility with surrounding uses. Among the restrictions are those which prohibit noxious or offensive activity, exterior storage, and erection of temporary buildings, antennas, and billboards. Warehousing and manufacturing uses which are compatible with these performance standards may operate within the center. Protective covenants drawn up for the entire Irvine Industrial Complex are also in force. Actual zoning is specified in the Irvine Industrial Complex Planned Community District Regulations, approved before the area was annexed by Costa Mesa.

Marketing

The industrial condominium concept is particularly appealing to the successful small businessman who is renting his present space and consequently is interested in becoming an owner. The marketing program emphasized strongly the benefits inherent in property ownership from both the financial and psychological points of view. Both warehousing

PROJECT DATA
Koll/Irvine Center

LAND USE INFORMATION

Site Area	51.4 acres
Floor Area Ratio (FAR)	.29
Parking	3.1 spaces per 1,000 sq. ft. of GBA
Space per vehicle	322 sq. ft.

BUILDINGS

Number of Individual Units	101
Gross Building Area (GBA)	596,100 sq. ft.

Land Use Plan	Acres	Percentage of Project
Building Coverage	13.7	26.7
Landscaping	9.0	17.5
Parking	14.7	28.6
Loading Zones and Local Access	9.0	17.5
Sub Total	(46.4)	(90.3)
Dedicated Streets	5.0	9.7
Total	51.4	100.0

UNIT CHARACTERISTICS

Type	Number of Units	Sq. Footage per Unit	Price per Unit	Building Utilization: Manu-facturing Area	Building Utilization: Office Area	Building Utilization: Percent of Structure Available for Manufacturing
Plan 1	27	4,800 sq. ft.	$109,000	4,384 sq. ft.	266 sq. ft.	91.3
Plan 2	19	6,400 sq. ft.	$139,000	5,986 sq. ft.	266 sq. ft.	93.5
Plan 3	20	6,000 sq. ft.	$128,000	5,586 sq. ft.	266 sq. ft.	93.1
Plan 4	21	8,000 sq. ft.	$158,000	7,586 sq. ft.	266 sq. ft.	94.8
Plan 5a	7	4,000 sq. ft.	$ 92,000	3,734 sq. ft.	266 sq. ft.	90.2
Plan 5b	7	4,150 sq. ft.	$ 95,000	3,884 sq. ft.	266 sq. ft.	90.0

2-35

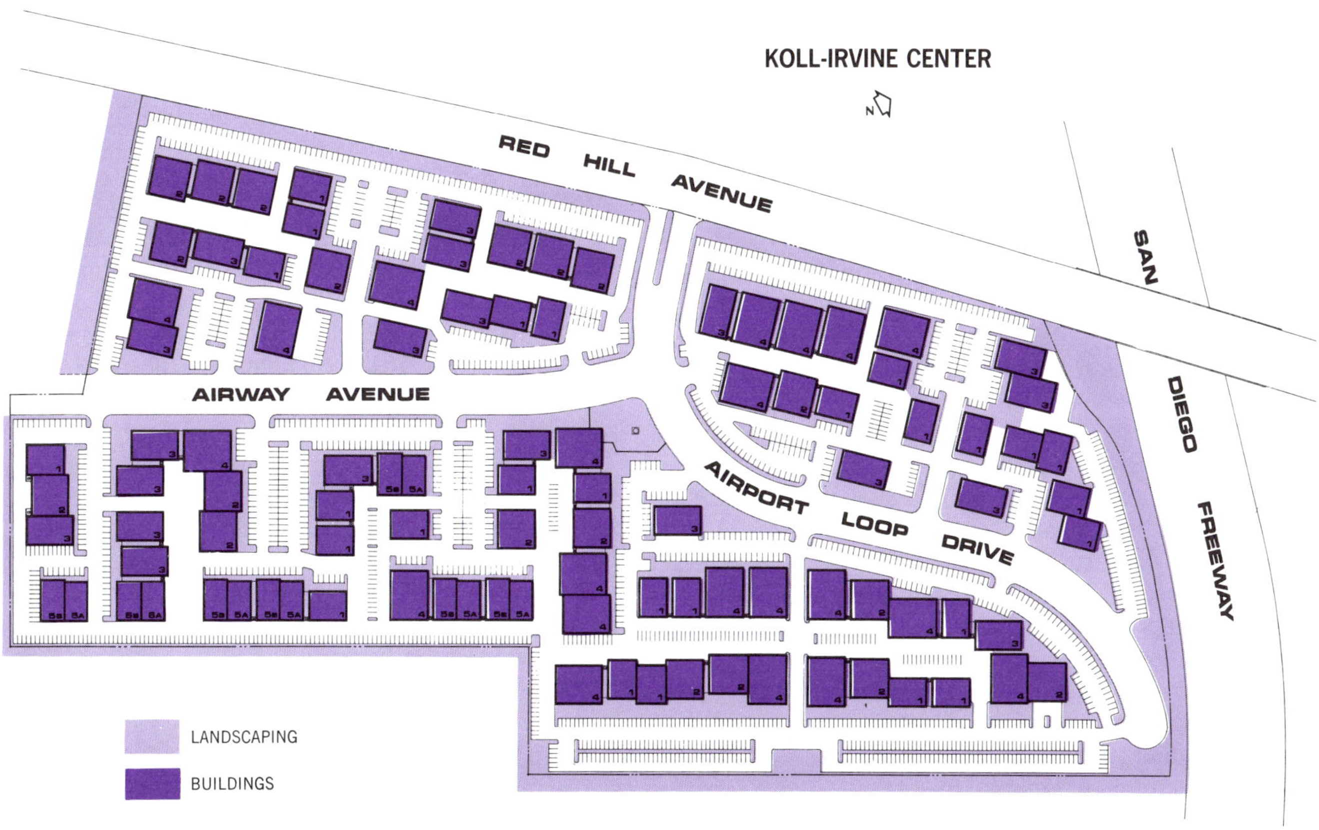

2-36 Koll/Irvine Center site plan.

and manufacturing uses have been attracted to the Center. The developers had anticipated drawing most purchasers from a ten-mile radius in Orange County; however, many more than expected have come from Los Angeles County.

Developer Observations

The developer, in evaluating his project, offers the following observations which might benefit others involved in the development of similar projects:

- The developer considered including a service center with sandwich shop, central copying facilities, and office supply store, but chose to eliminate such a facility in the absence of any clearly identified demand. Likewise, a basketball court shown in the original plan has yet to be built.
- The marketing effort has relied heavily on techniques more commonly associated with residential sales programs. These include use of a model complex, extensive landscaping at the outset, and an open listing policy with outside realtors who receive five percent of each sale.
- The condominium concept is a more efficient way of providing space for small industrial users than the traditional approach in which an entrepreneur erects a single freestanding building in an industrially zoned area. At Koll/Irvine Center, it was possible to provide higher quality buildings at lower cost by taking advantage of the efficiencies of scale. Likewise, the cost of landscaping can be spread over a number of owners which enables parking and circulation areas, which are used in common, to be provided more efficiently.
- Success may be attributed to several different factors. First, is a location within a highly successful and carefully planned industrial environment. Second, are the efficiencies inherent in this type of development. Third, financing is available from the Irvine Company which offers terms with a 15 percent down payment.
- Small units sold better than the larger ones. The number of 6,400 and 8,000 square foot units in Phase III was reduced to take this into account.

Burns Harbor

(Single-Industry Site)

Certain industries—notably primary metals, chemicals and petroleum—must develop their production facilities as independent, freestanding units. This is true for a number of reasons. For example, these industries require extensive sites to accommodate a multiplicity of industrial processes. Also necessary are specialized transportation links which frequently include both water and rail access. Often, plant locations are dictated by the availability of natural resources and an adequate power supply. The manufacturing processes themselves require the establishment of self-contained utility and pollution control systems. Because of these factors, investments for extensive facilities of this kind tend to be far greater than required for other industries. When a corporation must maintain complete control over every aspect of the development process, a single-industry site becomes essential.

Location Considerations

The decision of Bethlehem Steel Corporation to move into the Midwest steel market illustrates many aspects of developing a single-industry site. During the 1950s, a 3,300-acre tract was assembled along the shores of Lake Michigan in Porter County, Indiana. This site is located just east of long-established steel mills in Gary and East Chicago. The general location was chosen for the same reasons the earlier ones had been: a large and growing market, and ready access to basic resources of steel manufacture via highway, rail, and Great Lakes vessels. The site itself met predetermined transportation criteria and afforded an adequate supply of power. Availability of a trained labor force was a secondary consideration as the company chose to bring in a cadre of skilled specialists and train the remainder of the personnel needed.

Located on the western edge of the Burns Harbor site is the Port of Indiana, owned and operated by the state. This deep water harbor is intended to

2-37 An aerial view of Bethlehem Steel Corporation's Burns Harbor plant shows the size of the growing steel complex located on the shores of Lake Michigan in Porter County, Indiana. Bethlehem Steel has invested approximately one billion dollars in Burns Harbor since construction began in 1962.

provide Indiana-based industries with an alternative to the use of Chicago port facilities. The port was opened for operation in 1969. Bethlehem Steel paid for its east harbor area docking facilities, while the outer breakwater, public dockage facilities, and warehouses were financed by a combination of state and federal funds. The U.S. Army Corps of Engineers is charged with the responsibility of dredging the channel to an authorized depth of 27 feet and performing other ongoing maintenance.

The Plant

Construction began in December 1962. By 1975 total investment exceeded one billion dollars. Completed facilities include two blast furnaces, a basic oxygen furnace shop, a coke plant, a 160-inch plate mill, an 80-inch hot strip mill, and sheet and tin mill facilities. Expansion is continuing and a sintering plant, a third basic oxygen furnace vessel, a continuous slab casting facility, and a 110-inch plate mill are now under consideration. When in operation, these facilities will bring plant capacity to 5.3 million tons of raw steel per year.

The original development plan has been modified many times in response to changing steel technology and market demand. This circumstance was anticipated at the outset, and the property acquired was sufficiently large to accommodate virtually any conceivable adjustments necessary. Originally, plans for 10 million net tons of raw steel capacity called for the installation of eight blast furnaces.

At this point two furnaces, which will support over five million net tons of raw steel, are in operation.

As a fully integrated steelmaking complex, Burns Harbor converts basic raw materials, including coal, limestone, and iron ore, into finished steel products. Many of these products are utilized by the heavy construction industry and by automotive, appliance, and beer and soft drink container industries.

Built in several phases, the first portion of the mill to be constructed was the finishing end which was completed in the mid-1960s. Since the plant at that time had no steelmaking capacity (was not fully integrated), slabs required for the rolling mills were shipped to Burns Harbor from other Bethlehem plants located in the East. Then, in December 1969, phase two was completed, giving the Burns Harbor plant the capability of making steel from raw materials.

Geographical proximity of these two processes —steelmaking and finishing—provides important transportation and energy efficiencies, and contrasts with many older plants in the steel industry, where such facilities are isolated from one another. Product flow on the site is generally west to east, with the unloading of raw materials at the harbor on the western edge of the plant and customer shipment of finished steel products from the eastern edge of the plant site.

Total employment at the Burns Harbor facility stands at approximately 7,000. Most of the workforce resides in the Gary-Hammond-East Chicago SMSA (1970 population—633,367) which includes Porter County and LaPorte County (1970 population—106,200). Since most primary metals jobs fall within the semi-skilled range, providing the workforce with requisite skills is not overly difficult. Of far greater importance was locating such a facility in an area with sufficient population to permit assembling and maintaining the large workforce required. Actual impact of the plant on surrounding communities has been ameliorated by the phased development process which has seen slow but steady increases in both plant employment and capacity since the start-up of the first production unit in 1964.

Development Considerations

The site itself is bisected by a limited access freeway, U.S. Route 12, and trackage of the Penn Central Railroad and the Chicago South Shore and South Bend Railroad. Existing plant facilities and those under construction are located mainly in the northern half of the property, which has approximately a mile of shoreline on Lake Michigan. The southern section is earmarked for future Bethlehem Steel facilities and for metal using industries seeking to locate near a primary source of steel.

2-38 Raw materials handling area. Great Lakes ore boats discharge more than 4 million tons of raw ore and limestone each year at the deep water docking facilities to feed the plant's two blast furnaces.

For years prior to development, the site was zoned for industrial uses. The corporate decision to locate in the area met with environmental opposition. The U.S. government has been seeking to acquire nearby properties, including a portion of the site, for inclusion in the Indiana Dunes National Lakeshore. The pre-existing natural environment of sand dunes, marshes, and woodlands, has been preserved in the National Lakeshore, located predominantly to the east of the plant site and extending 13 miles eastward to Michigan City.

Recognizing the environmental concerns, Bethlehem Steel has made a concerted effort to minimize this aspect of the mill's impact on surrounding areas. The task has been vastly simplified by the fact that Burns Harbor has been planned and developed from the outset to be a freestanding, integrated steel plant. Most of the environmental and pollution control features were incorporated in the original designs, unlike older steel mills where such facilities must be added to operational units. Older plants are typically found in fully built up urban areas which impose great problems of space limitation. Further, improvements at these older plants from an aesthetic point of view can be only minimal at best.

At Burns Harbor, a substantial investment has been made for landscaping, including spreading peat on the sand dunes to promote revegetation and stabilization. The peat was carefully saved at the time the site was initially graded. Building exteriors feature a sand beige, chocolate brown, and green color scheme to blend with the surrounding terrain. The retention of a wooded greenbelt between U.S. 12 and the mills provides a visual screen of the heavy industrial uses. The plant office building is located in a wooded area just south of the highway.

Pollution controls include a secondary waste treatment plant which chemically treats water used in the steelmaking and finishing processes before it is discharged into a holding lagoon and ultimately returned to Lake Michigan. Actual volumes of water treated and discharged are minimized since much of the process water is cooled, cleaned, and recirculated within partially closed systems. A sewage treatment plant sufficiently large to accommodate a community of 7,500 persons has been provided.

Investment in air pollution control devices, including baghouses, high-energy scrubbers, and enclosed conveyor belt systems, have minimized particulate discharges from plant stacks and from the movement of raw material from open storage areas. Older plants now are being required to install similar equipment.

2-39 Blast furnaces.

Observations

When discussing the Burns Harbor project, one must deal in superlatives. When started it was the largest privately financed construction job in the world and the newest fully integrated steel mill in the United States; and, it included the largest blast furnaces in the Western Hemisphere. In substantial measure, these superlatives also define many of the difficulties inherent in developing a single-industry site. The magnitude of the undertaking, with respect to cost and complexity, vastly increases its difficulty. Accommodation of heavy industry to the environment has become a major issue and the environmental controls installed at Burns Harbor are now required of all industrial developments. Site acquisition and obtaining necessary approvals for major industrial projects are increasingly difficult. These limitations tend to suggest that fewer freestanding heavy industrial complexes will be developed in the future.

Combined public/private investment, as evidenced by development of the Port of Indiana, has worked very well. Both the state and the company have benefited from the development of a deepwater port facility. Traffic to the port, much of it iron ore pellets destined for Bethlehem Steel, is growing steadily.

To date, only one company strongly linked to steel production, Continental Can, has located within the Burns Harbor site. The economic rationale for locating major steel users in proximity to the primary materials source, however, is well established and derives from reduction in transportation costs gained through adjacent locations. The company, therefore, expects to provide additional sites for steel users in the future.

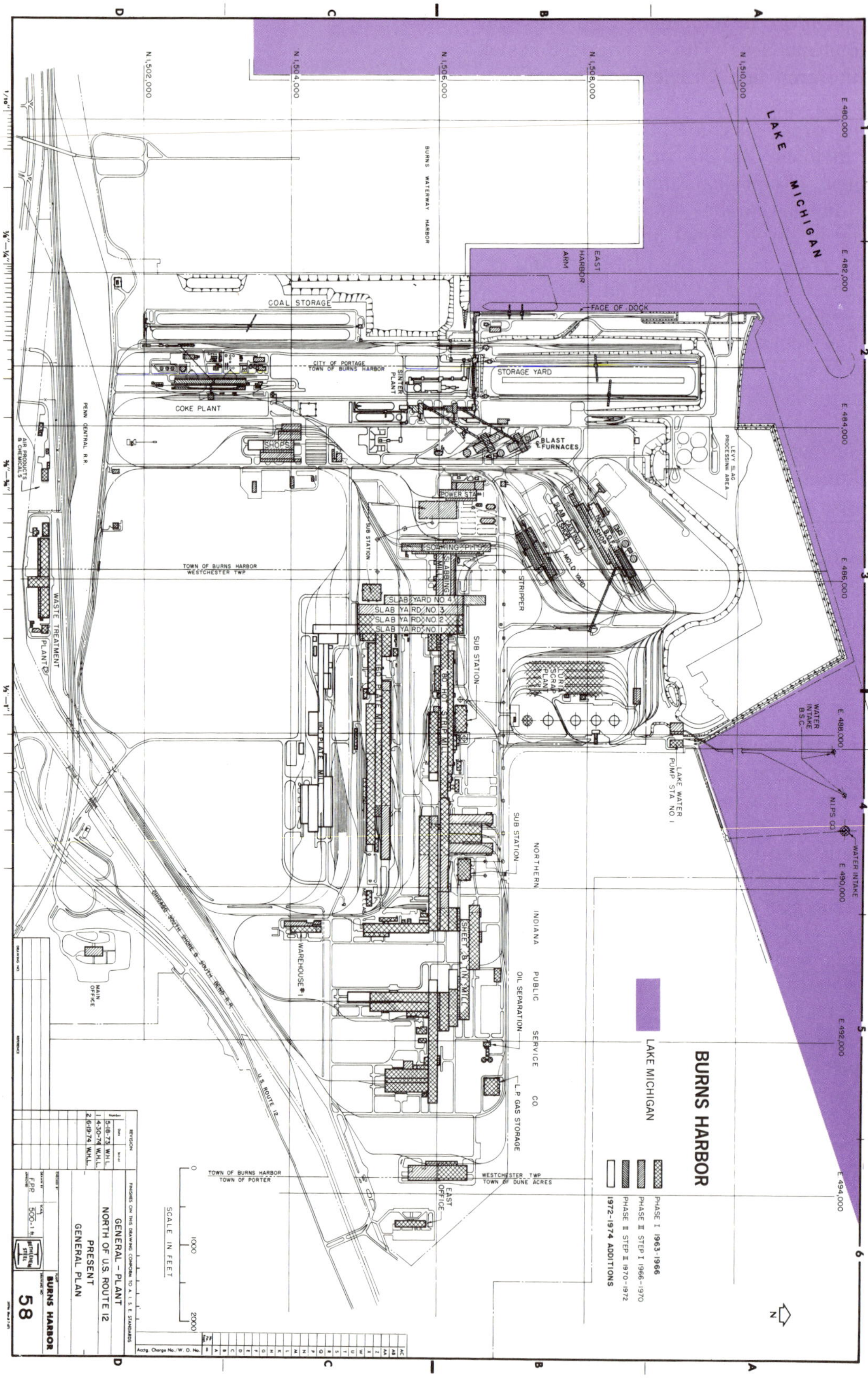

2-40 General plan—Burns Harbor. The plan is flexible and designed to accommodate shifts in steel manufacturing technology and market demand as the facility is developed over several decades.

Bayport

(Port)

General Description

Bayport is a 10,500-acre industrial project located adjacent to Galveston Bay, 25 miles southeast of downtown Houston, Texas. The land is part of 30,000 acres owned by Exxon Company U.S.A. since 1938. The property was acquired originally for its mineral value. Subsequent exploration defined some 7,000 acres of oil and gas production. In the 1950s, Exxon initiated studies of use of the remaining 23,000 acres. Lehman Brothers of New York recommended two kinds of development: (1) a new town development, now known as Clear Lake City, and (2) an industrial development, now known as Bayport.

Basic reasons for developing Bayport for industry, as defined by the initial planning, were twofold: (1) to generate a job base to help provide a market for Clear Lake homes, and (2) to provide a long-term, substantial market for energy and related products for the nearby Exxon Baytown refinery. In the mid-1960s, energy market conditions caused a change of emphasis from energy consumption to real estate development. This change of direction did not affect the original planning concept.

From the outset, emphasis was placed on maintaining air and water quality in the recreation-oriented area surrounding Bayport. Environmental standards were adopted as part of Bayport's initial planning, acting as deed restrictions and enforceable in civil court. Bayport was the first entity in Texas and, at the time, one of few in the United States, to voluntarily adopt such standards. The standards were drawn before the current high level of public interest in environmental and ecological matters, and were implemented without public suggestion or pressure. The original standards remain in force.

Bayport originally contained 7,250 acres. In 1967, an additional 1,500 acres were purchased and added to Bayport to provide more acreage with direct access to the port (Bayport Division, Port of Houston). In 1974, a third addition—1,750 acres—was made to Bayport from land already owned by Friendswood Development Company to achieve the current gross acreage total.

2-41 Celanese Chemical Company occupies a 1,100-acre site at Bayport, a 10,500-acre industrial development located on Galveston Bay 25 miles southeast of Houston, Texas.

2-42 The Bayport Ship Channel, extending inland 1.8 miles from Galveston Bay, provides deep water access for occupants of this development for heavy industry. The view is westward toward the turning basin. Plant construction at Bayport is in its initial stages.

Bayport Division, Port of Houston

Original planning called for development, in two stages, of a publicly owned and operated port. Stage one was construction of a barge channel and turning basin and stage two was enlargement and deepening of a port designed to accommodate ocean-going shipping.

Approximately 700 acres of land were transferred to the Port of Houston Authority. In 1964, construction began on the barge phase of the channel. This phase was completed in 1966. Expansion of the channel to create a deepwater facility was begun in 1972 and was completed in October 1974. The new channel dimensions are a 40-foot water depth, a 300-foot bottom width, and a turning basin of 1,600 feet square. The channel extends 1.8 miles inland from the Galveston Bay.

Plant sites with direct access to Bayport Division are available. Pipeline, rail, and road access are also provided.

Significant aspects of the Bayport Division include:

- a location only two hours up the Houston Ship Channel from the Gulf of Mexico entrance to the channel (which compares to an up-channel, one-way passage of up to six hours from the Gulf entry to the original turning basin);
- a larger turning basin than the Port's original with water sheltered from wave action and currents;
- room for expansion to 19 ship berths including 12 for liquid cargo, and numerous barge berths; and,
- ability to accommodate ships in the 70,000 to 80,000 dead-weight-tons class.

Environmental Factors

A major role in enforcing environmental standards is vested in the Bayport Industrial Association (BIA) which is composed of representatives from Bayport landowners and service companies (railroads, etc.). This self-policing body has been very effective in upholding standards and has resisted attempts by other groups to dilute their effectiveness. The BIA also establishes procedures for helping its members in times of emergency.

The standards administered by BIA cover:

- emissions of smoke, suspended matter, and dust;
- odors;
- discharges of toxic materials;
- airborne discharges of nine specific substances such as carbon disulfide, carbon monoxide, and others;
- radioactive materials;
- explosive and flammable materials;
- noise levels, vibrations, and glare;
- liquid wastes, surface drainage, and solid wastes;
- property controls, including off-street parking, off-street loading, building setback lines, and area maintenance; and,
- acceptability criteria for materials sent to Central Waste Treating Plant (CWTP).

The Bayport Central Waste Treating Plant treats and disposes of both domestic and industrial process wastes. Companies have the option of subscribing to CWTP services or making other arrangements for treatment of wastes. To date, all plants have subscribed to CWTP services because of two main advantages: (1) ease in obtaining necessary permits and exercising necessary controls; and (2) economies of scale inherent in one central operation versus individualized treatment at or near each plant site. Companies are charged for treatment services on basis of volume and nature of effluent. In some cases, firms pretreat wastes before sending them to CWTP in order to meet ac-

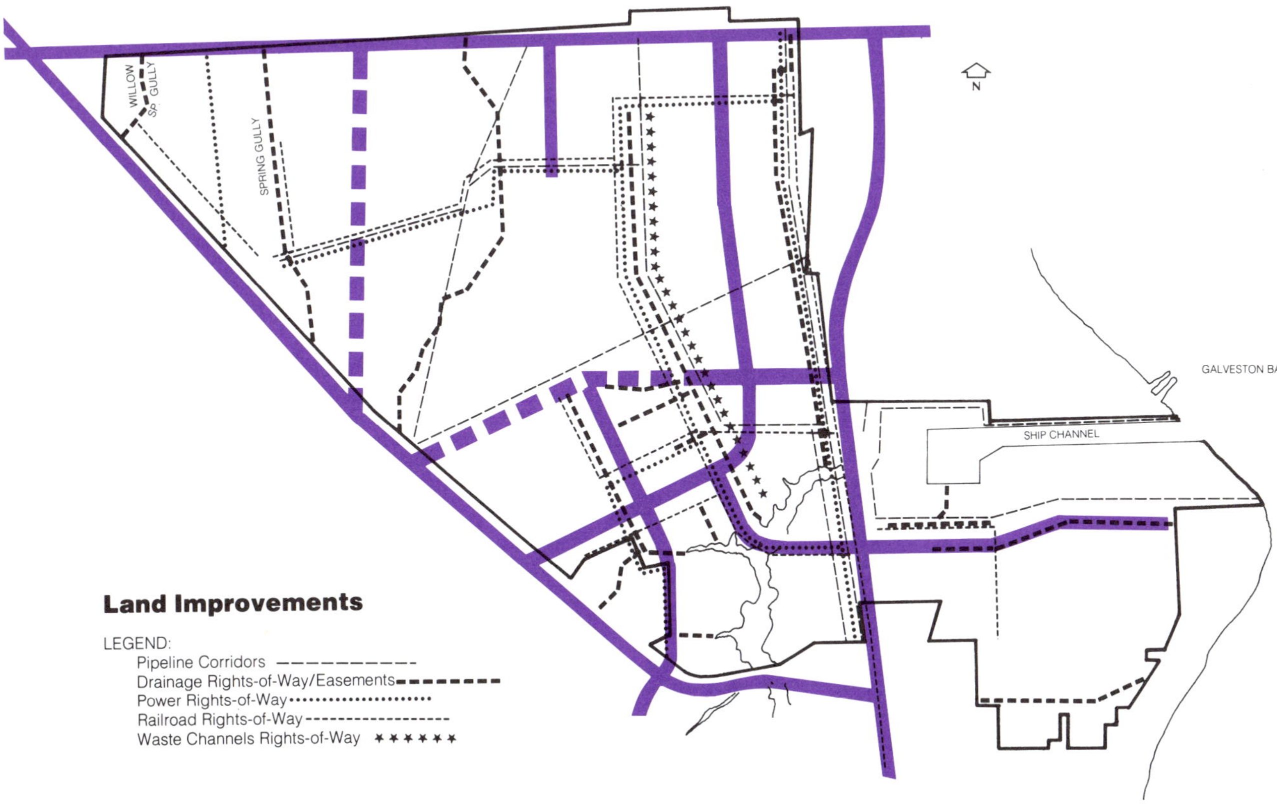

2-43 Land improvements at Bayport include provision of a modern arterial street network, pipeline corridors, and rights-of-way for railroad lines, power lines, drainage canals, and waste water channels.

ceptability criteria. This step helps assure continuity of operations in the biological process plant.

Two other aspects of environmental planning at Bayport can be cited. During construction of the port, safeguards were taken to guard against silting of waterways important to marine life breeding. Due to the recreational nature of Galveston Bay, no new spoil banks were created because of their possible interference with sailing and fishing activities.

Planning Considerations

Following publication of the Lehman Brothers report recommending development of an industrial project at Bayport, Exxon Chemical Company and the Texas A&M University Industrial Economics Research Division were hired to develop a list of the most likely prospects. Prime prospects were identified via Standard Industrial Code (SIC) classifications. An examination of the needs of such prospects, together with initial planning, provided the basis for developing overall plans for the port, pipelines, roads and highways, rail, water, waste, and electric and utility systems. For example, all roads were designed for heavy loads, and the plan allowed for rail spur access to all plant sites within Bayport. Much of this planning was done by Friendswood Development Company with assistance from Engineers of the Southwest, a consortium of Houston consulting engineers.

Important planning considerations included the prohibition of overhead road/highway pipeline crossings, provision of flexible plant site sizes in the original layout of sites, and reservation of adequate service corridors to meet various utility and pipeline needs. The original plan, with additions as Bayport has been enlarged, remains in force with only minor modifications. Provision has been made for the extension of pipeline corridors and a central pipeline corridor has been combined with railroad, drainage, and power rights-of-way.

The original 7,250-acre size was sufficient to permit a comprehensive master plan to guide development, both quantitatively and qualitatively, while retaining flexibility to meet market demand for sites of all sizes (from one acre to more than 1,000). The plan also was compatible with adding land to the development.

Government/Industry Cooperation

Bayport illustrates some of the advantages which can accrue through cooperation between public and private sectors. Examples include several major changes and improvements in access and peripheral roads. Developer involvement included donations of rights-of-way and an agreement to build one side of major boulevards with the other side to be built by the appropriate government entity.

A great deal of public/private cooperation was necessary for construction of the port. The developer donated the land along with the necessary rights-of-way and easements, and also financed construction costs and operating deficits. The Port of Houston provided engineering, management, and secured all permits and licenses required for construction and operation.

In general, good rapport with governmental entities was maintained throughout the formulative stages of development. Bayport has resulted in significant tax roll additions without a commensurate demand on governmental services.

Occupants of Bayport

Thirty-six companies hold 5,706 acres (including the 1,057-acre Hudson option), accounting for 71 percent of the land available for sale. Twenty-one plants are operational, two others are under construction, and six are announced for construction.

As of January 1975, the permanent work force, excluding construction workers, is 1,600 with an annual payroll of 20 million dollars. Total investment by all parties is approximately 400 million dollars. As announced plans are implemented, payroll is expected to increase to more than $30 million and total investment will approach $1 billion.

2-44

PROJECT DATA

Bayport

Site Plan	Acres	Percent
Development Area	8,000 (±)	76.2
Sold or Optioned	5,706 acres (71 percent of net salable)	
The Port	831	7.9
Street System	425	4.0
Utilities (includes waste plant)	1,192	11.4
Other	52	.5
Total Area	10,500	100.0

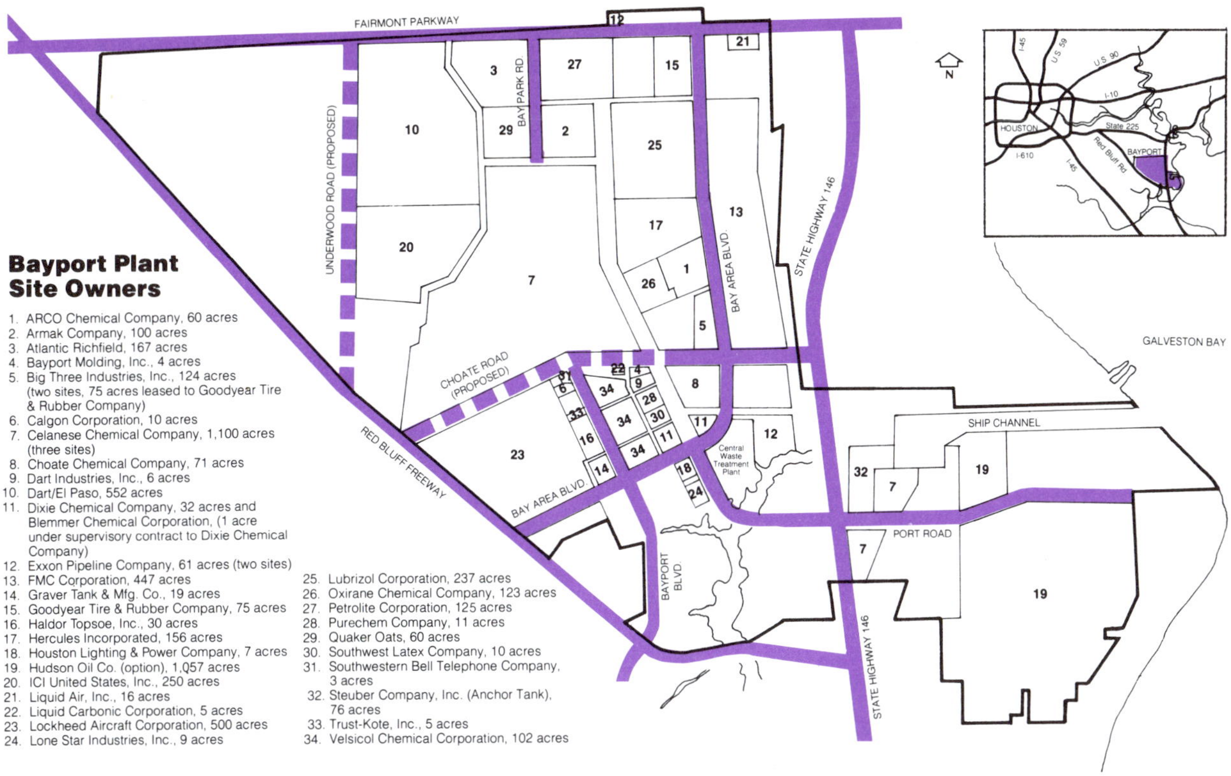

2-45 Approximately three-quarters of the land at Bayport has already been sold. Some parcels are being held for future development, while construction is underway on many others.

2-46 DEVELOPMENT AT BAYPORT AS OF JANUARY 1975

COMPANY	SITE SIZE IN ACRES	PROJECT, STATUS
Anchortank/Houston, Inc.	76	Liquid terminal, storage
Arco Chemical Company (Division of Atlantic Richfield Company)	60	No plans announced
Armak Company	100	Mid-1975 completion of peroxydicarbonates plant scheduled
Atlantic Richfield Company	167	Crude oil terminal
Bayport Molding, Inc.	4	Molded plastic products plant in operation
Big Three Industries, Inc.	200	Steam and air separation plant doubled in past year. Lease 75 acres to Goodyear Tire & Rubber Company
Blemmer Chemical Corporation	1	Glycidyl methacrylate plant in operation
Calgon Corporation	10	Industrial water treatment chemicals plant in operation
Celanese Chemical Company	1,100	3 billion pound per year plant in operation; completed new deepwater terminal at port
Choate Chemical Company	71	Polyether polyol resins plant in operation
Dart Industries, Inc.	6	No plans announced
Dart Industries, Inc., and El Paso Products Company	552	150 million pound per year polypropylene plant to begin operations in 1976
Dixie Chemical Company	32	Custom chemical manufacturing plant in operation
Exxon Pipeline Company	5	Office/warehouse in use
Exxon Pipeline Company	54	Site for future liquid cargo terminal
FNC Corporation	447	Plant produces synthetic glycerine and other chemicals

Goodyear Tire & Rubber Company	75	Hydroquinone plant in operation
Graver Tank & Manufacturing Company	19	Large-scale metal fabricating plant in operation
Haldor Topsoe. Inc.	30	Ammonia catalyst plant in operation
Hercules Inc.	156	Completing 400 million pound per year polypropylene plant
Hudson Oil Company	1,057 option	200,000 barrel per day refinery announced. $500 million project scheduled for completion in 1978
ICI United States, Inc. (Subsidiary of British-based Imperial Chemical Industries, Ltd.)	250	Construction begins in 1975 on plant to produce various chemical products
Liquid Air, Inc.	16	Air separation plant in operation
Liquid Carbonic Corporation	5	Industrial and specialty gases plant in operation
Lockheed Aircraft Corporation	500	Lockheed has no building plans. Their land is for sale
Lone Star Industries, Inc.	9	Ready-mix concrete plant in operation
Lubrizol Corporation	237	Petroleum additives and specialty chemicals plant in operation
Oxirane Chemical Company	123	900 million pound per year propylene oxide and propylene glycol plant in operation
Petrolite Corporation	125	No plans announced
Purechem Company	11	Titanium trichloride catalyst plant in operation
Quaker Oats Company	60	1,000 tons per day furfural and furfuryl alcohol plant under construction
Southwest Latex Company	10	Styrene butadiene latexes plant in operation
Southwestern Bell Telephone Company	3	Equipment station operational
Trust-Kote, Inc.	5	Industrial paint and sandblasting plant in operation
Velsicol Chemical Corporation	102	Agricultural chemicals plant in operation
(Company has not authorized use of name)	21	No plans announced

Centex
(Industrial District)

One of the largest industrial districts in the United States is located in the Village of Elk Grove, Illinois—18 miles northwest of downtown Chicago. The 2,853-acre district, 2,250 of which is the Centex Industrial Park, has been under development since 1957, and consists of eight separate industrial parks and several parcels being developed for industrial use. The district is projected to reach an ultimate size of 3,273 acres as additional parcels destined for industrial use are annexed to the community. More than 1,000 firms have located within the district and total employment as of 1975 exceeds 30,000. Since 1968, the annual value of industrial building permits have ranged as high as $30.2 million and never dropped below $12.1 million.

Development in this area was initiated by the Centex Construction Company of Dallas which planned the original 4,500-acre tract for a balance of residential, industrial, and commercial uses. Actual development has proceeded according to the original development plan. Centex Industrial Park, planned and marketed by Bennett and Kahnweiler of Chicago, dominates the district. Centex is almost entirely sold out and the company has established the 675-acre Centex-Schaumburg park several miles to the west in Schaumburg. Subsequently, other industrial developers established industrial parks adjacent to Centex. These projects are moving ahead rapidly as well. Community population, predominantly housed in a single-family detached suburban setting, has risen from 6,608 in 1960 to 25,303 (special census dated October 1974).

An advantageous location relative to transportation is perhaps the most important single factor relating to the success of an industrial district. This consideration—coupled with the availability of large, contiguous parcels of land—was of primary importance in motivating Centex to make its initial land acquisitions in the 1950s.

Immediately to the east of the site was Orchard Place, a war-surplus airfield which the city of Chicago was in the process of converting into O'Hare International Airport. The area was also served by two railroads, the Chicago Northwestern Railway and the Chicago, Milwaukee, St. Paul and Pacific Railroad. Various highway improvements either were being planned or already complete. Two tollways were built by the state of Illinois—the Northwest and Tri-State insured ready highway access throughout Chicago and the region. At no other point in the Chicago area could such a fortuitous combination of locational advantage be found.

The actual size of the area made available for industrial use was dictated by the size of the Chicago industrial market. Since the property was divided into a relatively small number of large-scale agricultural holdings, acquisition of extensive areas

2-47 Aerial view of Centex and portions of Elk Grove Village. O'Hare International Airport can be seen at the top of the picture.

was not overly difficult. Centex was thus able to preplan an area of more than three square miles for industrial use.

Development of Centex and the rest of the district has proceeded apace with suburbanization. The preponderant share of metropolitan population and industrial growth over the last two decades has taken place outside the city limits of Chicago. Because of its size, Centex has been able to attract a broad array of industries. Primary reasons cited for locating in the district include relocation from congested city locations, provision for plant expansion, and establishment of regional facilities by national corporations wishing to serve the Chicago market.

An estimated 65 percent of the district occupants are warehousing and distribution activities, and the remainder are industrial uses which meet both developer and community performance standards. At Centex, food processors, steel fabricators, auto parts distributors, warehouses, contractors, and printers are represented heavily. The skilled blue collar labor force has become a primary attraction for industries seeking to locate in the district. The skilled workforce has followed the movement of industry to the suburbs. Semi-skilled and unskilled workers, however, are drawn primarily from the city of Chicago.

Elk Grove Village

Elk Grove Village differs from most suburban municipalities in that almost half of its current and projected ultimate land area has been set aside for industrial use. The benefits derived from the preplanning are readily apparent in both residential and industrial sections.

Within the 3,228-acre residential area, single-family homes are situated along curvilinear streets. Multifamily areas, schools, and the municipal center are clustered along arterial and collector streets. The floodplain of Salt Creek, which traverses the area, has been set aside as parkland. Residential and industrial sections of the community are separated from one another by means of a 200-foot right-of-way occupied by a four-lane divided arterial and a power line easement. Deleterious impact of industry upon residences is further mitigated by a 400-foot-wide strip zoned Restricted Industrial, located across from the single-family district.

Benefits accruing to the community in pre-planning its industrial areas include the provision of public facilities—water, sewers, and streets. It was possible to plan these systems for projected ultimate demand rather than by responding to development decisions on an individual basis. Arterial street rights-of-way were protected until they could be upgraded to ultimate standards. Likewise, police and fire systems have been geared to meet the specialized needs of industry. For instance, Elk Grove Village has been divided into two police districts, one of which has the industrial district as its primary responsibility.

The Village has established two industrial zones, an I-1 Restricted Industrial District and an I-1 Generalized Industrial District. Developers seeking to annex prospective industrial areas to the municipality have been required to conform to the provisions of these districts. In this manner, it has been possible to expand the district incrementally while maintaining established developmental standards.

The Industrial District

Eight separate but contiguous industrial parks are under development in the Elk Grove Industrial District. The largest of these—and the original catalyst for development of the district—is Centex. Its 2,250 acres represents 76.1 percent of the district's current size. Most other developers acquired 30- to 100-acre tracts contiguous to Centex and are developing in conformity with Centex standards.

Current industrial acreage in the Village amounts to 2,853 acres. Elk Grove Village, however, has reached boundary agreements with surrounding municipalities and an additional 420 acres are to be added before the district reaches its ultimate projected limits. These boundaries will include the Northwest Tollway to the north, O'Hare International Airport to the east, the 200-foot right-of-way which separates the industrial and residential portions of the village on the west, and prearranged boundaries with neighboring municipalities to the south.

The resulting industrial area is an irregular rectangle roughly three miles north-south by two miles east-west. Within this area, pre-existing highways and section line roads have been upgraded to the status of major arterials. The industrial property itself has been platted in a predominantly gridiron pattern, with numerous rail crossings at street grade, similar to the Los Angeles Central Manufacturing Districts.

Centex Industrial Park

At Centex, block dimensions vary from 300 to 600 feet in depth by 1,500 to 4,200 feet in length, ex-

PROJECT DATA

Centex

INDUSTRIAL BUILDING PERMITS
ELK GROVE VILLAGE

Year	New Construction[1]	Building Expansion[1]	Total[1]
1968	$10.6	$5.3	$15.9
1969	9.6	4.7	14.3
1970	10.6	3.8	14.4
1971	7.4	4.7	12.1
1972	10.0	6.4	16.4
1973	26.1	4.1	30.2
1974	14.9	3.5	12.4

[1] Value in millions of dollars.

ELK GROVE INDUSTRIAL
DISTRICT PROJECTS

Projects	Acreage
Centex Industrial Park	2,250
Davis—Brickvale	10
De Bruyne—Mayfair	41
McLennar Devon—O'Hare	114
McLennan Oakton	10
Reintz Tritan	30
Rogers	66
Trammel Crow—Elk Grove	208
Miscellaneous Industrial Areas	124
Industrial District as of 1975	2,853
Future Annexation	420
Anticipated Total Area	3,273

LAND USE—ELK GROVE VILLAGE

	Residential & Other		Industrial District		Total	
	Acres	Percent	Acres	Percent	Acres	Percent
Incorporated	3,228	53.1	2,853	46.9	6,081	100.0
Future Annexations	977	70.0	420	30.0	1,397	100.0
Grand Total	4,205	56.2	3,273	43.8	7,478	100.0

2-48

cept for a few small blocks not served by rail. For the small sites, lot depths are usually 200 feet. Most lots range from two to eight acres in size, but there are parcels as small as 10,000 square feet. The largest occupied site in Centex is the 50-acre location of the Halo Light division of McGraw Edison although Sears is building its one million square foot midwest fashion warehouse on a 90-acre parcel. Initial employment in the Sears facility will be over 1,000 persons, a figure which is expected to rise as the facility is ultimately expanded to 5 million square feet. Of the 2,250 acres in Centex, 1,600 are rail served, with approximately 40 percent of the rail-served occupancy actively using rail. Freight traffic amounts to about 15,000 cars per year.

Service and commercial development within the district includes motels, restaurants, small office buildings, and gas stations. These uses, intended to meet employee needs, are located on both developer sites and on out-parcels not acquired during the site assembly phase.

Developer Observations

The developer, in evaluating his project, offers the following observations which might benefit others involved in the development of similar projects:

- Because of the size and variety of tenants accommodated, industrial districts present unique problems to the developer. Perhaps most serious is the question of proper handling of pollutant emissions. Even when stringent controls are enacted and enforced vigorously, a concentration of industry on the scale of Elk Grove can be expected to generate air pollution. This situation is compounded by the necessity of accommodating the heavy traffic volumes required to serve the district.

- Associated with this are a variety of difficulties which size brings to the developer. Projects on the scale of Centex involve a long-term financial commitment and marketing strategy. Problems relating to provision of public utilities and vehicular traffic tend to be on a massive scale as well. Marshall Bennett points out that Centex Industrial Park, because of its unique proximity to the world's busiest airport, and its size and success (over 900 buildings), provides jobs for 30,000 employees, an annual tax income of millions of dollars, and a substantial profit to the developer.

• Minor admixture of incompatible uses is a common outgrowth of an industrial area being developed by several different organizations because economics rarely justify acquisition of small subdivisions and miscellaneous commercial uses to round out extensive industrial districts. The result in no way hinders the efficient operation of the industrial district, but the isolated uses themselves are usually adversely affected. Their presence detracts from the overall appearance of the district as well. This has generally been avoided at Centex.

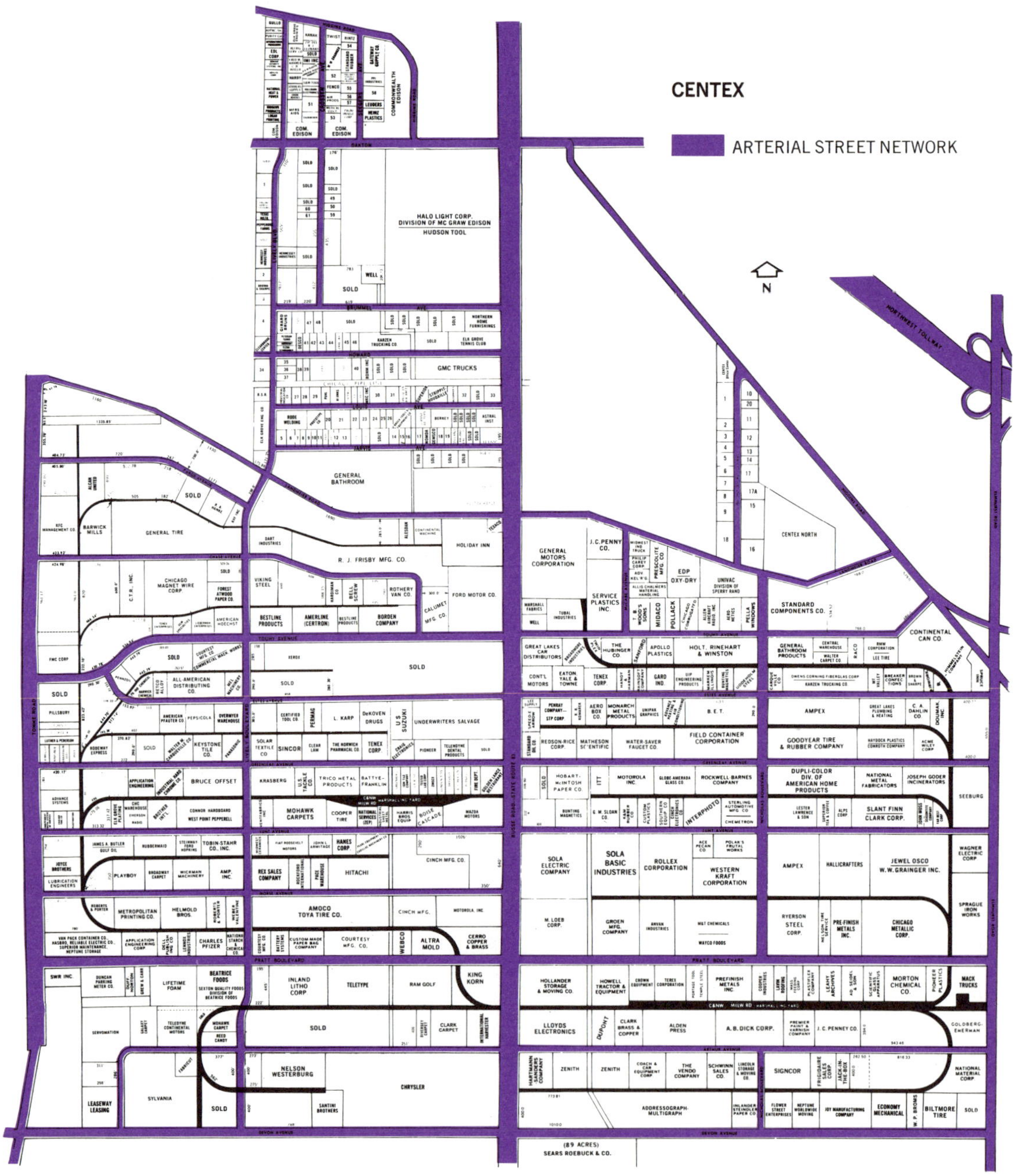

2-49 Centex Industrial Park site plan.

Kent Industrial Center

(Adaptive Reuse/Public Involvement in Industrial Development)

Kent Industrial Center combines two important aspects of industrial development: it illustrates the opportunity to convert obsolete facilities to industrial use, and exemplifies the potential for public involvement in the field of industrial development.

By 1960, the Kent County Airport serving Grand Rapids, Michigan and adjacent areas was obsolete. Having served as the community airport since the inception of commercial aviation, the facility had been hemmed in by commercial, industrial, and residential development. Given the relatively constricted 405-acre site, it was no longer feasible to extend the 5,000-foot runway to accommodate large jets. Therefore, the owner and operator of the airport, Kent County, decided to build a new facility.

The question then remained, what could be done with the existing site? For several reasons, industrial reuse appeared to be the most feasible option open to the community. Deciding factors included the nature of surrounding development, a desire to strengthen the local tax base and obtain a maximum return for the land, the need to provide modern sites for industries seeking to locate in Grand Rapids, and the wish of local planning officials to round out their proposed master plan for development with a single, large site for industry.

Initially, the property was offered for sale because the County had pledged to retire a portion of the bonds for the new airport with land sale receipts. Offers submitted were inadequate so Kent County decided to develop the property on its own. Williams and Works, a local consulting firm, was hired to prepare both a development plan and a strategy for disposal of the site. From the outset, the County committed itself to creation of a quality

2-50 Kent Industrial Center—light industrial firms are located along either side of a landscaped boulevard which follows the location of the former main runway of the Kent County Airport. The airport has been closed and the site converted to industrial use. The view is looking south.

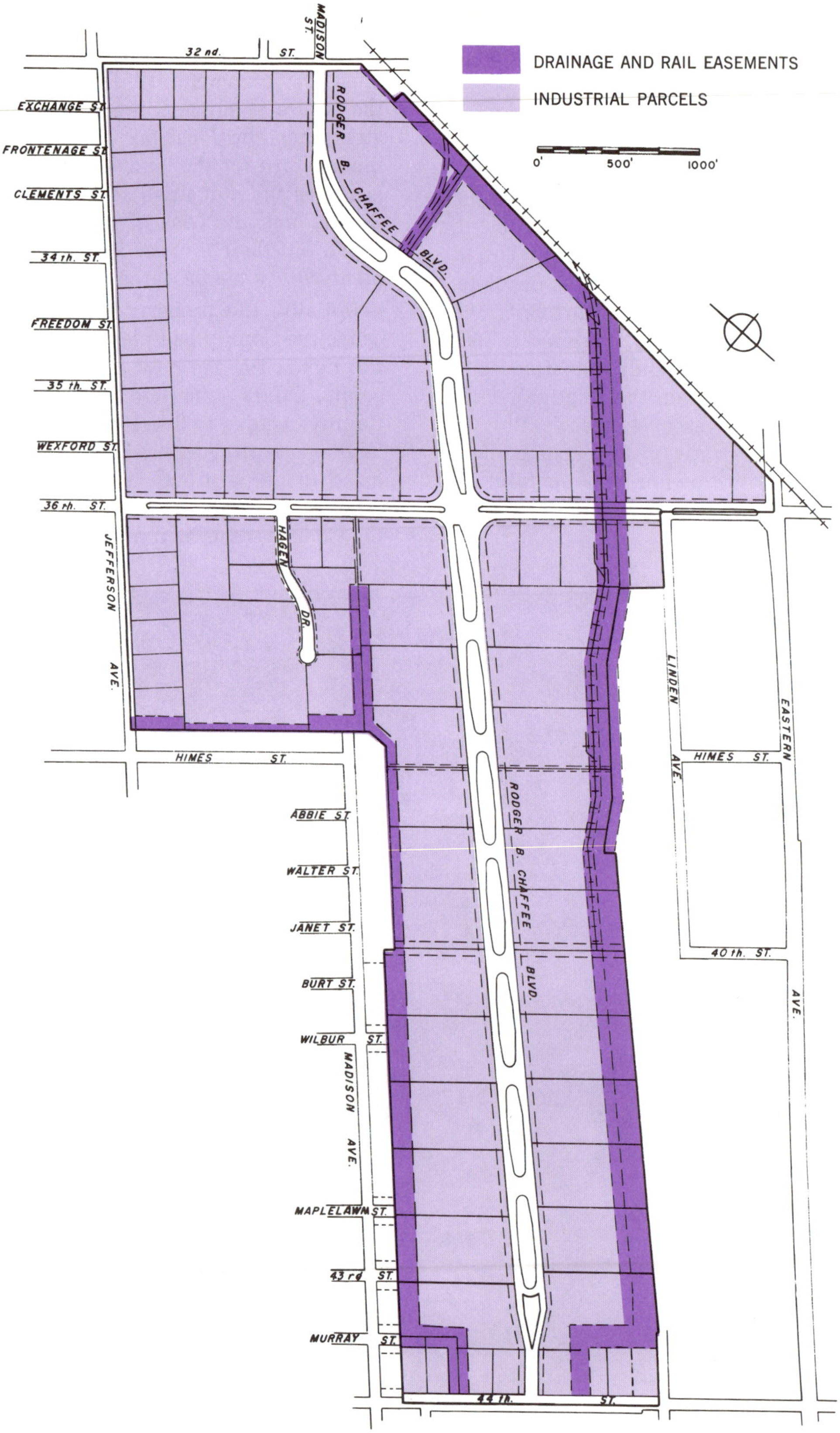

2-51 Kent Industrial Center site plan. The 50-acre tract in the northwestern quadrant of the park was re-subdivided into 21 separate parcels when it became apparent that demand for very large sites was limited.

environment for industry, fully recognizing the high capital costs this would entail. Funds to redevelop the airport were generated through sale of bonds.

The property was platted into 121 rectangular parcels with access provided by a landscaped boulevard following the path of the old main runway constructed along the length of the site. Total capital investment required to prepare the property for industrial use was $2.4 million. Improved sites were offered for sale beginning in 1966. As of 1975, 77 percent of total acreage had been sold. All improvement costs have been recovered and sales now represent a net gain to the local government. Fifty-three firms employing 2,000 persons now are located within the Kent Industrial Center.[25]

The Site

Kent Industrial Center is located in the southeast quadrant of the Grand Rapids metropolitan area. It is situated in the suburban community of Wyoming, an incorporated township which grew rapidly during the 1950s and 1960s. The Penn Central Railroad forms the northeastern boundary of the property. Predominately residential and industrial uses are found to the east and west of the park, respectively. Traffic circulation in this area is accommodated by at-grade arterials which interconnect with interstate highways and freeways located several miles away. Some additional contiguous acreage forming part of the original airport site was disposed of separately.

Engineering and Design Considerations

Redevelopment of the site presented formidable obstacles and necessitated an imaginative solution on the part of the design consultants. Specifically, the site plan had to be devised carefully to permit maximum use of the 5,000-foot by 150-foot main runway. Removal of the 750,000-square foot runway, built of reinforced concrete six to ten inches thick, would be prohibitively expensive. The solution adopted was to use the runway as a concrete

[25] *Project Reference File*, vol. 5, no. 19 (Washington, D.C.: Urban Land Institute, 1975).

2-52 Landscaped islands were placed on top of the original reinforced concrete runway surface. This strategy substantially reduced the cost of converting the property to industrial use, since the concrete could be retained as the base for the industrial boulevard and did not have to be removed.

base course for a boulevard extending the length of the property. Landscaped islands were placed upon the concrete base which was removed only where necessary to accommodate utilities and trees. Originally, the islands were intended to become the location of a variety of service-commercial uses which would complement the industrial park. Now that these areas have been landscaped, occupants have become accustomed to considering the medians as common open space. It seems unlikely such a shift will ever take place.

Provision of an adequate surface drainage system was necessary also. Virtually no topographic variation existed within the site, and the water table was only four feet below the surface. Such conditions present major problems for industrial parks if more than half of total site acreage is covered by roofs and paving. To accommodate stormwater runoff, a system of canals draining into a nearby stream was located in easements along the edges of the property. The system also lowered the water table to 15 feet below the surface. Since runways are flat and streets must have a slight gradient to remove surface water, the thickness of the bituminous surface applied over existing runways was varied and storm drains placed at 200-foot intervals.

Taxiways, aprons, and other paved elements of the old airport were built of asphalt and could be removed for new construction without difficulty. Nevertheless, many park occupants have designed their facilities to maximize use of original paving for parking lots and internal circulation. The terminal building and all but one of the ancillary airport structures were demolished because they had no potential for industrial reuse.

The site plan was designed for maximum flexibility and created a variety of square and rectangular parcels appealing to a wide range of industrial land users. The largest single parcel—subsequently replatted—was 50 acres. The smallest sites are 50,000 square feet. Minimum lot frontage and depth is 200 feet. The arterial street plan internalizes traffic on two at-grade arterials which intersect at the

2-53 Kent County Airport, circa 1960.

center of the property. These arterials, Roger B. Chaffee Boulevard and 36th Street, have rights-of-way of 180 feet and 100 feet, respectively. They incorporate extensive landscaped medians which create a spacious setting for industry, and they emphasize the high level of exterior maintenance required by park occupants.

Approximately 30 percent of the Kent Industrial Center is served by rail. A spur of the Penn Central was extended halfway down the eastern edge of the property and occupies a 20-foot easement. Further extension of the track is provided, should demand warrant it. Currently, rail carloadings average 200 per year, with one firm accounting for almost 90 percent of the total.

Protective Covenants

A set of covenants and restrictions covering the Kent Industrial Center, to which all occupants must subscribe, was drawn up at the outset. These covenants and restrictions specify uses permitted within the park, building lines and setbacks, building design and construction, landscaping, and other design controls. In addition, the covenants provide for establishment of the Kent Industrial Center Association of Property Owners which is responsible for maintenance, enforcement of covenants, operation of an information office, and performance of "other functions in the best interests of the Kent Industrial Center."

The Association has a five-member board of directors, the make-up and responsibilities of which are outlined in the bylaws. Currently, two of the members are Kent County representatives who are serving at the request of the property owners. When all lots are sold, the board of directors will be solely the responsibility of park occupants. Other controls in force within the park include the provisions of the city of Wyoming's Planned Industrial District.

Marketing

Land absorption at Kent Industrial Center has averaged 30 acres per year. Sales patterns have corresponded closely to fluctuations in the national economy. If a sellout is accomplished by 1980, as anticipated, disposal of the property will have taken 13 years.

Pace of development is slower than feasible for many privately developed industrial parks, but has been sufficient for the County to attain its original financial goals. With redevelopment having cost $2.4 million, and sales projected to total $4.6 million, the County will have experienced a $2.2 million net gain excluding the increase in the tax base. Sales prices average $12,000 to $14,000 per acre which is comparable to privately developed industrial parks in the area.

Most companies located within the park conduct distribution, warehousing, and light manufacturing

2-54

PROJECT DATA
Kent Industrial Center

LAND USE INFORMATION

Site Area: 405 acres

Parking: One space per 1,000 sq. ft. of floor area or, one 300-sq.-ft. space for each two employees.[1]

Employment: 2,000

Development Acreage: 158 acres

Acreage Sold: 271 acres

Employees per Developed Acre: 12.6

LAND USE ALLOCATION	ACRES	PERCENTAGE OF PROJECT
Platted Area[2]	352	86.9
Rights-of-way	43	10.6
Landscaped Medians	10	2.5
	405	100.0

ECONOMIC INFORMATION

Land Sale Price:[3] 33¢ to 47¢ per sq. ft.

Site Improvement Cost:

Construction[4]	$1,836,182
Engineering	$ 166,004
Inspection	$ 14,476
Other	$ 412,845
Total	$2,429,507

Construction Cost: (Representative examples)

BUILDING TYPE	SQUARE FOOTAGE	YEAR BUILT	COST PER SQUARE FOOT
a) Office and Warehouse	159,000 sq. ft.	1974	$12.58
b) Office	3,000 sq. ft.	1972	$38.33
c) Warehouse	45,000 sq. ft.	1972	$ 8.46
d) Office and Warehouse	30,240 sq. ft.	1972	$ 8.50

[1] City of Wyoming zoning ordinance.

[2] Includes 35 acres of drainage easements and 2 acres of railroad easements.

[3] Firm offering prices. Figures as of 1975 constitute a 30 percent increase above original offering prices.

[4] Includes streets, watermains, storm and sanitary sewers, and $138,651 in landscaping. Landscape maintenance amounts to $14,000 per year.

operations. The employment level is somewhat less than would be the case if most occupants were engaged in manufacturing. An estimated 85 percent have relocated from elsewhere in the metropolitan area—primarily from congested sites in the older industrial areas of Grand Rapids. While no apparent net gain occurs to the community from shifts of this kind, provision of modern sites helps strengthen the local economic base by permitting orderly plant expansion and corporate growth.

Public/Private Cooperation

To a considerable degree, the Kent Industrial Center has been the result of a cooperative effort by public and private sectors of the Grand Rapids community. Groups which have contributed expertise and perspective have included local government, the local Chamber of Commerce, engineering and design professionals, and the real estate community. The smooth conversion of the project can be attributed to several factors. The County made it clear that while it intended to maximize its return through conversion of the airport to an industrial park, there was no plan to undertake any additional such projects in the future. At the time the project was initiated, no private industrial parks were under development in Grand Rapids. It has been contended that a pioneering effort by the public in this area paved the way for private initiative.

Provision for sale of parcels was accomplished in the following manner: an agreement was drawn up with 15 local realtors, each of whom paid $1,000 for the privilege of representing Kent Industrial Center. Under this multiple listing arrangement, other realtors can sell property, but must turn over a portion of their fee to the contributing realtors.

Developer Observations

- Acquisition of a prime occupant at the outset has been a major factor in the success of the park. The covenant provisions alone do not explain the level of architectural quality and landscaping at Kent Industrial Center. It would appear that an internally generated consensus on the overall appearance of the park rather than elaborate regulations has been responsible for at least a part of the superior industrial environment created at the Center.
- Once sale of industrial parcels was underway, land speculation in nearby areas had the effect of pushing up land values within the Center also. This has proven to be an unexpected benefit and the return to the County from this project has been greater than anticipated originally.
- It is important to keep in mind that in industrial park development, the size of parcels required by prospective occupants is impossible to predict. Therefore, it is important to design for as much flexibility as possible so that parcels can be split or combined to create the sizes in demand. At Kent Industrial Center, a 50-acre tract was set aside for a single, large industry. When no such industry came along after eight years, the County decided to extend a cul-de-sac into the tract and subdivide it into 21 smaller parcels. Provision for such a step had been made during the initial planning phase.
- The original purpose for designating the median islands for service-commercial use was to internalize traffic within the Center. This design also had the advantage of reducing the amount of lower value, rear-lot property and of expanding the amount of higher value property fronting on the major north-south arterial.

3. Planning

The planning process for industrial parks and employment centers should be continuous. Improvements in transportation, shifts in tenant mix, technological changes in processing or communication, changes in the lifestyle of employees, and other factors may modify the layout, the structures, and the optimum mix of uses—even in the final stages of marketing. An ongoing planning process is important to insure completion of a balanced and viable project which can be expected to have a long and productive life. However, the concentrated planning effort comes during the initial stages of development.

Selecting the Site

Few persons are aware of the local and regional impacts involved in the choice of a location for an employment center.

A single industrial park directly affects several thousand families, and indirectly many thousands more. The location of an industrial employment center determines the length and mode of transport time required for the trip to work. The choice of a site indirectly increases or reduces social and economic costs to the families. The nature of freight transport facilities available at the site—such as access to rail or proximity to the interstate highway system—will determine the types of industries which can operate efficiently in the development. The variety of industries which can be attracted to the location will affect future local employment opportunities. A community's capital budget program can be substantially affected by the water and sewer facilities, and the street and public transportation system improvements needed by the future industries.

Numerous existing industrial parks and individual large industrial plants throughout the nation provide excellent proof that planned location and development have enhanced the community's appearance. Local employment, tax revenues and sales are also increased. Air and water quality, and the natural environment, need not be violated when employment centers are properly located, designed, and serviced in accordance with a comprehensive community development plan.

The selection of future sites for planned employment centers involves the following considerations:

- physical and economic factors,
- technological, social, and aesthetic standards supported by industry, and
- social, ecological, natural resource, and energy issues.

Location Factors

Experienced industrial developers list the following major factors which determine industrial location: materials, labor, market, transport, energy, and public policy. However, industrial parks and employment centers discussed in this Handbook are primarily market oriented; or, in the case of rural development areas, they are centers for agricultural, or mineral, processing and distribution. Therefore, location is considered the more constricted sense of accessibility (transportation), prestige, and potential for aggregation of related or interdependent industries.

CABOT, CABOT AND FORBES' GUIDELINES:

R. John Griefen who is senior vice president of Cabot, Cabot and Forbes, an industrial park development authority, and member of ULI's Industrial Council, has listed the criteria used by his firm in choosing both general locations and specific sites for its industrial park properties. While oriented toward the needs of large industrial developers, these guidelines reflect a development strategy which will be of interest to small and medium-size developers as well.

When selecting the general location, it is important to:

- Select only major metropolitan areas which have a population of one million or more and have the expectation of substantial population and economic growth. (A CC&F policy related to their strategy as a national industrial development firm and not meant to imply that industrial parks cannot be successfully located in smaller metropolitan areas.)

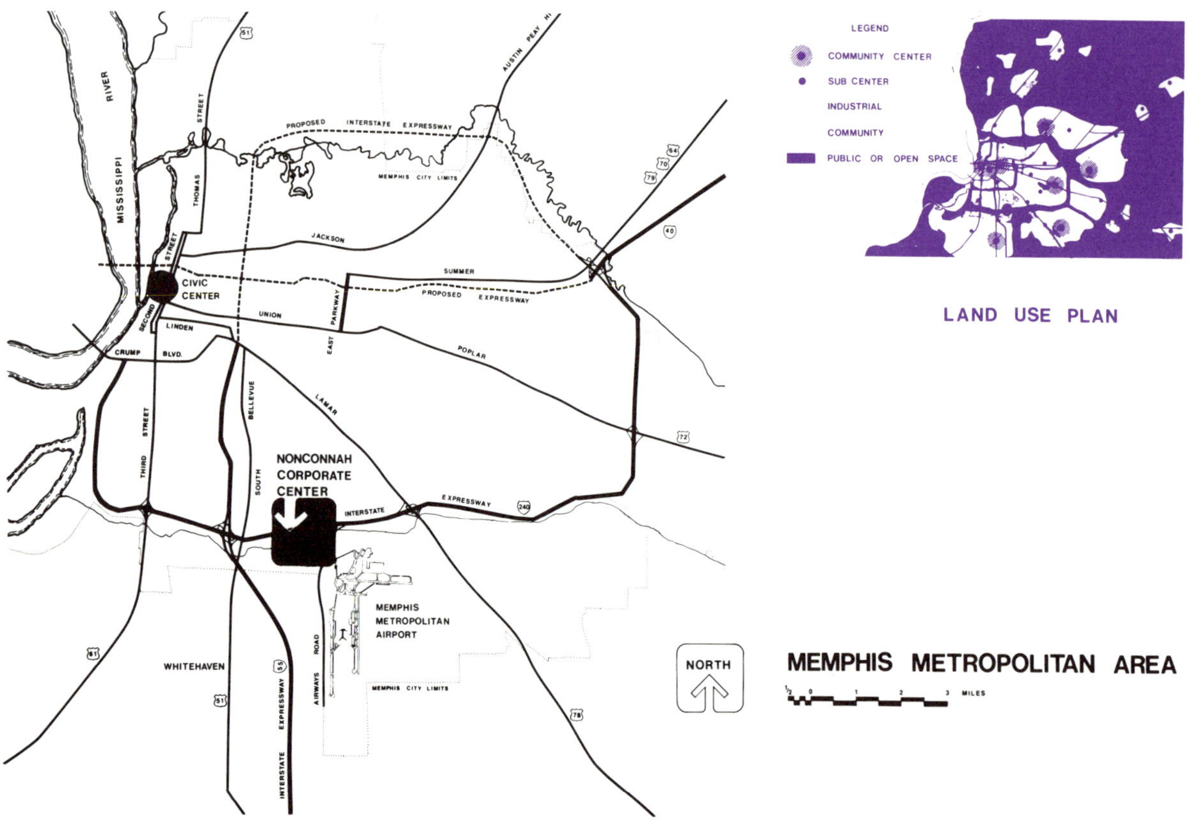

NONCONNAH CORPORATE CENTER

3-1 Proper location is a critical factor in successful industrial development. In the case of Nonconnah Corporate Center, primary locational advantages include a site at the Airways Road intersection of Interstate 240, and ready access to the Memphis Metropolitan Airport just to the south.

3-2 The Crawford Tract, one of several Central Manufacturing District developments in Chicago, places primary emphasis on rail access. Note spur lines serving individual firms throughout the site.

- Be certain the site is served by either an existing expressway system or one slated for construction.
- Analyze direction of population growth and its composition.
- Assess direction and type of industrial growth near such areas as highways, airports, or seaports.
- Ascertain both community attitudes toward industry and economic pressures in a given area.
- Determine types of industries which are expanding and might be moving into the region.

When selecting the specific site it is further important to:

- Estimate amount of land required through a study of local absorption rates for a five-year period (acres absorbed per year by type of industry typical to industrial parks, and number of transactions).
- Seek sites which are either immediately accessible to major highway routes or have highway frontage; are adjacent to the main line of a railroad, or near an airport or, if appropriate, near a harbor.
- Watch topography: acquire land with minimum of ledge rock, water, and peat or soft ground; however, land which is easy to develop may not be the best location for industry.
- Ascertain that water, gas, electricity, telephone, and, if possible, sewer should serve the site, at competitive rates with appropriate capacity.[1]

[1] R. John Griefen, "Creating an Industrial Park," *Lawyer's Title News*, May/June (Richmond, Virginia: Lawyers Title Insurance Corp., 1972). Also reprinted in *SIR News*, August (Washington, D.C.: Society of Industrial Realtors, 1972).

Development Criteria

In selecting his site, the industrial developer is thinking in terms of the resources of the specific community, the characteristics of the site and its environs, and how adequately and rapidly essential community facilities can function.

Most industrial parks include some or all of the following participants:

- distributors requiring warehouses with office space,
- light manufacturing,
- related commercial facilities, and
- small to medium-size research and development laboratories.

In contrast, an industry seeking a large single site undertakes complex studies which will vary by type of industry. Studies which might be done include labor quality, water supply characteristics, waste disposal, comparative transportation rates, distribution costs, and power rates. Because of the freestanding nature of many large industrial sites, companies seeking locations for these facilities tend to pay less attention than would the industrial land developer to land use trends and community planning proposals. Frequently, deleterious development has occurred at the boundaries of multimillion dollar industrial plants in a manner not anticipated by corporate management. Anheuser-Busch has taken the step of acquiring sites much larger than required for its breweries in order to develop the remainder of the property as an industrial park or primary recreation destination. The industrial park, well located and with strict land use controls, can frequently provide a protected environment for firms which previously sought an individual site.

The most important location consideration for an industrial park, or a large single industrial site, is the array of transport services available at the site. Rail is considered essential in most industrial parks over 500 acres. Proximity to airports is increasingly sought by many industries. However, the major difference between planned industrial districts before World War II, and the modern industrial parks dating from the mid-1940s, has been the new freedom in choice of locations made possible by improved roads, expanded auto ownership, and an expanding trucking industry. Developers who have become leaders in the industrial park movement began acquisition of highway-oriented sites years before the first new freeways were completed.

3-3 Construction of Interstate 83, linking Baltimore and Harrisburg, Pennsylvania, created a valuable industrial location in a previously rural area. Hunt Valley Business Community is in the foreground. Downtown Baltimore is at the top of the picture.

Highway Orientation

The Interstate Highway System has strongly influenced the location of modern industrial parks, industrial plants, and laboratories.[2] It has also provided customers for the new parks from industries displaced by clearance of older industrial areas in central cities for freeway rights-of-way. Both slum clearance and urban renewal—frequently done in concert with highway system development—also brought about the relocation of many small and medium-size industries whose old plants were demolished to make way for redevelopment.

---

[2] For a more detailed discussion of freeway impact on industrial location, see: A. J. Bone and Martin Wohl, *Economic Impact Study: Massachusetts Route 128, Preliminary Report, Industrial Development Survey* (Cambridge: Massachusetts Institute of Technology Press, 1958); and, Dorothy A. Muncy, "Influence of New Highway Program Upon Industrial Location," *Proceedings of the American Industrial Development Council*, Twenty-third Annual Conference (Boston, 1957).

Advantages of Freeway Locations

- *Transport savings:* the speed of transport brings economies in transport costs and savings in capital investment, and permits measurement of distances in terms of time rather than miles.
- *Expanded labor market:* the labor market area serving a plant is vastly expanded. The employee who spends 30 minutes getting to work driving five to ten miles on traffic-jammed local streets, will consider traveling 25 miles to work if it takes the same half hour.
- *Expanded market area:* plant or warehouse near the freeway can serve the central city, the expanding suburbs, and often other nearby metropolitan areas. Trucks also can serve a wider overnight delivery area.

With the advent of the interstate highway system, prime industrial land—previously defined by availability of rail service—was redefined in terms of frontage on, or proximity to, the freeways.

The tendency of industrial parks to cluster near freeway interchanges is illustrated in statistics taken from an inventory of industrial parks and vacant industrial land made in 1970 by the Port of New York-New Jersey Authority. This study was the first detailed industrial land inventory ever made in the 18 counties of the New York-New Jersey metropolitan area and was part of a comprehensive study of industries, labor supply, markets, industrial land, transportation, and educational resources.[3]

[3] Saverio Cappello, *The New York-New Jersey Metropolitan Area Industrial Development Guide* (New York: The Port of New York-New Jersey Authority, Planning and Development Department, 1971).

3-4 A glass wall building in the Waltham Research and Development Park, Waltham, Massachusetts, reflects the vehicles passing by on adjacent Route 128.

Of the 157 industrial parks in the New York-New Jersey metropolitan area which met the standards of the National Industrial Zoning Committee's definition of an industrial park: 82 percent were within two and one-half miles of an interchange of a controlled access freeway; and 38 percent had on-site rail, or had an available right-of-way for a spur to a main line within three miles of the industrial park.

In the inventory of individual vacant sites zoned for industry not located in industrial parks, the survey showed that of the 195,000 acres in the 18-county New York-New Jersey metropolitan area: 74 percent were freeway-oriented within two and one-half miles of an interchange of a controlled access highway; and 63 percent had rail on site, or had an available right-of-way for a spur to a main line within three miles.

The move toward freeway locations, for individual industrial plants and for planned industrial parks, has substantially increased land requirements because of the large parking lots needed for employees and visitors and because of off-street truck facilities required to accommodate inbound and outbound shipments and vehicular storage areas.

Prestige Locations, Visibility

Visibility is one advantage which freeways provide to industry and to industrial parks. Prominent sites with attractive buildings and extensive landscaping, which can be seen by the passing public, constitute an important form of free advertising for park occupants and are sought by firms wishing to establish or maintain a progressive and forward thinking public image. One freeway which has attracted a multiplicity of individual research laboratories and industrial parks seeking highly visible sites is the Massachusetts circumferential Route 128.[4] Another is the Dallas Central Expressway which has become the location of laboratories, industrial operations, and service facilities including Texas Instruments' 300-acre campus for its Semiconductor Division, and the Collins Radio Research Laboratory.

The Research Park

For research and development laboratories and other industries with professional staff, locations near a major university can provide opportunities for professional enrichment, use of library facilities, and consultations with professorial experts. Industries without large research or testing facilities can frequently contract for special services if the research problems are of interest to the appropriate university department.[5] These sites have prestige value as well.

The state of North Carolina recognized the importance of these considerations by locating the 5,000-acre Research Triangle on a site near Duke University, North Carolina State University, and the University of North Carolina. Limited to organizations engaged in research, development, and scientifically oriented production, occupants now include International Business Machines Corp., National Environmental Research Center of the Environmental Protection Agency, Forest Sciences Laboratory of the U.S. Forest Service, and Chemstrand Research Center of the Monsanto Company. Developed under the auspices of the Research Triangle Institute, a not-for-profit corporation, the Research Triangle has been in operation since 1959.

The industrial developer is urged to approach research parks with caution. These developments require an extended period to be built out; and the economic viability of many tenants is closely tied to the uncertain source of federal government funding.

The unique and specialized nature of many research parks is illustrated by Technology Square, located adjacent to the campus of the Massachusetts Institute of Technology in Cambridge, Massachusetts. This project, which began as a joint venture of Cabot, Cabot and Forbes with the Institute, occupies a 14-acre parcel served by mass transit. Previously occupied by an old soap factory, the site was cleared by the Cambridge Redevelopment Authority. Five high and low-rise buildings clustered around an open plaza are now complete. Outwardly, the development resembles an office park. Tenants are seeking the benefits of a location in proximity to the Massachusetts Institute of Technology.

Work now underway at Technology Square includes construction of approximately 450,000 square feet of additional low-rise laboratory and office space including landscaped plaza areas. This space is on a long lease to the Draper Laboratory, an independent nonprofit corporation whose research and instrumentation activities include atmospheric and oceanographic sciences, bio-medicine, transportation, energy, and industrial process controls.

---

[4] See New England Industrial Center—case study p. 35.

[5] See Stanford Industrial Park—case study p. 41.

3-5 Technology Square, Cambridge, Massachusetts. This research park occupies a highly urban site close to the Massachusetts Institute of Technology and Harvard University.

Family Oriented Recreation, Theme Parks

Selection and development of the site can have an important public relations value if the plans encourage sharing some part of the industrial facility with the public. For example, much of the traveling public is now familiar with the Busch Garden attractions at several of the Anheuser-Busch breweries. Gardens, aviaries, rathskeller hospitality houses, re-creations of natural landscapes such as the African veldt, an aerial tramway, boat tours and train rides, have made the breweries in Los Angeles and the Tampa Industrial Park into major tourist attractions. For its most recent brewery site, Anheuser-Busch chose the historic Williamsburg, Virginia area. The site was selected to serve the growing beer market in the southeastern United States; however, the plant is being developed in conjunction with a public attraction called The Old Country consisting of English, French, and German villages. Overnite tourist accommodations have been added as well.

3-6 A fountain is located in the German Village created by Anheuser-Busch, Inc. as part of The Old Country theme park in Williamsburg, Virginia.

3-7 This 92-foot tower, created by sculptor Aristides Demetrios, overlooks San Francisco Bay in the South San Francisco Industrial Park.

Other Special Features

A prestige location can be achieved by taking advantages of the evirons, or by designing special features into the employment center. A location overlooking a public park provides neighboring industry with both prestige sites and protected open space. In Don Mills, the planned community near Toronto, one of the attractive industrial areas features small industrial plants on a hillside overlooking nature trails in the public park, a favorite lunch time walking area. If no opportunity to utilize the existing natural environment exists, some developers have gone beyond the usual flowering shrubs and grass landscaping by installing contemporary art in their industrial parks. Atlanta Gateway Park's original developer, Great Southwest Corporation, has created a modern outdoor art gallery with sculpture, fountains, and ponds scattered throughout a 3,000-acre industrial park.[6] Outdoor art has been installed also in smaller industrial parks, including Evans-Pitcairn Corporation's Mid-Atlantic Park in West Deptford, New Jersey.

Airport Orientation

Proximity to an airport has become a locational factor for an increasing number of industries. Some companies have built research, engineering, production facilities, and headquarters offices adjacent to airports. Such a location appeals to industries with the following characteristics:

- Headquarters offices for firms with widely dispersed locations for their production plants, warehouses, research facilities, and regional offices;
- National and international engineering service companies which custom-build special equipment, supply special parts, provide maintenance service and operator training for unique installations (computers, power sources, environmental treatment facilities), or provide engineers for clients with dispersed locations;
- Warehouse and distribution centers for parts and perishable goods (food, flowers, hospital equipment, drugs); fragile or seasonal merchandise (fashion apparel/accessories, sport goods); and lightweight and high-value merchandise; and,
- Service industries related to air freight.

[6] Atlanta Gateway is currently a project of Bennett and Kahnweiler.

3-8 A light plane adjacent to an industrial building at Hayward Park, Hayward, California. A taxiway connecting with the Hayward Air Terminal has been extended through the park.

For example, Westinghouse Air Arm Division built a large factory and development facility to produce airborne radar equipment. The factory is located at the edge of the Baltimore-Washington International Airport with a taxiway to the main field. This location enabled Westinghouse engineers to supervise installation of airborne radar into planes at the point of production. The equipped plane then can be taxied from the plant to the main runway for take-off for flight testing. Not only does this proximity save travel time for the engineers who designed the equipment and supervised its installation, but controlled conditions for installation of precise instrumentation are provided also.

By 1973, the immediate environs of the Baltimore-Washington International Airport had attracted several manufacturing facilities which employ over 8,000 persons engaged in producing environmental control instrumentation; meteorological data loggers; corrugated shipping containers; fabricated and portion-controlled meat products; airborne electronics; and space and surface systems. Also included are engineering, research, and development facilities. In addition, several office headquarters buildings have been constructed on the parkway entrance road to the main terminal.

A partial list of industrial parks located adjacent to or a short distance from major airports includes:

South San Francisco Industrial Park South San Francisco, California	San Francisco International Airport
Crocker Ind. Park Brisbane, California	San Francisco International Airport
Great Southwest Industrial Park Arlington, Texas	Dallas/Fort Worth International Airport
Centex Ind. Park Elk Grove Village, Illinois	O'Hare International Airport
O'Hare Ind. Center Schiller Park, Illinois	O'Hare International Airport
Montbello Ind. Park Denver, Colorado	Stapleton International Airport
Irvine Ind. Complex Orange County, California	Orange County Airport
Arizona Interstate Industrial Center Phoenix, Arizona	Sky Harbor International Airport
Orlando Central Park, Inc. Orlando, Florida	Orlando Jetport

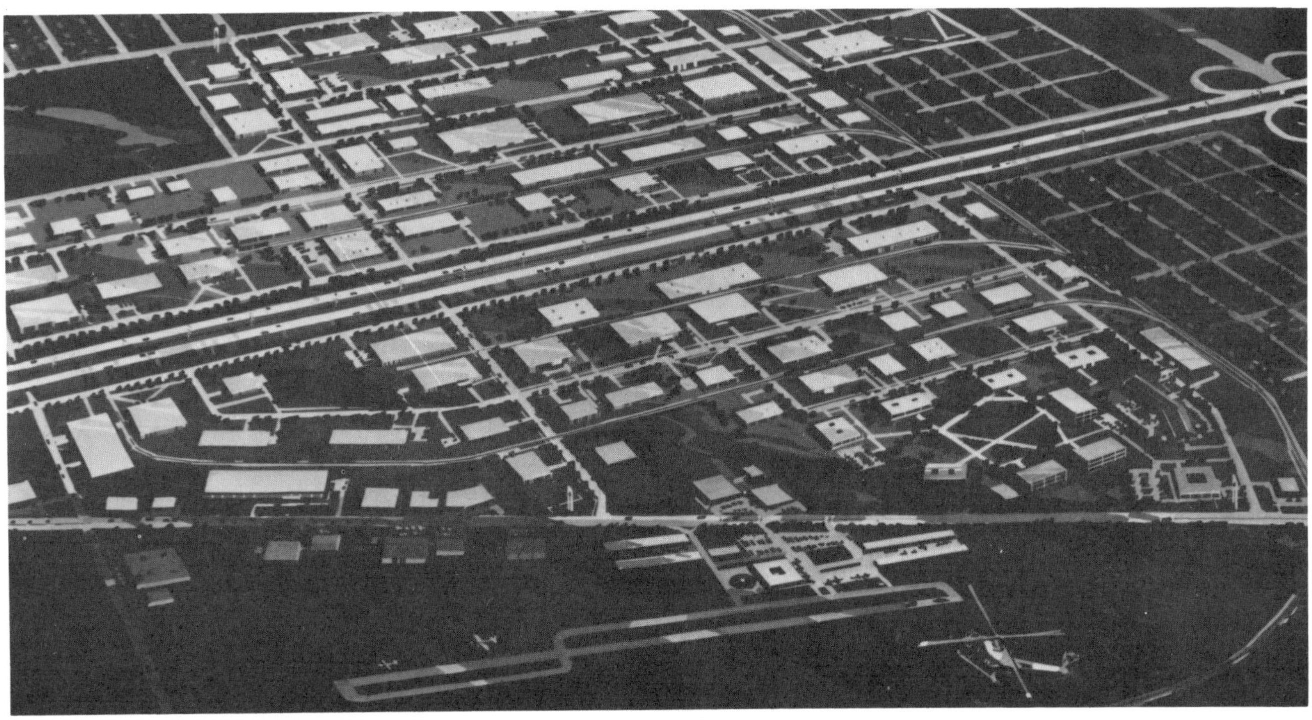

3-9 Centex—Schaumburg Industrial Airpark. This suburban Chicago industrial park is located adjacent to the village of Roselle Airport.

In some of these instances, an airport was not the primary site selection factor, but rather a matter of circumstance. However some park developers—for example, Cabot, Cabot and Forbes; and, Bennett and Kahnweiler—tend to seek proximity to airports for their major parks. For example, Marshall Bennett, chairman of ULI's Industrial Council, states that the site selections for Centex and the O'Hare Industrial Center were based very strongly on their proximity to O'Hare. Among industrial parks located adjacent to or in proximity to smaller air fields or general aviation airports are:

Industrial Center and Industrial Center West Hayward, California	Hayward Air Terminal
Hayward Park Hayward, California	Hayward Air Terminal
Miami Lakes Industrial Park Miami Lakes, Florida	Ocala Field
Atlanta Gateway Park Atlanta, Georgia	Fulton County Airport
Inverness Arapahoe County, Colorado	Arapahoe County Airport
Centex-Schaumburg Industrial Air Park Schaumburg, Illinois	Schaumburg Airport
Spirit West Industrial Airpark St. Louis County, Missouri	Spirit of St. Louis Airport

Although proximity of some major industrial parks to airports does not appear to be a primary attraction for a majority of industrial tenants, it is certainly not a drawback. The more transport services available to an industrial park, the better the location.

Airports can be expected to increase in importance as a factor in industrial location. Certainly the nature of airports makes industrial neighbors compatible. Many airports are ringed by a variety of industrial uses, some of which benefit from ready access to air transport.

Subways and Mass Transit Facilities

Programs to install or extend mass transit in urban areas will provide opportunities for industrial parks to locate in inner areas with redevelopment potential, or on larger vacant sites in outer areas to be served by the new lines. However, more likely uses for these sites will be office parks, highrise housing, and shopping/commercial centers.

The New York-New Jersey Metropolitan Area Industrial Development Guide prepared by the Port of New York-New Jersey Authority is an example of a positive strategy. Included in its 18-county survey of vacant industrial land is a count of vacant land zoned for industry and served by subway. The percent of vacant industrial land within one-half mile of a subway was high, for four of the boroughs. According to the survey, however, additional land near subways will have to be cleared to meet the expansion and modernization needs of industries now located in New York city, if the city is to retain this job and tax base.

THE PORT OF NEW YORK-NEW JERSEY AUTHORITY
1970 SURVEY OF VACANT INDUSTRIAL LAND IN THE 18-COUNTY AREA
3-10 Subway-Oriented Land †

COUNTY	SUBWAY-ORIENTED*	TOTAL VACANT INDUSTRIAL LAND	PERCENT VACANT INDUSTRIAL LAND SUBWAY-ORIENTED
Bronx	318 acres	647 acres	49%
Brooklyn	676	894	76
Manhattan	115	172	67
Queens	176	738	24
Subtotal for New York city	1,285 acres	2,451 acres	52%
Hudson, N.J.	406	7,547	7
Total	1,691 acres	9,998 acres	17%

* Vacant industrial land located within one-half mile of subway.

† Saverio Cappello, *The New York-New Jersey Metropolitan Area Industrial Development Guide* (New York: The Port of New York-New Jersey Authority, Planning and Development Department, 1971).

3-11 MODE OF TRIP TO WORK BY NEW YORK CITY BLUE-COLLAR INDUSTRIAL EMPLOYEES †

MODE OF TRANSPORT	NUMBER OF EMPLOYEES	BLUE-COLLAR INDUSTRIAL EMPLOYEES PERCENT OF TOTAL	CUMULATIVE PERCENT
local public transit	845,900	68.2%	68.2%
walk	43,800	4.6	72.8
suburban commute (public transit)	59,900	6.3	79.1
drive	198,400	20.9	100.0%

The Planning Department of New York City has recommended that the industrial renewal program be speeded up to provide 40 million square feet of new floor space for 100,000 new jobs in the city by 1980. The nine point program to increase manufacturing jobs had the following objectives:

- Protect the city's industrial land;
- Rebuild the city's older accessible industrial districts through staged renewal;
- Maintain mixed-use areas where mixed uses are compatible;
- Use vest pocket renewal to assist manufacturers to expand on sites currently in use;
- Develop new industrial parks on vacant land at the city's outer edges;
- Change zoning policy to encourage multi-story development in industrial areas where it is now discouraged;
- Use city financing to permit the construction of new multi-story buildings at rents most firms can afford;
- Encourage vertically zoned mixed-use development in suitable areas along the waterfront and over railroad cuts; and
- Allocate space in industrial renewal areas to companies which can contribute most to the city in terms of more and better jobs and higher wages.

3-12 BART System—San Francisco Bay Area.

The Port of New York-New Jersey Authority's industrial survey also showed drive-to-work patterns are related to transit access at the work place. Employees tend to patronize public transit which serves industrial locations. Among the few industrial parks served by rapid mass transit are:

- Technology Square, adjacent to the Massachusetts Institute of Technology, which is served by Massachusetts Bay Transit Authority's Kendall Square station.
- Hayward Park, Hayward, California, which is 0.6 mile from a Bay Area Rapid Transit station.
- Governor's Gateway—a mass transit terminal planned in the middle of the 1,000-acre Governor's Park, situated 25 miles south of Chicago in Park Forest South. The Illinois Central Suburban train makes this employment center readily accessible to a large labor force. It is the only major industrial park location in the Chicago metropolitan area to be served directly by rapid mass transit.

3-13

PERCENTAGE OF WORK FORCE COMPARED WITH JOB DENSITY AND SUBWAY ACCESS †

NUMBER OF JOBS PER ACRE	PERCENT DRIVING TO WORK IN ZONES WITH GOOD SUBWAY ACCESS	POOR SUBWAY ACCESS
10	47%	64%
30	37	54
50	36	52
70	35	52
90	33	52

† Saverio Cappello, *The New York-New Jersey Metropolitan Area Industrial Development Guide* (New York: The Port of New York-New Jersey Authority, Planning and Development Department, 1971).

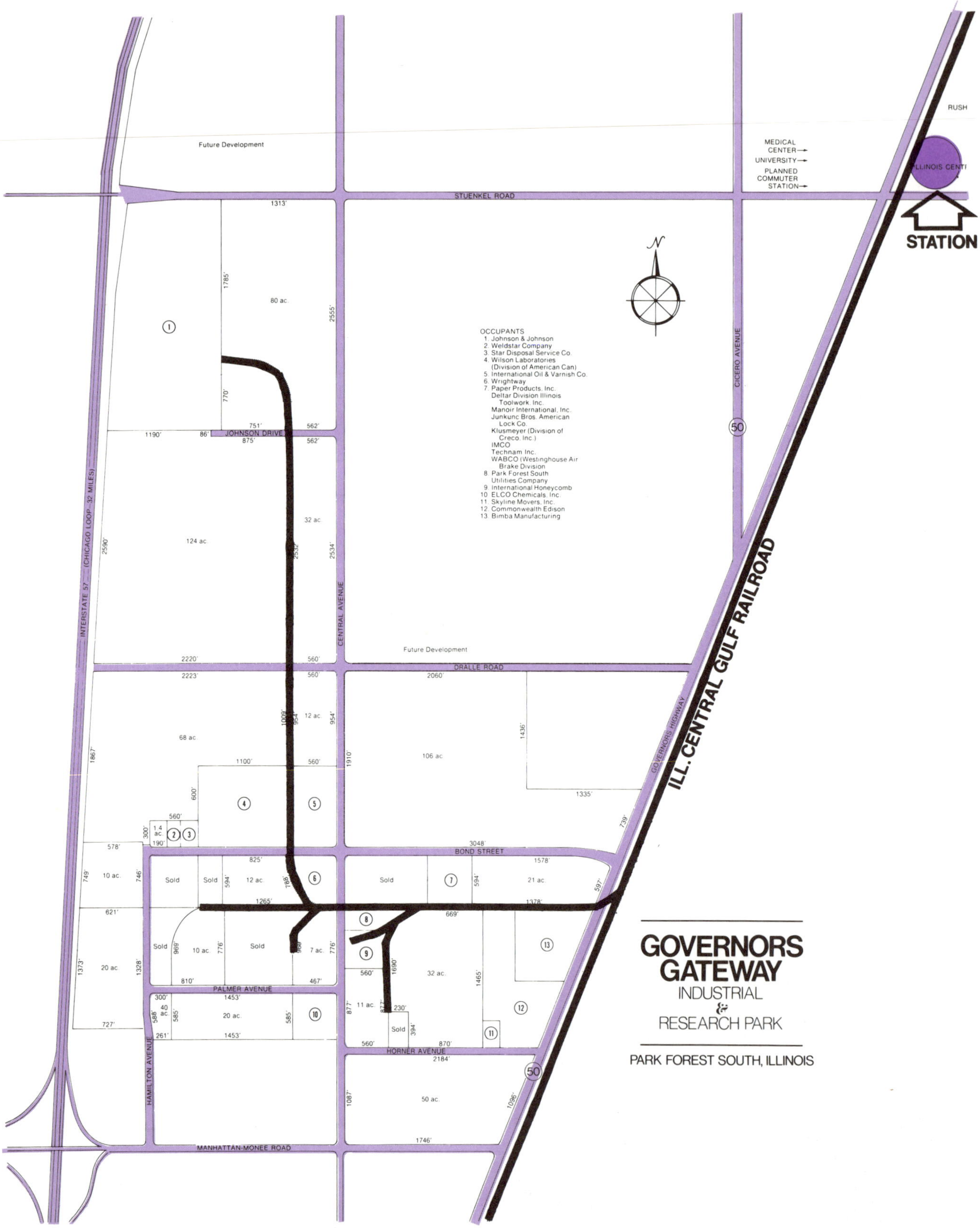

3-14 Site plan—Governors Gateway Industrial and Research Park, Park Forest South, Illinois. Plans call for the extension of commuter service on the Illinois Central Gulf Railroad to this industrial park, which is now under development. Note proposed location of passenger station on Stuenkel Road.

Linkages and Industrial Park Location

Many industries benefit from the proximity of several functions closely related to, or relying upon, the production process. The close proximity of linked industrial activities can reduce production costs by eliminating unnecessary movement of goods and personnel.

Industries are becoming increasingly aware of the interrelationship between transportation costs and industrial location, including the increasing cost of transporting service personnel, equipment, and supplies *within* a region. The freeway and interstate highway program is almost complete. Few extensions will be added. Emphasis in policy and in funding has shifted to mass transit. The national rail picture is one of consolidation, with little-used branch lines being abandoned and overall mileage shrinking. Many transport costs not emphasized in recent years will come under closer scrutiny, including:

- Distance in miles to be traveled within the metropolitan area;
- Reduced speed and longer travel time;
- Production time lost during travel;
- Employee time (on payroll) required for travel in the metropolitan area;
- Cost of vehicular fuel;
- Availability of vehicular fuel;
- Cost of vehicle, including maintenance, per mile traveled within metropolitan area;
- Reduced use of personal vehicles; and
- Air quality regulations and number of vehicle trips in metropolitan area.

Thus, the spatial clustering of interdependent linked industries, once a common phenomenon in central cities, is receiving increasing attention from both the urban planner and the developer of planned industrial land.

Typical examples of natural clustering of interdependent industries are the specialized districts which developed in large metropolitan centers. Often located in loft buildings at the edge of the central business district; these concentrations of small firms are typified by such names as the garment district, the fur district, and printers row. Other examples are wholesale flower and food markets, machinery parts and repair, commercial groceries, commercial kitchen equipment/supplies, and jewelry manufacturers. Within a given industrial cluster, each firm performs a process or service, or provides a product essential to the next stage of production or distribution, usually in an industry in which time is a major cost. In the case of the garment manufacturing trade, fashion is the perishable factor. In custom printing, the interrelationships between typesetting, photo plate making, binding, etc., are critical. Deadlines require close proximity of the various activities.

A manufacturing example of interdependent or linked industries is found at Tampa Industrial Park. Two breweries—Anheuser-Busch and Joseph Schlitz—are served by a glass manufacturing plant and a dried brewer's grain supplier, located in the same industrial park. The management of Tampa Industrial Park, and the Committee of 100 of Greater Tampa Chamber of Commerce, are to be credited for their studies of the location requirements of a glass bottle manufacturer. Availability and quality of raw materials, availability of gas and electricity, and the existence of a potential market, resulted in the location of the two breweries in the park. Subsequently, a carefully planned program to interest various glass firms to locate in Tampa was undertaken.[7]

The potential of industrial parks to serve the need of some industries to cluster, is worth further analysis. Clearance of old buildings on the fringe areas of the central business district—to create more space for commercial office centers, government buildings, housing, and sometimes urban university or hospital expansion—can be expected to continue. Attendant with this clearance will be the displacement of small, linked industries. In the case of small industries, government relocation programs and similar assistance is often inadequate. In the past, too many small firms—cleared out of the urban center—have died on the way to new locations.

Before this problem arises, however, the planning department, the industrial development agency, and experienced industrial developers must establish a cooperative program to examine:

- Potential areas for demolition;
- Potential locations for future concentration of industry;
- Needs of industries which will be moved; and
- Needs of industries which may remain after many firms functionally related to them have moved.

[7] Robert E. Boley, *Industrial Districts: Principles in Practice,* Technical Bulletin no. 44 (Washington, D.C.: Urban Land Institute, 1962), p. 93.

A second reason for considering the possibility of industry clusters is the opportunity to save transport costs and time. Fuel, especially, can be conserved through the minimization of intra-urban movement of parts, supplies, and service personnel between related industries. In a vertical industrial development, this can be accomplished by elevator or other materials handling systems. Within the more typical horizontally developed industrial park, special areas can be planned for the use of linked industries. While potential for developing sections within industrial parks for interrelated industries will not exist at many locations, this opportunity needs more attention and analysis than it typically receives in the usual planning, sales, and marketing programs of most industrial parks.

Bypassed Parcels

In every metropolitan area, pieces of left-over land are interspersed among the residential developments, commercial centers, and industrial facilities. While some of these vacant parcels can be seen by auto inspection, or by study of up-to-date land use maps, bypassed areas are most noticeable from a low-flying plane. Varying in size from a few acres to more than a hundred, many of these bypassed parcels are potential small industrial park sites or vest pocket employment centers—if vehicular access is either available or can be developed. Intrusion of industrial traffic into narrow residential streets is the foremost problem in development of these parcels.

Benefits of utilizing this left-over land for small planned industrial uses include:

- Industry is not handicapped by delayed construction of streets or other public services;
- Existing sewer, water, and streets do not require additional public investment;
- Travel time is shortened for many employees;
- Tax base is expanded; and
- Walk-to-work opportunities are available for many workers.

Frequently, zoning is a problem in utilizing bypassed vacant land for industrial purposes. In many instances, existing zoning will be residential or light commercial. If the project is rezoned industrial, using an older and less restrictive industrial zone typical of many ordinances, the ordinance controls will not incorporate development standards which protect either the neighboring uses or the future industrial occupant. The new and more restrictive industrial park-type zones usually require spatial standards which cannot be accommodated on these smaller sites. Thus, a new zoning district may be needed which does not have excessive yard requirements because these are not justified at locations where walk-to-work housing and public transit are available. Compatibility between the small industrial park and its neighbors can be achieved through careful design review and site plan approval.

Community and Regional Land Use Plans

The search for potential locations for future industrial parks should be initiated by contacting the operating land use agencies and by referring to its publications. If the sites under consideration are within an incorporated city—the municipal planning department, county planning department, or regional planning council probably will be in authority. Some states have land use control legislation and, consequently, a state agency may be concerned about large employment centers and development at critical locations.

The Comprehensive Plan

At the local level, the developer should refer to the comprehensive plan or master plan for the community, county, or region. This will identify locations proposed by the planning agency for future industrial areas. Early identification of potential industrial areas is essential for orderly development of a community's economic base. Therefore, an economic study and industrial land use plan are basic elements in most comprehensive planning efforts.

The purpose of these studies is to quantify the existing nature of the local economy, and to measure its strengths, weaknesses, and future potentials. The findings of these studies will serve as a means of determining the probable magnitude of future industrial growth and of developing strategies to effect those changes. In order to accomplish these objectives, a wide variety of techniques are available to the public planner—including industrial land use surveys, economic base analyses, employment studies, and input-output analyses.[8] Actual techniques used in any given situation will depend upon the nature of the community and the purpose for which the planning program is being conducted. The work itself is carried out either by

[8] See also Section 5 p. 162 for discussion of these studies in relationship to the developer's market analysis.

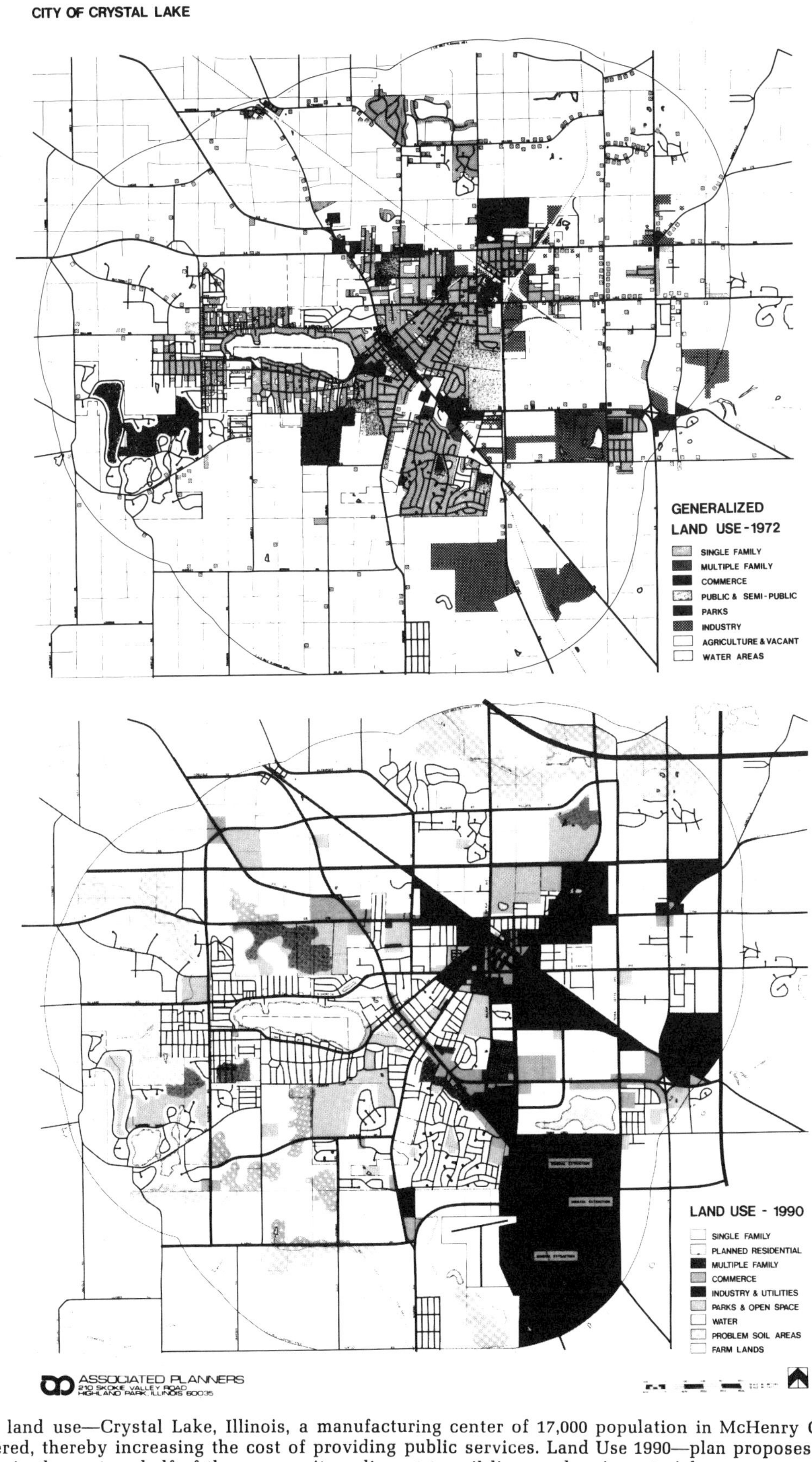

3-15 Existing land use—Crystal Lake, Illinois, a manufacturing center of 17,000 population in McHenry County. Industrial areas are scattered, thereby increasing the cost of providing public services. Land Use 1990—plan proposes concentrating industry generally in the eastern half of the community, adjacent to rail lines and major arterials.

a municipal planning staff or by consultants. Based upon these studies, it is possible to project alternative future levels of employment and industrial land absorption rates—factors which are critical in determining what provisions need to be made to accommodate future expansion of industrial activity in the community.

Normally, the comprehensive planning program is either a designation of those areas deemed most suitable for industrial use, or a listing of goals, objectives, and criteria upon which site location decisions will be based, or both. The plan also incorporates mechanisms—most notably zoning—for the community to preserve areas for future use.

An example is the comprehensive plan prepared by Harland Bartholomew and Associates for the Racine (Wis.) Urban Planning District. Initial surveys quantified the heavily industrial makeup of the community and identified 2,678 acreas devoted to this use.[9] The nature and location of this industrial land, and prospects for future change, were evaluated in detail. These steps provided the basis for projecting a 41.5 percent increase in industrial land area by 1990. Despite this substantial increase, acreage of industrial land is expected to decrease from 1.89 acres to 1.69 acres per person.

Areas proposed for conversion to industrial use were then located on the 1990 land use plan. Arrangement of these areas was based upon 1) existing land use pattern; 2) suitability for development; 3) arterial street systems; 4) existing zoning; and 5) amount of land estimated to be needed.

Both the nature and utility of such studies are aptly illustrated by the Baltimore Regional Industrial Land Study.[10] This study was made for the Baltimore Regional Council and covered the city and four adjacent counties. Existing industrial areas and vacant land zoned for industry were evaluated. Projections of future industrial employment were made and locations for future industry recommended.

Existing zoning was not a criteria for recommending locations for future industry. Most important was transportation potential—including expansion of the port, of the planned system of regional highways, and of regional plans for the extension of water and sewer transportation.

Usually proposed locations for future industry were adopted by each county planning commission, together with other recommendations for upgrading zoning standards and land use policy, as part of a comprehensive plan for future industrial land use. Thus, when land owners applied for rezoning of land for industrial development, the local legislative body would consider the adopted plan for industrial land use in making its decisions. Published studies of this type often guide developers in selecting land for future industrial park development.

Today, most of the potential industrial locations recommended in this 1959 study are now developed as industrial parks or occupied by major industrial plants. One of the largest is the Hunt Valley Business Community, developed by McCormick and Company, Inc. on a 250-acre site in Baltimore County, Maryland. The property had been designated industrial because it afforded both rail and freeway access.

The owners of the future site of Hunt Valley Business Community were experienced residential developers. They planned to subdivide into one-acre lots; and the property was zoned residential. The chief planner of Baltimore County objected to this subdivision proposal, urging that this unique location—identified in the 1959 regional industrial land study—be rezoned from residential to industrial to protect it for the county's future economic growth. By 1962, approximately 250 acres were rezoned to M-L (Manufacturing-Light).

At this time, McCormick and Company, Inc., spice manufacturers with a multi-story factory facing the Inner Harbor in Baltimore city, was looking for a site to expand its manufacturing and distribution facilities. Its involvement led to the development of the Hunt Valley Business Community. As of 1975, the Hunt Valley Business Community has approximately 16,000 employees and is the third largest suburban employment center in the Baltimore region.

Clearly, the industrial developer should encourage comprehensive planning for industrial development. The existence of such plans will assist him in his site selection process, and will reduce the risk of rejection of a proposal for nonconformance with the community's plan. Likewise, the community should elicit the participation of the industrial developer in the development of a comprehensive plan for the area so that areas identified for future industrial use are consistent with the needs of industry.

[9] *Comprehensive Plan for the Racine Urban Planning District*, Vol. 1, Inventory Findings & Forecasts; Southeastern Wisconsin Regional Planning Commission, Harland Bartholomew and Associates, December 1970, Planning Report No. 14.

[10] Dorothy A. Muncy, *Industrial Land Development*, Technical Report no. 2 (Baltimore: Baltimore Regional Planning Council, 1959).

Analysis of a community's comprehensive plan will allow a developer to make the following basic determinations:

- If a location for the proposed industrial park has been purchased already, the developer can determine if the planning agency has officially recognized its industrial use potential.
- If the land is under option, and is shown as future industrial on the comprehensive plan, the developer can evaluate his land in relation to other proposed land uses in the vicinity, with regard to such concerns as the availability of existing or proposed transport facilities; proposed public works such as water, sewers, schools; and other essential public services. All of these will affect the quality of the location, development timing, site planning, and to some extent the types of industry which can be attracted.
- If the location is not shown as future industrial on the master or guide plan, then the developer will need to take the initiative, and convince the citizens, planning commission, and city council, that his proposed undertaking will constitute a desirable addition to the community.

Other studies made by the planning department staff and other public agencies will be useful to the industrial park developer throughout the process of development. These include, as discussed earlier, special economic and social studies, projections of population and income, economic growth studies, public open space and recreation programs, and special area studies focusing on subareas of the community which provide more detailed land use recommendations.

Special Planning Studies

City and county planning—with large jurisdictions and substantial amounts of undeveloped land—update sections of the master land use plan for those areas which soon may be slated for extension of water, storm, and sanitary sewers, and other major public works such as bridges, highway widening, or freeway extensions. In the urbanized sections of a county, a proposed mass transit route, or intensive commercial and office development, also may activate a sectional study to update the planning agency's proposals for land use changes.

If the future industrial park site is to be located in a section under special study by the planning department, the zoning application for the industrial park should be held in abeyance. The developer should have a preliminary study made of the site by a planning consultant, and then should discuss the proposed industrial park with the chief public planner. These steps should be taken before the planning department is ready to publish a preliminary sectional land use plan and prior to public hearings on the sectional land use plan. While this approach will delay the application for rezoning, the legislative body responsible for zoning changes usually would postpone such decisions anyway pending receipt, or even adoption, of the final sectional land use plan.

Preliminary Planning Analysis

The prezoning stage of the industrial development process involves formation of the planning/zoning team; a detailed analysis of the character of the site and its environs; an evaluation of the effect of the official land use plan upon the site development; a calculation of the effects of zoning ordinance standards upon the utilization of the site; and the impact of the development on the community. ULI Industrial Council member Ben Dyer notes that a key preliminary analysis consideration should be the economic impact of the project on the local community. He observes that unsureness as to the value of the proposal to a community has frequently been the basis for withholding or delaying approvals. The developer must be able to quantify these benefits to the community. In addition, the developer should begin to put his concept of the park on paper. Finally, the relationship of the industrial park to the community goals should be thought through objectively.

The Planning and Zoning Team

By the time the preliminary planning analysis has begun, the developer already should have formed his planning and zoning team, the key members of which are:

- Urban planner: trained and experienced in community planning, industrial land use, and zoning for industry, the planner's role includes evaluating project impacts, preparation of design alternatives, and serving as an expert witness at zoning and other approval hearings.
- Engineer: the engineer is responsible for evaluation of site characteristics including topography, soils, utilities planning, drainage, and other areas of engineering expertise. The engi-

neer relates these to development alternatives and prepares development cost estimates.

- Attorney: the attorney is responsible for development of all legal documents related to acquisition, development controls, and conformance to legislative requirements. Experience in zoning and other land use control matters is required.
- Other specialists: a variety of specialized personnel also may be required to attend to matters such as environmental standards, air quality, noise control, traffic engineering, seismology, economics, etc.

This team should be assembled early in the development process. Information essential for preliminary meetings with the public agency planning staff, and at the zoning hearing, is also needed by the developer during every phase of the development process including the sales program. However, the developer must realize the desirability of keeping permanent staff to a minimum in order to reduce overhead. Maximum use of consultants should be made either on a full or part-time basis.

Character of Site and Environs

Every person involved in the planning of the site, or testifying at the zoning hearing, should be familiar with the physical characteristics of the site and of adjacent areas. Much of this information can be obtained from aerial photos, topographic maps, land use plans, highway plans, flood plain maps, and other source materials. It is also essential that the whole development team—developer, planner, engineer, attorney, and other specialists—walk the site in order to become as personally familiar with it as possible. This exercise often can lead to development decisions which make the difference between a great industrial park and a mediocre assembly of industrial structures. The best developer is one who understands his land, and recognizes his caretaker responsibility for the natural environment as well as the economic health of the community.

ULI Industrial Council member R. John Griefen recounts an interesting example of advance environmental planning for Cabot, Cabot and Forbes' (CC&F) Arizona Interstate Industrial Center:

CC&F decided that the construction of a new interstate highway joining Tucson and Phoenix, plus the expansion of the Sky Harbor International Airport, would change the trend of industrial expansion from the north and west sectors of the city to the south and east sectors. Accordingly, CC&F purchased a substantial amount of land to the southeast of downtown Phoenix, adjacent to the proposed new highway and in close proximity to the airport.

There was a long delay in the completion of the highway construction. The property was located in the bed of the Salt River, an area that had a history of flooding. A government flood control project, installed many years previous to the purchase, had actually diminished the danger of major flooding; but as an added precaution, CC&F installed its own dike on the property.

Initially, the industrial community offered substantial resistance to the location because it was not in the accepted expansion pattern for the community, the highway was delayed, and the flood problem remained a question mark. CC&F was confronted with some challenging marketing conditions for the first three years. However, as a result of the completion of the highway, a flood along the Salt River in which the property remained bone dry, and a groundbreaking for several nationally prominent firms, the park began to gain acceptance. The wisdom of the location now has been proven with over one million square feet under roof for 60 firms in a six-year period.

Griefen, in describing CC&F's development land policy, reports:

> The natural landscaping is preserved as much as possible in preparing the property. Wherever possible, trees are marked and carefully saved during the construction period to retain the natural environment. In the 750-acre, Interstate 95 Industrial Center at 495, in Eastern Massachusetts, landscape architects have gone over the entire property to determine the value of all growth and to mark the appropriate trees and shrubs that will be preserved.

Effects of Official Land Use Plans and Public Works Programs Upon Site Development

The land use plan, the master plan of highways, plans for mass transit, scheduling of water and sewer facilities, and recent rezoning decisions in the vicinity, will all affect the development pattern.

Even though the land use plan may designate a general area as a future employment center, the amount of land held by the developer may be in excess of the acreage recommended by the planning agency staff study. The official highway plans may not provide adequate access for the size of the site. Programming for water or sewer extension

3-16 Arizona Interstate Industrial Center, Phoenix. Location and development of this industrial park was based on a carefully-devised series of criteria. See pages 76-77. Downtown Phoenix is in the top of the picture.

may need to be advanced. Adjustment of the differences in the public plans and the private development proposal, will need planning and engineering analysis.

Effects of Zoning Standards Upon Site Development

Developers logically view zoning requirements in the light of land preparation costs, net acres, number of sites, and current market price for industrial acreage. Unless zoning standards result in very high subdivision costs, the developer usually will accept them. The industrial park entrepreneur, however, may discover that some standards in the zoning ordinance will seriously circumscribe his development plans. The types of industries appropriate to his location, together with ancillary commercial uses intended to benefit the occupants, may be restricted. Rigid space standards may preclude an efficient or an imaginative site plan.

Usually, the industrial builder will find his property falling within one of three industrial zoning categories: (1) an industrial park zone (most restricted); (2) a light industrial zone (seldom requires a park-like environment); and (3) a heavy industry zone (permits virtually all industrial uses). Although the light industrial zone might seem to be the easiest to live with the legislative body would usually prefer to grant industrial park zoning to an application which is proposing to develop an industrial park. The problem is many industrial park zoning standards may no longer fit the changing needs of employment centers.

Enhancing Ordinance Flexibility

Although industrial zoning standards in many communities have caught up with, or frequently exceed, modern industrial park development standards, it is appropriate to evaluate industrial zoning in light of recent land use planning innovations which will be changing the location and development pattern of industrial park employment centers. Some potential changes include:

Land areas required for off-street parking. For some industries, more land is required to park the employees' car than to provide and service his work space. Land area required for parking can be reduced by:

- walk-to-work housing;
- use of special bus lanes on Interstate and regional freeways to provide fast public transit for employees to freeway-oriented industrial parks;
- reservation of land near rapid transit terminals for employment centers; and parking structures or parking under, or on top of, buildings.

Yard requirements. Reduce large front and side yard requirements on individual lots in industrial parks by using the saved open space for one or more shared recreational areas for employees and families in the employment center.

Provision of recreation space:

- use roof tops for recreation areas;
- use parking lots during nonworking hours for temporary recreation activities (tennis, basketball, etc.).

Approval procedure. Consolidate local performance standards and state or regional air and water quality standards into a simplified one-stop procedure for approval of control equipment and continued monitoring and enforcement.

Expansion of permitted uses. Permit the range of uses in employment centers, which are needed by the occupants, to reduce the many costs of transport (time, money, space, and pollution):

- encourage linked industries to cluster in a section of an employment center;
- permit commercial services (hotels and meeting facilities, offices, shopping, etc.) in employment centers to meet industry's operating requirements and accommodate employees' work and personal needs;
- design industrial parks to provide walk-to shops, banks, restaurants, housing, and recreation; and
- link bike and pedestrian paths in employment centers with regional or community path systems.

Height limitations. Encourage use of multi-story structures to save land for recreation and conservation areas. The developer may decide that he would like to propose a new zone to be added to the existing zoning ordinance, contributing the professional services of his organization to the local planning staff and zoning administrator in drafting an industrial zone to incorporate some of the changes suggested above.

Developing the Industrial Park Concept

The developer and his team must begin by identifying the unique location characteristics of the site; the types of occupants seeking this location and their space needs; the special physical qualities of the site to be preserved or enhanced; and the services to accommodate future occupants. With these steps, the industrial park concept begins to form. A design concept graphically interprets the planning recommendations for the site's potential development. Based upon physical, economic, and social factors; and an analysis of the site—its environs and the community; these recommendations are translated into spatial relationships of functions and facilities to be distributed upon the specific piece of real estate. The plan should reflect unique factors inherent in the site, and special requirements of types of industries which might be attracted to this kind of location. At this stage, the developer will be able to make judgments about the overall character of the park, including tenant makeup and land space and service needs of prospective occupants. The orientation of the site to freeways or major roads, to the landscape within or nearby, and to the potential of man-made features as focal points, are major considerations for the industrial park layout. Possibilities for functional sub-areas within the development become apparent. Preliminary road and access points can be sketched. Rail service, if available, should be considered.

Against these tentative outlines, existing impediments may indicate the need for additional land purchases or changes required in the space or use standards in the zoning ordinance.

3-17 Artist's conception of Pacific Gateway Center, a 278-acre industrial development now under way at the intersection of the San Diego and Harbor Freeways in Los Angeles. The park occupies the former site of a Shell Chemical Co. synthetic rubber manufacturing plant. Development concept anticipates a combination of commercial and industrial uses.

The Approval Process

The approval process includes finalizing the design concept; conducting work sessions with public planners and related agencies as needed; preparing covenants if required; and completing graphics and final written reports, which are to be submitted for approval. With these steps completed, the developer and his team are ready for the public hearing.

ULI Industrial Council member Lowell Eastman makes the point that obtaining approval for an industrial development is a cooperative undertaking which requires effort on the part of both the developer and the community. "Normally it takes at least one year for a developer to reach a working rapport with a community in which he is undertaking a project. The city too must become familiar with its responsibilities in meeting requirements imposed by other governmental jurisdictions, in order to properly respond to the developer. Among these requirements to be understood as they relate to local conditions are environmental impact reports, and the various requirements of the Environmental Protection Agency."

Preparation of a successful application for rezoning requires competent professional teamwork, adequate time to prepare the case, and a developer who understands the community and his site in relation to that community.

Advance Preparation

Given the many tasks to be performed in preparation for the approval hearing, it should be clear that the best strategy is for the consulting team to prepare its materials sufficiently in advance of the hearing date to be able to resolve any unanticipated problems which may arise. Several months may be required to conduct studies, complete work sessions with public agencies, and fully document testimony to be presented. Conferences with public agencies produce better results if scheduled and prepared in advance, rather than under pressure and with interruptions. Lead time is also needed for developer, attorney, planner and engineer to develop a basis for working effectively with one another. Most important, a reasonable preparation period will give the developer time to question, assimilate, partially test alternative development strategies presented by the planner and engineer, and to insure that all potential assets of the land will have been identified and utilized in the chosen design scheme.

Preparation of the design concept should be accomplished during this stage of the development process even though it may not be actually necessary for the zoning hearing or other approvals. The design concept is a graphic summation of the developer's objectives and of all the planner's observations and analyses. It will be useful for the orientation of public officials, as well as for future site design.

The design concept will help the developer retain his objectives while his plan is first modeled, then modified in the approval process, and later modified in detailed design and marketing programs. It has the further advantage of demonstrating to the zoning examiner and other public officials the sincerity of developer intent and the care he has taken in evaluating the sites potential.

Coordinating the Approval Process

The first hurdle in the approval process is the report and recommendation to be made by the local public planning staff. Community planners can be conservative. Frequently they recommend against or heavily condition, industrial rezoning applications or other approvals. Community planners are concerned with both the immediate and the long-range impact of sizable developments upon the community. The planning agency and the public often contend that too much land has been zoned for industry which highlights the needs for having a completely detailed comparative inventory of industrial land. The local legislators may be worried that present zoning standards are not sufficient to protect the neighborhood. Deficiencies should be noted if they exist and proposals made to make needed modifications in the regulations. Furthermore, public officials are deeply concerned that premature development may result in excessive public costs in utilities and streets, creating an imbalance in budget programming and in design plans for public facilities. A cost/benefit analysis may be required.

The planning staff report is an important document in the record of the approval hearings. Therefore, the cooperative analysis of problems before the hearing is very beneficial, and often results in a more favorable report. Assuming the developer receives basic approval of the project, the detailed engineering and design is undertaken. The next section discusses these aspects as well as describing in greater detail the preliminary work which must proceed simultaneously with the broader planning and approval process discussed in this section.

4.
Engineering and Design

Introduction

This section provides the developer and his staff with insight to engineering and site design considerations which are vital to industrial land development. The intent of this section is to provide the developer with a checklist—rather than specific engineering design methods or criteria—which can be applied to a parcel of land under consideration for development.

The goal of a preliminary engineering analysis of a parcel of land is to develop an accurate preliminary estimate of the cost of improvements and maintenance. With this cost information, the developer can prepare a credible financial projection of development costs for the entire project. Project engineering and design is not an end in itself. Rather, it should be considered an integral part of a developer's proforma analysis of the development costs, expenses, sales, income, and profit projections comprising the totality of the project.

Preliminary Engineering and Design Considerations

Project feasibility evaluation begins early in the development process with a preliminary engineering study of the site. By such procedure, physical site characteristics can be studied and evaluated. Any problems which may be too costly to remedy can be uncovered before a major financial commitment is made to proceed. Such preliminary work is usually done on a limited budget and, therefore, cannot be expected to determine and evaluate all problems. However, major problems should be evaluated during this phase of the work. Ideally, preliminary work is completed before site acquisition, as is discussed later in the section on development strategy.

Topographic Survey

The topographic survey is basic to site evaluation. Survey data are displayed on a map form which shows the ground surface contours of the site. Contours of perimeter property should be shown also, if available—particularly if they relate to watershed areas. This base map is used for terrain analysis studies which indicate such characteristics as ground slope and surface drainage patterns which provide important clues to the presence of sinkholes in limestone terrain, fault lines in earthquake prone areas, and susceptibility to flooding.

USGS Maps

Important and relatively inexpensive tools for preliminary engineering evaluation of a site are the quadrangle maps published by the United States Geological Survey (USGS). These maps cover four-sided areas bounded by parallels of latitude and meridians of longitude. The most useful maps for land development studies are the 7.5 by 7.5 minutes (one inch = 2,000 feet), and the 15 by 15 minutes (one inch = one mile).[1]

The slope of the land portrayed by contours is the distinctive characteristic of these maps. Contours are imaginary lines following the ground surface at a constant elevation above or below sea level. Where contours are close together, the ground surface is relatively steep. Where contours tend to spread apart, the ground surface is more gently sloping or flat.

[1] United States Geological Survey (USGS) maps can be bought over the counter at nine public inquiry offices located in Anchorage, Dallas, Denver, Los Angeles, Salt Lake City, San Francisco, Spokane, Washington, D.C., and Reston, Virginia. The two major public inquiry offices are located in Washington, D.C. (202) 343-8073 and 8074, and Reston (703) 860-6167 and 6168.

4-1 A difficult site was accommodated to industrial use by building a four-lane bridge across a stream which separates the property into two parts. Nonconnah Corporate Center, Memphis, Tennessee.

While such published topographic maps are useful in preliminary investigations, their use is limited due primarily to the large contour interval—generally 10 or 20 feet. They may become outdated quickly in urbanizing areas where man-made improvements are being made frequently. The USGS has a program of updating maps in urban areas as funds permit. A careful check should be made for the most current update available and for any updating in process. These maps are excellent preliminary planning tools, but they are not substitutes for field instrument or aerial surveys which must be conducted as part of a detailed engineering site analysis. In addition to USGS maps, the U.S. Corps of Engineers has other maps and studies of value—particularly for areas of flooding or near navigable waters. For example, the Corps publishes a series titled Flood Plain Information Studies, covering many urban areas.

Field Instrument Surveys

Topographic surveys with field instruments should be run on a certified plat of the property to be developed. The work should be done by a registered surveyor. In the typical survey, the property is laid out and staked in a grid pattern using a transit and chain. The grid, usually rectangular or square, may be any conventional dimension from 50 feet to hundreds of feet, depending upon the slope of the terrain and the topographic detail required. Elevations are determined at the location of the stakes and at other locations in between where abrupt changes in elevation of the terrain occur. A topographic map is then prepared from this field data.

Existing physical features such as springs, creeks, and the perimeter of wooded areas, also should be located in the topographic survey. It is desirable to show the density of wooded areas as well. Large specimen trees suitable for retention should be singled out for more precise location. The location of existing man-made features such as roads; railroads; utility lines—including poles and appurtenant structures, sewer lines, culverts, and other drainage facilities; and buildings, also should be determined. It is costly to send a survey crew back into the field for additional data; therefore, the developer and engineer should determine carefully what data ultimately will be needed before proceeding with a field survey. For example, the field location of nonphysical encumbrances such as easements and underground rights might be appropriate for surveying and mapping.

Topographic maps prepared from ground instrument surveys are generally limited to relatively small developments. Where developments of considerable size are contemplated, it is more economical to have topographic maps prepared by aerial survey using photogrammetric methods and augmented by ground survey work only where necessary.

Aerial Surveys

Topographic maps prepared by aerial photogrammetric methods are in wide use today. These maps can be produced quickly, economically, and accurately, with the modern equipment currently available. An added benefit of the aerial survey is the availability of an aerial photograph series which extends beyond the limits of the topographic map and may be used in the planning process to study off-site influences including access, vehicular circulation, and recent construction in adjacent areas.

Prior to proceeding with new aerial photography, a check should be made for available aerial maps from sources such as local public agencies, state and county highway departments, railroads serving the area, and adjacent property owners. While not necessarily suitable for final survey map preparation, these maps may be very acceptable for preliminary analysis.

The aerial survey is an alternative to the ground survey, but does not totally eliminate the need for ground surveying. A limited amount of field work is required to establish control (distances between, and elevations of, identifiable points on the aerial survey). The decision to use aerial methods rather than ground instrument surveys is primarily one of economics. An important factor affecting the cost of the field survey is the slope of the terrain. Ground surveys in flat open terrain are less costly than those in steep terrain with heavy foliage. Survey time is increased in steep terrain or terrain with heavy foliage due to limitations of the equipment. Therefore, aerial surveys have a definite advantage in such locations. Nevertheless, these surveys also are limited by heavy foliage because the camera must be able to record the ground surface in order to obtain the information from which the ground elevation can be determined. If possible, surveys should be run during the winter, or early spring season, when the deciduous vegetation is at a minimum.

Soils and Foundation Conditions

Preliminary evaluation of the soils and foundation conditions at the proposed site should be made at the earliest practicable time by a competent soils or geological engineer. Early analysis often can determine potentially troublesome conditions such as soils of low bearing capacity, and swelling or shrinking soils which can affect floor slabs and pavements. Also, the presence of shallow depth to bedrock can increase the cost of excavation, pinnacled bedrock can interfere with grading operations, and sinkholes in terrain underlain by limestone may require expensive stabilization. High groundwater levels requiring special subdrainage systems around foundation and basement walls, or requiring costly spreading and drying of excavated materials prior to placing in fills during rough grading operations, can also be identified. Many of these problems can be determined by the preliminary soils and foundation evaluation. A preliminary study of soil characteristics will determine suitability for industrial development. Most important will be identification of site areas where soils characteristics render the land economically unsuitable for building or other surface or subsurface improvements. The presence of significant amounts of land with these impediments must be ascertained as early in the site analysis as possible because this condition will affect the developer's feasibility analysis.

The state geologist often can provide valuable preliminary soils information useful to the developer in evaluation of his property. For example, the Missouri Division of Geological Survey and Water Resources has published a series of geological studies of the St. Louis area.[2] The report and accompanying geologic map describe the geological features of St. Louis County, including an engineering interpretation of these features.

[2] *Engineering Geology of St. Louis County, Missouri*, Engineering Geology Series no. 4 (Rolla, Missouri: Missouri Division of Geological Survey and Water Resources, 1971).

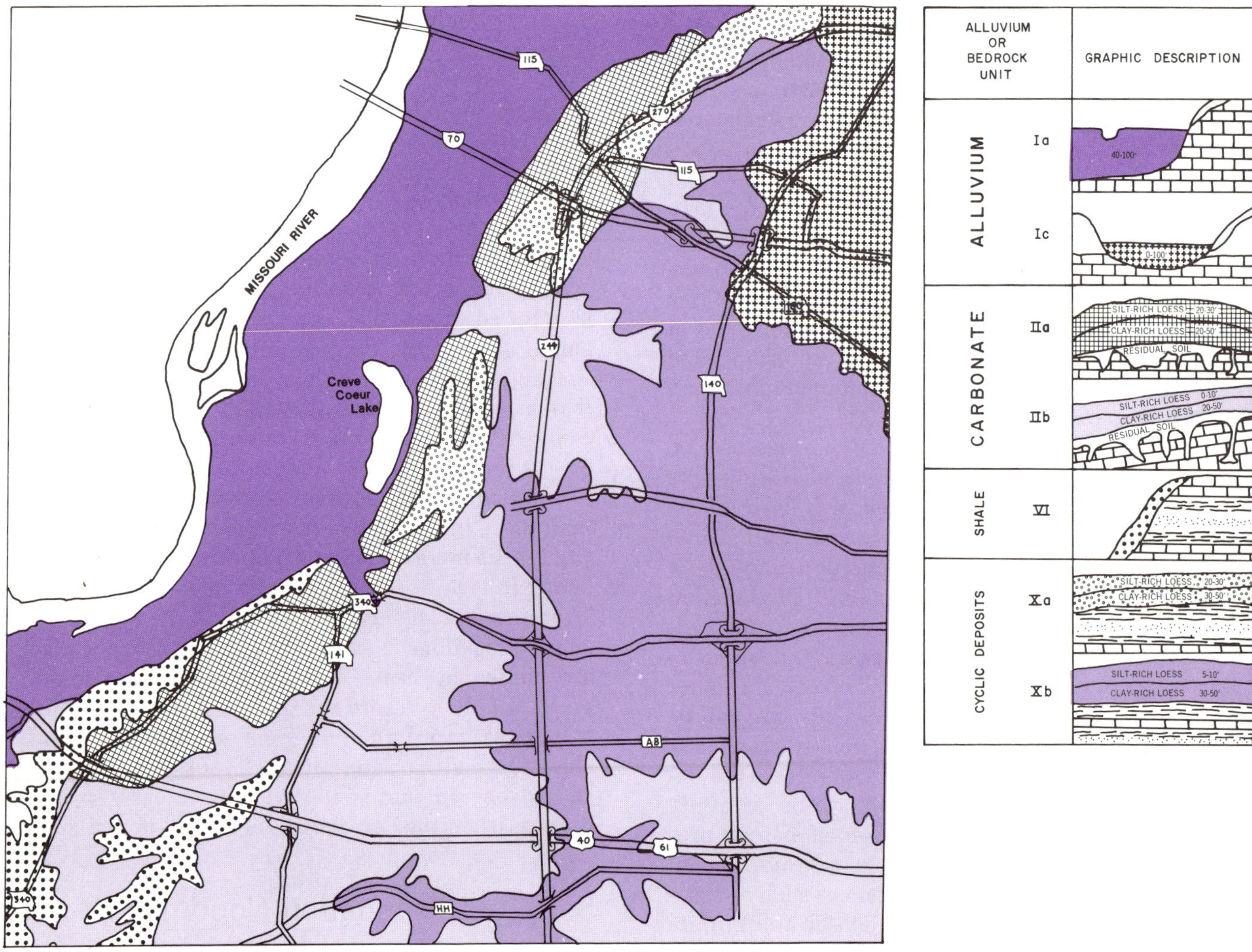

4-2 Geological map, St. Louis County, Missouri. Maps of this kind reveal important information about the development potential of varying land areas. This map shows a portion of the flood plain of the Missouri River and adjacent hilly areas.

The report evaluates the area with respect to waste disposal, water pollution, slope stability, excavation properties, bearing capacity, conservation of mineral resources, and special problems. Such reports provide a sound basis upon which to begin geologic investigation of a specific site. The developer should contact government agencies in his locale regarding availability of similar useful information pertaining to the potential development area.

Another useful document for preliminary engineering evaluation of a site is the agricultural soils map. These maps are prepared by the U.S. Department of Agriculture and by various states. The maps show the extent of soils of similar characteristics across a site. Soil surveys ordinarily extend to a depth of three to six feet and are of such design purposes as roadways and other paved areas.

The Geological Report

The soils or geological engineer will determine site conditions and select sites for borings using as a guide the topographic maps, published geological information, and tentative development plans.

Subsurface investigations normally are made with a truckmounted rotary drilling rig. Since this process is expensive, the number of preliminary borings is kept to the minimum needed for an adequate evaluation. Consequently, care must be taken to locate the borings around the property in such a manner that full coverage of various soil types is obtained. Borings should extend to strata of adequate bearing capacity and should penetrate all deposits which are unsuitable for foundation purposes such as unconsolidated fill, peat, organic silt, and very soft compressible clay.[3]

Samples will be taken from the drilled holes. Testing will be done in the boreholes as drilling progresses and in the laboratory using soil samples taken from the drilled hole. Groundwater levels generally can be determined by observation in the drilled hole. Typical tests conducted in the drill holes include the standard penetration test and some shear tests. These soil tests are used to determine in-place strength and safe bearing capacity. The standard penetration test is generally reliable in soils other than clays. Strength and bearing capacity of clay soils are commonly determined by laboratory testing or by vane shear testing in the borehole. Settlement potential of the foundation soils due to loading from structures can be predicted by tests performed during preliminary investigations.

The geologic report may include the following items:

- A plat showing borehole locations.
- A profile and log of each borehole describing materials encountered. The log usually will be presented in the report as an exhibit including standard penetration test results, moisture content, and other pertinent test data.
- Discussion of the general nature of surface and subsurface soil and rock conditions.
- Results of laboratory tests of soil samples.
- Range of soil bearing capacities.
- Settlement considerations applicable to foundation footings.
- Evaluation of groundwater conditions.
- Generalized recommendations relative to applicable foundation treatments including anticipated foundation supporting elements.
- Generalized information relative to roadway and parking lot subgrade treatment.
- Anticipated conditions to be encountered during utility construction, particularly deeper facilities such as sanitary sewers.
- Suitability of on-site soil for use as fill.
- Suitability of site for ponding of stormwater (potential leakage).
- Thickness design recommendations for rigid (concrete) and flexible (asphalt) pavements for streets and parking lots. (Design of streets to be dedicated for public use will normally comply with the standards of the public agency having jurisdiction but frequently variations are possible if based on well conceived engineering recommendations.)
- Stripping requirements for vegetation and organic material.
- Rippability of bedrock (as opposed to blasting).
- Depth to bedrock.

A preliminary geological investigation and report necessarily will be general in nature. A foundation investigation should be performed for each building to be designed and constructed on individual sites within the tract. A common practice is to take a test boring at the location of each corner of a proposed building. Intermediate borings are frequently taken depending upon the building's size and geometric configuration. As a spot check of the analysis of test borings, a relatively inex-

[3] M. Juul Huorsler, *Subsurface Exploration and Sampling of Soils for Civil Engineering Purposes* (Vicksburg, Mississippi: Waterways Experiment Station, 1949), p. 16.

pensive method of visually and physically checking subsurface conditions at selected locations of a site is by use of a backhoe. This equipment can excavate a trench to a depth of 12 to 14 feet or more as may be required. The observer may enter the excavation and check the soil in its natural state. Some undesirable conditions such as land fill sites may be discovered in this manner. However, this technique in no way substitues for a thorough professional investigation.

Availability of Utilities

Early investigation to determine availability and capacity, should be made of all utility services which are normally required by anticipated industrial users. These include sanitary sewers and treatment facilities, water supply, treatment and distribution facilities, and natural gas, electric, and telephone service. Industrial parks in outlying suburban areas frequently require extension of utility services. These services may be provided either by a private company or by a governmental agency. In large rural or suburban developments, sewer and water services may have to be provided by the developer. If services will be provided by an existing agency, the developer should meet with these organization and obtain written agreements for extension of services which indicate when these services would be constructed, the capacity of services to be provided, and the share of cost to be borne by the developer if any.

Water Service

An important consideration is the extent of expansion of the existing water distribution system which may be required to provide adequate flows to the industrial development to meet fire protection requirements. Preliminary determination of flow and pressure of the public main can be obtained from the utility for planning purposes. While the extension of an existing public main may be adequate for initial phases of development, industrial districts of several hundred acres in outlying areas will often require additional sources of distribution from other parts of the existing system as the area grows. When a master plan of the development has been prepared, an analysis of the proposed water distribution system will be necessary so that adequate pipe sizes can be selected based on anticipated water main extensions. This matter will be covered in more detail later. In preliminary analysis, however, it is important to be aware of the extent of the existing system and to come to some agreement with the water company about location of water supply main tie-ins to the total park layout.

Such early discussions with the water company will be beneficial to both the developer and the utility. The developer will have a more definite basis on which to plan water distribution system loops and the phasing of development, and the utility will have justification for including extensions, in its future plans, which coincide with the developer's master plan. Such cooperation will result in savings to both parties. As pointed out by ULI Industrial Council member Ben Dyer, in many localities, the mains are sized and built within the dedicated streets by the utility company. Costs are assessed against the abutting properties. Preplanning is important to avoid over or underdesign. In additon, Council member Sherman Eubanks stresses the need to determine the quality of the available water particularly if such potential users as a brewery might have special requirements for water quality.

Sanitary Sewer Service

Sanitary sewer service typically is provided by a public agency and sometimes by a private sewer company. The developer should meet with the authorities to determine: 1) the capacity of the existing sewer system to accept estimated flows from the proposed development; 2) the capacity of the sewage treatment plant to accept the estimated flows; 3) whether discharge by gravity is possible or pumping is required; 4) the policy of the sewer authority on sewer extension to pick up the proposed development; 5) who pays for the sewer extension; 6) the charge for connecting the park sewers to the authority sewers; 7) the periodic service charge for use of the sewers and sewage treatment facility; and 8) quality restrictions on sewage effluents.

While policies vary widely, connection charges usually are made by private companies, and frequently by public agencies, to allow recovery of the cost of construction of facilities to serve new customers. These connection charges typically are made either on an acreage basis or on a per fixture unit charge basis. These charges can be substantial and need to be determined in advance even though such advance decisions may be based on an estimation of anticipated flows. Even though these charges may not be a cost to the developer, a good marketing program must be able to identify these costs for the potential user.

Connection charges made by municipally owned sewerage systems are usually quite small compared to those of private companies. This is because municipal systems are eligible for federal grants to help finance construction of new facilities. Even so, specific main extensions, or enlargements of treatment facilities to serve new industrial projects or industries, are generally excluded from these grant programs. The ground rules under the new revenue sharing provisions of The Housing and Community Development Act of 1974 are still unclear, but presumably, the community will allocate funds only to projects of community wide impact. Private sewer companies are currently ineligible for these funds and must make other arrangements for construction of sewage treatment facilities.

Federal and state agencies will have a continuing and growing concern for the adequacy of collection and capacity of treatment facilities, and also for the quality of treatment. Pre-treatment requirements are becoming very common for many industrial uses. These requirements should be identified during preliminary engineering evaluations.

Periodic service charges are made to occupants of the industrial park by the owner of the sewerage system. The service charge may be made monthly, quarterly, or some other convenient period, and is made either on a flat rate or based upon water consumed. The latter basis is more common because it is generally more equitable. The service charge enables the sewer authority to recover operating and maintenance costs. The service charge rate structure, while not directly affecting the developer, will be very important to prospective purchasers of property—particularly those who are heavy water users such as bottling plants or vehicle maintenance facilities with washing operations. Accordingly, the developer should be thoroughly familiar with service charges as well as with connection charge practices of the sewer authority. ULI Industrial Council member Otto Pongrace suggests that in discussing service charges, the basis for negotiation should be actual cost of treatment of anticipated discharge rather than an arbitrary fee tied to water consumption which for high water users has no relation to discharge in many instances. Paul Shepherd, a vice president of Cabot, Cabot and Forbes, points out that heavy use of water for irrigation of landscaping can result in a penalty if water consumption is used as the basis for sewer system service charges.

Electricity, Telephone, Gas Service

Other local utility companies should be contacted during preliminary work to determine the availability of services. Utility rate structures may include either partial or total cost of placing overhead electrical and telephone service, and of placing gas mains underground. Due to environmental and aesthetic considerations, it is becoming more common to bury electrical and telephone lines. The cost of an underground system throughout the development may be substantially greater than the cost of a typical overhead system. As demand increases for underground utility services, however, the additional cost is likely to be borne by the utility company and incorporated into future rate structures. In recent years, there has been a marked reduction in the cost differential between overhead and underground systems.[4] Curvilinear street layouts of modern industrial parks favor underground systems which need no guys at each turn and are unaffected by changes in street grade. Equipment improvements are also responsible for reduced underground costs. A further factor affecting the relative cost of overhead and underground services is the policy adopted by many units of the Bell Telephone System of automatically placing lines underground. The Bell System has found, in most instances, it is easier for them to install and maintain underground facilities. Such a policy leaves power companies, which normally enjoy joint pole agreements, with the full cost of overhead construction and thus encourages them to pursue an underground service policy as well.

Electrical and telephone services are generally in ample supply. However, in outlying areas, the developer may have a choice of obtaining service from more than one source for these utilities. These choices should be evaluated carefully in the preliminary engineering stages. Small power companies frequently have records of frequent and lengthy power outages and voltage fluctuations which industrial users may find unacceptable, Service from independent telephone systems may be totally inadequate to potential users requiring an excellent level of communication with customers or a tie-in to computer services.

Many industries rely on, and expect, ample supplies of natural gas in industrial developments. As the supply of natural gas has diminished, its availability to industry has become more restricted.

[4] *Buried Cables*, Technical Bulletin no. 48 (Washington, D.C.: Urban Land Institute, 1964).

Priorities recently established by gas companies give new residential customers first priority. When curtailment is necessary, industry is the first to be affected. Curtailments occur during winter months when home heating exerts heavy demand on supplies. Some areas which have favorable geologic conditions have developed underground storage space and are affected less severely. Early engineering studies should determine the present availability of natural gas, the likelihood of curtailments, and should evaluate alternatives.

Design Standards and Criteria of Public Agencies

Engineering for industrial development requires early determination of the standard practices of the reviewing authorities from whom approvals will have to be obtained prior to construction. While ample room exists for design innovation, certain improvements will have to be designed and constructed in accordance with established standards and criteria.

In larger municipalities and in metropolitan areas, typical street cross-sections and intersection details are available to the engineer. Design criteria for streets may include pavement widths, permissible grades, and centerline alignment, including maximum degree of curvature and stopping sight distance.

Standards for sewer construction may include pipe bedding, inlet, and manhole details; and open channel details. Engineering design requirements for sanitary and stormwater drainage facilities also are available from local agencies. Requirements for sanitary sewers may include type of pipe, gradients, depth, manhole spacing, population factors, and basis for flow design determination. Criteria for stormwater drainage facilities may include design method to be used, rainfall frequency, and runoff coefficients based upon percentage of imperviousness. Standard construction specifications also should be obtained by the engineer.

The engineer should be completely familiar with the requirements of those public agencies having jurisdiction. In the interest of the developer, the engineer should become acquainted with agency personnel so that necessary communication and processing of documents can occur expeditiously. Such a relationship will minimize the time and effort required for review of submittals and will help to assure that submittals are prepared in accordance with established criteria. While practices vary across the country, street rights-of-way generally are required to be dedicated to the municipality. Streets, and storm and sanitary sewers, are dedicated to the public agency having jurisdiction over land to be subdivided for sale to the public. Private companies typically require easements over the land and/or ownership of the facilities.

All of the above requirements have significant cost implications. The preliminary engineering analysis should note those requirements which appear excessive. Based on discussions with the local agencies involved, the analysis should recommend modifications which, if accepted, could have significant effect on development costs. While current local practices must be complied with, modifications will be considered. More frequently, local agencies are willing to reconsider standards if presented with reasonable alternatives. However, securing changes is always time consuming and, unless they are identified early in analysis and planning, delays in obtaining public approval may be more costly than proceeding with established standards even though excessive. This is unfortunate when it happens, because it benefits no one and ultimately the unnecessary costs are borne by the public.

Environmental and Ecological Considerations

A short time ago, industrial land developers were free to design and construct projects with relatively limited governmental regulation of their efforts. What regulation did exist was at the local level. This is no longer true. Several recently enacted federal laws are having, or will have, significant effect on many current industrial park projects due to federal requirements for protection of the environment. The National Environmental Policy Act (NEPA) of 1969, the Clean Air Act of 1970, the Federal Water Pollution Control Act Amendments of 1972, the National Flood Insurance Act of 1968, and the Noise Control Act of 1972, all have a potentially far-reaching effect on industrial land development. As Sidney Robbins, an industrial engineering consultant of Chicago points out, "the developer now must allow considerably more time for processing of approvals and must be prepared to accommodate changes which are required for conformance."

Under NEPA, proposals for many projects using federal funds must be accompanied by an environmental impact statement (EIS). Several states and many units of local government either have already or are considering legislation which would require the preparation of environmental impact statements for various kinds of development activity including such industrial development types as power plants and mining operations. In California, any new construction requires compliance with an environmental review procedure including the preparation of an environmental impact statement. This report, according to ULI Industrial Council member M. Sherman Eubanks, takes 60 to 90 days to prepare and may cost a developer from $1,000 to $20,000 or more.

The purpose of the EIS is to document the effect of a proposed land use change before the change is made. Nothing in NEPA requires states to adopt an EIS law. However, as stated earlier, there is a growing tendency for states to do just that.[5]

An example of such activity is Florida's legislation enacted in 1973 to protect the ecology of the state. In certain restricted areas, the state planning agency has defined categories of development which have regional impact and therefore, which should be subject to review at the state level. General guidelines have been prepared for determining which development has regional impact. When development in restricted areas (primarily coastal) is contemplated, the developer must fill out a lengthy questionnaire. The questionnaire which is subject to review and approval by the state, seeks to determine impact of the proposed development on environmental and natural resources of the region; impact on the regional economy; impact on public transportation; impact on facilities such as sanitary and storm sewers, water and power supply, solid waste, and impact on housing.

The coastal states have special ecological considerations not found in the interior of the country. It is apparent that developers in these coastal areas must be especially alert to the rapidly changing and increasingly restrictive land use regulations in their areas. A review of current regulations indicates the need for extensive input in the early stages of a development project from the engineer, planner, economist, environmental scientist, and geographer, when answers to environmental questions are required.

The Clean Air Act of 1970 required all states to submit their plans to implement the Act's provisions by August 15, 1973. This law requires information on anticipated traffic flow and vehicular usage from indirect sources and will affect industrial developers' future projects. Effects on industrial park development at the present time are not known. As of February 1974, only Florida and the territory of Guam had submitted plans for air quality regulation which had been approved by EPA. While implementation of the Act has been delayed, it remains law and presumably will be implemented in all 50 states.

A major feature of the Federal Water Pollution Control Act Amendments of 1972 is the enforcement of a new national permit system. This system, the National Pollutant Discharge Elimination System (NPDES), requires industry to obtain a permit for discharge of any pollutant into navigable and certain non-navigable waters of the United States. Should a developer in an outlying area propose to provide sewage treatment and the effluent were to discharge to waters of the United States used for commerce, recreation, or other purposes, his project would be subject to review by EPA and to open public hearings. The same permit regulation would not necessarily apply to industrial waste discharges into a municipal system since the municipality would be required to have a permit. However, where pretreatment is required, the states can require individual industries to obtain permits from the state in order to reduce the administrative load on the municipal agency.

The federal government also has taken steps to provide safe working conditions for employees. The Williams-Steiger Occupational Safety and Health Act of 1970 (OSHA) requires the employer to provide safe, hazard-free working conditions for employees. For example, sewer trenches in excess of five feet in depth must be adequately sloped or supported to prevent cave in. The U.S. Department of Labor may make an inspection if a violation is suspected.

Another environmental issue which has resulted in federal legislation is the problem of development occurring in floodplains. The National Flood Insurance Act of 1968 will have far reaching effects on the financing of all development in flood prone areas. This act came into effect on July 1, 1975.

It will probably take several years of administration of these laws, combined with extensive litigation, before there is a clear understanding of the responsibilities of various levels of government and private developers. This legislative area is new and is changing very rapidly. A developer needs to keep abreast as requirements and procedures

[5] James A. Roberts, "Just What Is An Environmental Impact Statement," *Urban Land*, vol. 32, no. 5 (Washington, D.C., 1973), p. 9; Robert W. Burchell and David Listokin, *The Environmental Impact Handbook* (New Brunswick, New Jersey: Rutgers—The State University, 1975), 234 pp.

are clarified.

In addition to complying with national environmental legislation, the developer and his engineer should be aware of local performance standards which may be already in the regulations. Frequently, the zoning ordinance for various industrial zones sets forth performance standards for noise, dust, smoke, odor, vibration, and lighting which will require the industrial developer to submit what amounts to a partial or mini-EIS. Where local performance standards exist, they should be evaluated carefully at the preliminary engineering stage. Performance standards can have positive value in that they can eliminate the need for some requirements in the covenants. They can also have significant impacts on the land use arrangements and may limit the suitability of certain portions of a site for certain types of industrial development.

Energy Constraints

The provision of ample supplies of energy in an industrial development is a fundamental prerequisite to the production of goods and services. From the point of view of the occupant, any prospective industrial site must offer certain essential energy sources.

Energy sources usually are considered to be natural gas, oil, and electricity. Due primarily to an ever increasing demand, each of these forms of energy is in relatively short supply. Natural gas, priced at less than its full value, has been consumed in large volume uses to such a degree that a shortage in this fuel accentuates shortages of other fuel forms. As large-volume use of natural gas is reduced, fuel oil, propane, and coal are being used to fill the gap. The 1974 foreign oil embargo vividly demonstrated the narrow margin between adequacy of supply and serious shortage. In spite of determined government efforts, the market price of propane and heating oil has increased because demand exceeds supply.

Electricity often has been thought of as offering the potential to meet the energy shortage because it is not derived entirely from liquid and gaseous hydrocarbons. For the past decade or so, the trend for new fossil fuel power plants has been away from coal toward either natural gas or fuel oil. The two main factors causing this shift were the underpricing of heat content in natural gas, and the implementation of stringent air pollution standards which precluded the use of high sulfur content coal. It also must be mentioned that concurrently, the coal industry was undergoing a fundamental shift with respect to mine safety and other factors which made it less price competitive. Nuclear and hydropower, which often are termed renewable, furnish less than 5 percent of total electricity and cannot help alleviate the problems in the short run. The net effect of an energy supply-demand imbalance is pressure to raise the price as users attempt to outbid one another. It is important to realize that little can be done by energy suppliers in the short run to reduce the impact of the energy problem. For instance, a new fossil fuel-fired power plant involves at least six years from initial design to project completion.

The American growth ethic has generally obscured the fact that much of the energy crisis can be alleviated by devising energy conservation techniques. Up to now, energy usage has grown proportionately with increases in the Gross National Product (GNP); and, some experts seriously question the ability to reduce energy consumption and still maintain adequate GNP growth. Recent work by the National Bureau of Standards, however, has shown that there is a trade-off between energy consumption and building capital cost. That is, certain features can be built into buildings at the time of construction which will provide a reduced level of energy consumption throughout the building's life.

Design tools are available to aid in predicting energy consumption in buildings. One such tool is an advanced computer program originally developed under contract to the natural gas industry by the Southwest Research Institute. It is called E Cube because it has three sections which examine energy requirements, equipment, and economics.[6] A thorough analysis of an industrial development must rely on computerized analysis to determine the amount of energy which probably would be required and what steps can be recommended to effect a rational balance between energy conservation and building design.

The E Cube Program simulates the annual operation of a building on an hourly basis for the 8760 hours in a year. Computer output allows judicious selection of building construction materials and optimization of building mechanical equipment. But energy conservation cannot reduce energy consumption to zero. There is still a need for the in-

[6] "E-Cube Computerized Energy Analysis: Energy, Equipment, Economics," *Journal* (New York: American Society of Heating, Refrigerating, and Airconditioning Engineers, September 1971), pp. 46-59.

dustrial developer to critically examine how the overall energy needs of prospective firms will be met.

Central energy plants could be the answer in some industrial developments but the central energy plant must be constructed before the manufacturing plant or office building to have the energy required when building construction is complete. The energy plant, if hot and chilled water, could be oversized or undersized depending upon the mix of plants, warehouses, and office buildings being constructed. In most central energy plants, expansion is planned to add future chillers or boilers. Anticipated energy demand will determine the feasibility of a central energy plant for an industrial development.

The engineer has methods available to him for computerized energy analysis and for economic feasibility studies to determine the payback period related to the increased initial cost versus the savings in operating cost.

The application of solar energy as a source of power is in its infancy. It would appear that industrial developments with single-story construction and large roof areas, plus extensive open space, present a real opportunity for the development of solar energy. While this may not prove to be a primary source of power, it could be a significant supplemental source, particularly for space heating and cooling.

In planning an industrial development, the developer must be alert to secure adequate forms of energy. Certainly the cost of energy will continue to increase. By using energy conservation techniques, the price rise can be mitigated somewhat. However, firms will still require adequate energy to produce a desired array of goods and services. Therefore, the developer should prepare an overall energy plan at project conception and should continue to modify and adapt the overall plan to the demands of the industrial development as it emerges from concept to maturity.[7]

Landscaping and Preservation of Existing Vegetation

Preliminary evaluation of the site should be made to determine whether desirable tree stands, ponds, or rock outcrops, or other existing features worthy of preservation, can be included in the master plan. Such items can be preserved frequently without a significant effect on project economics. For example, a depressed area of the site may have been historically avoided during clearing for agricultural pursuits due to its poor value as cropland. The area may be of questionable value to the developer due to peat or other poor soils in the depressed area. Such a setting may mean expensive excavation and backfill or special foundation preparation if the area is to be used for building purposes.

Further, local regulations in the zoning category under which the project is to be built may require significant open space. Preservation of such an area may then be justified due to open space requirements. Setting aside land for stormwater retention or detention may be required. If this is the case, otherwise unusable land may be extremely valuable as part of a drainage system. Judicious creative planning and engineering can, therefore, turn negative site features into visual and economic assets. Except in areas unsuitable for buildings or pavements, it is quite difficult to justify preservation of existing vegetation in an industrial development except when it will be located along roadways or in front yard setback areas. Thus, new landscaping becomes an important part of project planning. Plantings can be used to soften the appearance of the park by breaking up the continuous expanse of pavement and buildings.

Landscaping should always be used to give industrial developments a coordinated appearance. Whereas the architecture of particular buildings may be somewhat uncoordinated due to the individual desires of the various owners and the need to meet the functional requirements of a manufacturer, the landscaping program can be coordinated throughout the development by the introduction of parkways, or median planting along the streets. Frequently, street intersections may be landscaped as part of the theme. Care must be used in selecting and positioning intersection plantings to avoid obscuring vehicular visibility. The covenants of an industrial park may be a convenient place to specify landscaping requirements so that suitable plantings may occur on each lot. The zoning ordinance may specify the size and diameter of each tree or the number of trees per unit of area of the site. Many jurisdictions have adopted very specific landscaping requirements—particularly for parking areas, screening of outdoor storage, loading areas, and public rights-of-way.

[7] For a description of how one developer is dealing with energy limitations in his area, see the Busch Corporate Center, Columbus, Ohio case study in a later part of this section.

4-3 Landscaping for this single-story industrial building in Erin Mills, Ontario, Canada features birch trees, carefully-placed boulders, and clumps of bushes.

Responsibility for maintenance of landscaping in dedicated areas and other quasi-public areas should be explored and taken into consideration in the landscape plan.

In addition, local agencies frequently can offer excellent advice, based on their past experience, as to selecting suitable varieties of trees and shrubs, soil preparation, and other aspects of landscape engineering which may be of great value to a developer.

Landscaping is included in this manual under preliminary engineering considerations because of the importance of having the engineer, planner, and landscape architect evaluate together the potential for preserving worthwhile vegetation. A special consideration is the cost involved in preserving vegetation when such preservation would require radical adjustments in storm drainage, roadways, rail, and other improvements necessitating a substantial outlay of additional funds.

The determination of the value of natural site features is not always a simple economic analysis. Frequently, a higher development cost can be justified if it can be shown that preservation of an existing feature would enhance the marketability and quality of the project.

Conclusions

The preliminary engineering considerations discussed thus far form the basis, along with preliminary planning and financial studies, for the decision to proceed with site acquisition and *detailed* engineering and design studies. The more data which can be collected regarding a specific site, the less likely it is that an unforeseen cost or physical constraint will appear later in the development process.

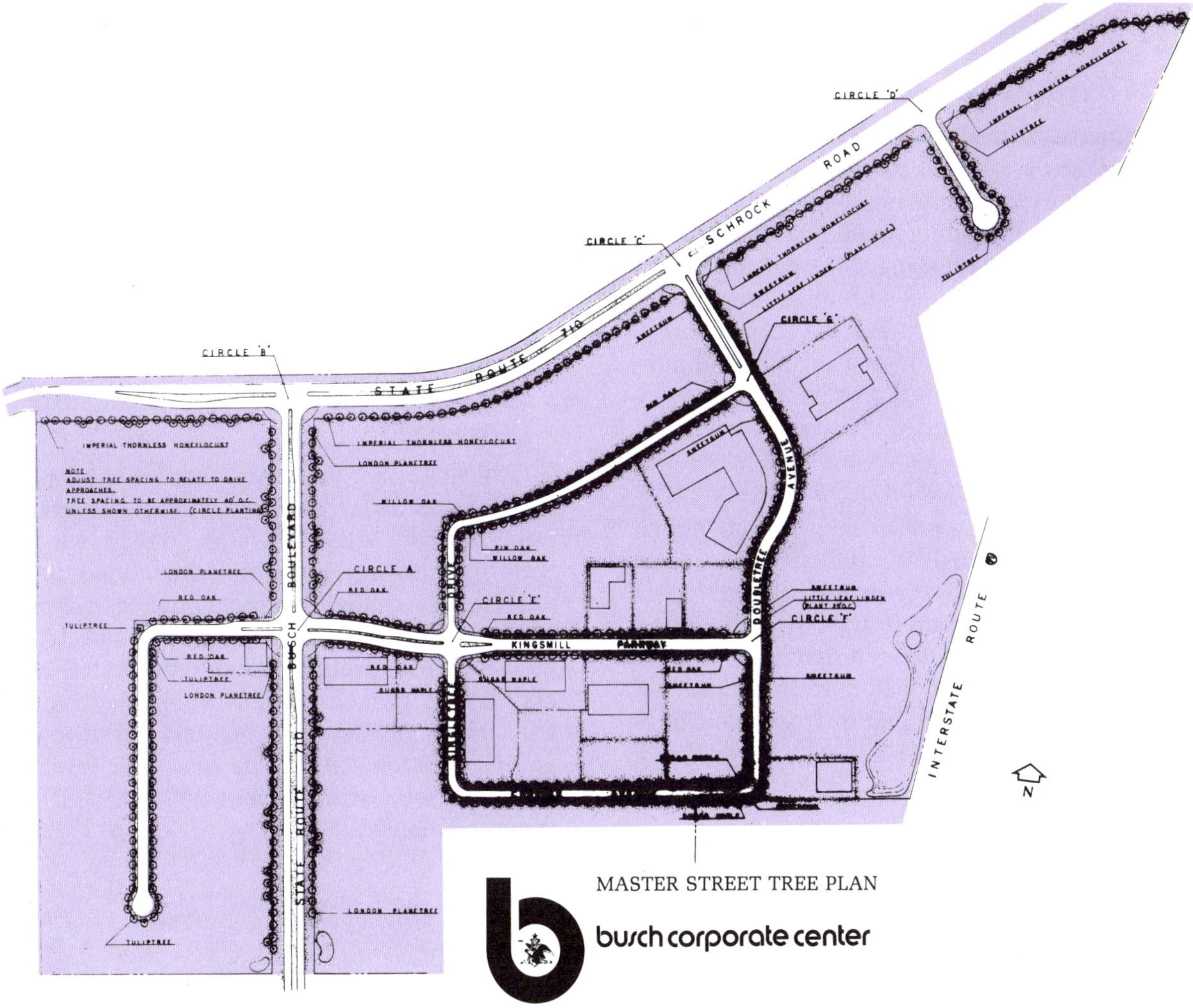

4-4 Master Street Tree Plan—Busch Corporate Center, Columbus, Ohio. Specific plant materials to be used are identified.

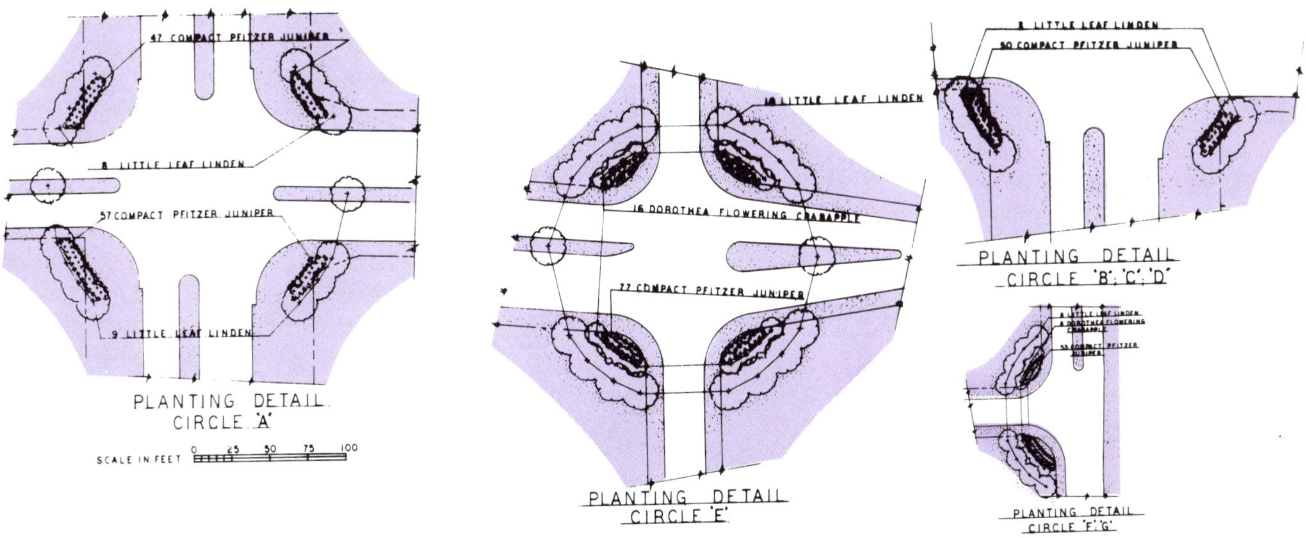

4-5 Planting details for street intersections at Busch Corporate Center. Vegetation is used to integrate the development and provide a unifying element. This is particularly important where most buildings are tenant designed.

Project Engineering and Design

Street Layout and Design

As the principal means of access for abutting property, proper street layout and design is most important in industrial development. The roads may be public or private, depending upon local regulations. Administrative and operational controls and procedures for planning, design construction, and operation of highways and streets, are the responsibility of authorized local, regional, state, or national public agencies. Streets, frontage roads, and intersections are normally designed and constructed by the developer at his own expense for dedication to the municipality or local road district or state.

A planned or existing major or principal public highway running through an industrial development is a mixed blessing to developer and occupants. Although such a public road will constitute a substantial portion of the internal road system at little cost to the developer, and enhances the value and increases the number of front sites, nevertheless disadvantages do exist:

- Intersections of internal streets with a major thoroughfare are accident prone;
- Unrelated heavy traffic constantly passing through the development on such a public road detracts from the appearance, order, and quiet environment which is characteristic of a well-planned industrial development, and,
- Such a highway tends to divide rather than unify the development. A through route would greatly handicap development of a planned employment center. The desirability of mixed uses, including residential and shopping/service areas, is predicated upon substantial pedestrian traffic.

Gridiron street patterns, traditional in industrial areas, have disadvantages which need consideration. Straight streets, which extend for several blocks within an industrial area, tend to attract unrelated fast-moving traffic which is avoiding congestion on nearby major routes. This unrelated traffic can speed through the industrial streets which are free of parked cars or unloaded trucks. The deleterious impact can be avoided through improved internal street design and through the avoidance of a street layout which would encourage unwanted traffic.

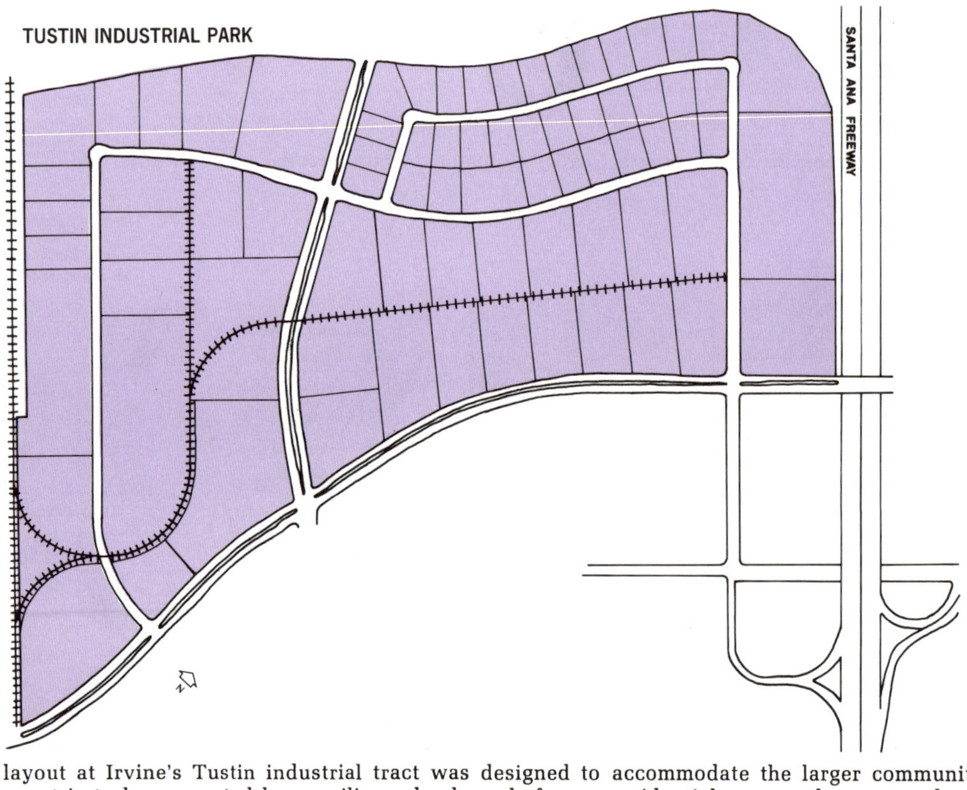

4-6 The street layout at Irvine's Tustin industrial tract was designed to accommodate the larger community's long-range plan. The development is to be separated by curvilinear boulevards from a residential area on the west and an "activity corridor" on the east.

General Considerations

The location of roads within the industrial development is frequently secondary to other layout features and concepts for the site. An early field reconnaissance should uncover any problems which would affect the feasibility of road locations suggested by other planning factors. Factors influencing the location of roads include the location of existing facilities, future expansion, grading and alignment relating to topography, rail service, soils, drainage, geological conditions, ground cover, maintenance considerations, desired lot depths, and local jurisdictional controls. Roads should be located on stable, well-drained soils of high bearing capacity. Surface and subsurface water may affect alignment, earthwork, and drainage features, as well as maintenance. Geological conditions may affect the excavation and stability of slopes. Excavation of rock formations is costly and should be avoided. Clearing and grubbing requirements, and the preservation of trees, are also considered in selecting the location of roads.

Another initial step in developing roadway improvement plans is obtaining soils information. Existing data from borings, soil profile maps, construction excavation information, and visual inspection may be available. Where there is rock, it is necessary to know the nature of the rock and the rock profile. Soils data will determine the nature and suitability of the on-site material for use in the roadway fill.

Land Use

As discussed in Sections I and III, most metropolitan areas are regulated by land use controls such as zoning ordinances, subdivision control, or private deed restrictions. Some communities have regulations providing for the location of future roads and rights-of-way, and for the future widening of existing roads. Regulations for the location and design of driveways by permit is a common practice used by official agencies.

Conditions for access may include an analysis of the traffic generated by the development and its effect on the safety, capacity, and level of service of the public road. Adequate ingress, parking, circulation, and egress may be required. Sometimes even provisions for access to other external roads are included.

Subdivision regulations provide for adequate streets in new areas through control of road patterns, alignment, lot size and configuration, drainage, grading, and utilities, as well as access to external roads.

Zoning ordinances normally regulate the type and intensity of land use developments without having any requirements governing the planning and design of roads. The land developer should recognize the influence of zoning upon roadway operation. Different types of land uses generate varying volumes of traffic, traffic peaks, and directional movement. By balancing different types of land uses, more uniform traffic load can be achieved and excessive peaks eliminated.

Ingress and Egress

While some industrial projects may be separated into halves or quarters by major public highways, thereby suffering from an excess of vehicular access, many otherwise well-developed industrial projects funnel all the cars within the project through one access to a major street or highway. Even staggered closing hours cannot prevent traffic jams within the development and on the highway if there is only one access point. Not only should more than one access exist along the development's longest highway frontage, also there should be access along at least one side, and also if possible, at the rear. To quote Paul P. Shepherd, of CC&F and currently president of the National Association of Industrial Parks: "In laying out the streets for an industrial park, you cannot go wrong in providing the maximum number of good access points." He provided the following example:

"Assume that you are developing a 160-acre industrial park, with an employment density of 20 persons per acre. With 3,200 employees at 1.1 persons per car, there would be approximately 2,900 cars. Three planned access points probably could take care of that volume in just over an hour.

However, if there is a half-hour traffic jam at our park nightly, and the competing industrial park down the road does not have a traffic jam, then we have a marketing problem."

Traffic Generation

Understanding the nature and extent of traffic generated by various uses of land is important to the industrial land developer in providing for adequate roads. Small land developments may not in themselves generate sufficient traffic to create an overload of the existing road network. On the other hand, industrial plants often generate substantial traffic which must be accommodated by the road system. Standard values for the amount of traffic generated for various uses of land are not available; however, typical values, based upon experience, can be used. The traffic generated is generally expressed in terms of the number of trips per acre, or per 1,000 square feet of gross floor area.

In traffic generation studies for industrial areas, the principal measure of density is employees per acre. The operations with lowest employee density are highly automated industries or warehouses supporting fewer than five employees per acre. A second class is light service industry, generally located on small parcels and having five to 20 employees per acre. An industrial tract in a larger development with more intensive employment may have from 20 to 100 employees per acre. One of the most intensive industrial classes is office industry (research and development). An industrial site of this type may support more than 100 employees per acre. A fifth class is mixed industrial development which has a variable employee density. Traffic generation rates are indicated in vehicle trips per acre and in vehicle trips per thousand square feet of floor area.[8]

As a general rule, one lane of pavement will handle from 800 to 1,200 trips per hour within the development. The actual number of cars accommodated within these limits is a function of several parameters including street layout, traffic control at intersections, and adequacy of highways serving the site. Therefore, all complex traffic movement should be analyzed by a traffic engineer to assure adequate design.

Street Planning and Feasibility

A preliminary street layout and cost estimate should be prepared as part of the master plan so the adequacy and cost of the plan can be evaluated.

Available information pertaining to the site and to the existing street and highway system should be obtained for planning purposes. When necessary information is not available, special engineering studies may be required. An inventory of existing roads serving the planned development should be made. Street function, cross section elements, traffic control measures, road condition, and special features are essential items which can be readily noted during a field inspection. Traffic counts, traffic projections, and intersecting traffic movements are often available from public agencies.

[8] Harold Marks, *Protection of Highway Utility,* Report no. 121 (Washington, D.C.: National Academy of Sciences, Transportation Research Board, National Cooperative Highway Research Program, 1971).

TRAFFIC GENERATION VOCABULARY

		Traffic Generation Rate (Vehicle Trips Per Day)			
		Number/Acre		Number/1,000 Sq. Ft. Floor Area	
Land Use	Density (Employees/Acre)	Range	Typical	Range	Typical
Highly automated industry low employee density (refinery, warehouse)	5	2-8	4	0.2-1.0	0.6
Light service industry single-lot industry (lumber yard)	5-20	6-30	16	0.4-1.2	0.8
Industrial tract (5 acres) (machinery factory)	20-100	30-160	70	0.6-4.0	2.0
Office campus research & development (research industry)	100	150-200	170	3-8	4
Mixed central industry small industrial plants	Varies	10-100		1-4	

SOURCE: National Cooperative Highway Research Program, Report 121, "Protection of Highway Utility."

4-7

Traffic desire-line plats from origin/destination studies, and home interviews and questionnaires conducted by official planning agencies, can be obtained from the agency or from master comprehensive plans prepared for municipalities, regional commissions, and the state or federal government. Proposed streets and highways also are shown on such comprehensive master plans. Plats and maps of existing streets are usually available locally.

Photographs should be taken at the proposed site for planning purposes. Topographical maps and aerial photos of the area are usually available. Geological data should be obtained during the planning stage. Flood prone area maps and flood studies should be acquired in areas subject to flooding. Physical improvements on the site observed during reconnaissance should be platted on an existing site plan if this information is not readily available.

A preliminary plat indicating street and parcel layout should be submitted to the appropriate public agencies for preliminary review and approval. Only after such a step will all of the public sector development requirements be known. Any public agency review requirements should be incorporated into the master plan. While an estimate of cost can be made from the proposed master plan, accurate development cost estimates can be prepared only after final plat approval has been obtained.

4-8 Street under construction at Busch Corporate Center.

Design Principles for Industrial Streets

Industrial streets have design requirements that are different than residential and commercial streets. All too frequently, however, local requirements treat all streets as being the same to the detriment of good engineering and planning design.[9]

Administrative and operational controls and procedures vary considerably from place to place and definite design criteria have not been adopted universally. Planning and design policies have been established by the American Association of State Highway Officials for urban and rural highways, local roads, and streets and highways other than freeways.[10] State and local highway agencies have controls and procedures for roads within their jurisdiction which must be accommodated. Since requirements for street design do vary, only general principles and procedures for industrial street design are presented herein, rather than specific design criteria.

General requirements of design principles for streets are based upon several roadway and operational controls. Typically, streets are classified in accordance with traffic volume ranges; and, the controls and policies governing design are divided accordingly.

Design traffic is expressed as Design Hourly Volume (DHV) or Average Daily Traffic (ADT). The Design Hourly Volume is the thirtieth maximum hourly traffic volume usually projected 20 years from the date of construction. The Average Daily Traffic is projected for future increases in traffic and is generally used for design purposes for roads with low volumes. It is the most suitable for use in design of industrial streets. It is important to anticipate the percentage of truck traffic and the directional distribution of all traffic.

---

[9] *Residential Streets: Objectives, Principles and Design Considerations* (Washington, D.C.: Urban Land Institute, National Association of Home Builders, and American Society of Civil Engineers, 1974), p. 47.

[10] *A Policy on Arterial Highways in Urban Areas* (Washington, D.C.: American Association of State Highway and Transportation Officials, 1973); *Geometric Design Guide for Local Roads and Streets* (Washington, D.C.: American Association of State Highway and Transportation Officials, 1971); *Geometric Design Standards for Highways Other Than Freeways* (Washington, D.C.: American Association of State Highway and Transportation Officials, 1969); *A Policy on Geometric Design of Rural Highways* (Washington, D.C.: American Association of State Highway and Transportation Officials, 1965).

A guide for relating pavement width and number of moving lanes to various ADT ranges in industrial areas is shown below:

GUIDES FOR DETERMINING WIDTHS OF NEW ROADWAYS

TYPE	NUMBER OF LANES	PAVEMENT WIDTH	APPROXIMATE ADT (ultimate development)
Local Access and Minor	2	26 feet (parking restricted)	less than 2500
Minor Collector	2	32 feet (parking one side)	2500
Major Collector and Arterial	4	47 feet (parking restricted)	5000/10000

Source: Adapted from St. Louis County, Missouri Department of Highway and Traffic Design Criteria, 1974.

Roadways constructed to Federal requirements may require additional width. The ADTs shown are approximate and the above criteria is intended as a general guide only. The actual need and width of collector and arterial type roads will be investigated for each development. Unlike traffic volumes for residential and multi-family developments which are based on the number of trips generated per unit, a detailed traffic report is normally required for commercial and industrial developments.

Roadway right-of-way widths are established by local jurisdictions and are sometimes subject to negotiation. In practice, a right-of-way width of not more than five feet beyond edge of pavement should be adequate in industrial developments. Easements adjoining each side of the street can be provided for roadway maintenance, drainage, and utilities.

On roads without curbs, minimum shoulder widths range from 10 feet on the left and 12 feet on the right for divided roads to as little as four feet for minor roads.

Reduced right-of-way width benefits the developer by increasing salable acreage and benefits the community by increasing taxable acreage. In some areas of the country, the practice is to sell land to the center of the street thus maximizing taxable acreage. Of course, land occupied by the roadway is not usable to the purchaser and an appropriate selling price adjustment is generally made.

As is noted above, the provision for all parking off-street is recommended in industrial developments. Off-street parking has many advantages over on-street parking including lower cost to the developer in terms of reduced roadway widths, better appearance enhancing land values, and greater safety for occupants and visitors. Parking along roadways requires additional minimum surface width—generally eight feet—with 10 to 12 feet preferable where parking of trucks is anticipated.

Design Standards

Many typical street features have been standardized and are available through official agencies. Most consultants have typical details for roadways that have been developed through their work experience.

The type of paving surface for industrial streets may range from rigid pavement (concrete) or flexible surface (bituminous), to gravel or crushed stone aggregate surfacing. This is typically the most rigidly applied local standard.

Pavement types and thicknesses are usually specified by the local authorities. Most street pavements are built to arbitrary thicknesses determined entirely upon the basis of past experience. Several structural design procedures have been developed for the different types of pavements, and should be followed where standards are not established. Pavements are divided into two broad categories—flexible and rigid. Flexible pavements are those containing a flexible base with an asphalt surface; and, rigid pavements are those made from Portland cement concrete. Structural design of rigid pavements is generally done by use of the theoretical stress analysis method or by soil classification methods. Structural design of flexible pavements generally consists of empirical methods or soil classification methods. The empirical methods include use of the California Bearing Ratio (CBR), or use of a plate bearing test whose values are correlated with pavement performance. Wheel loading, repetition of loads, and subgrade soil condition factors are used in pavement design. Frost action and subgrade drainage should be considered in design of pavements.

Other controls for design may be required. The developer should check with local authorities to see what controls govern the design of the roads that are to be constructed. Where no controls exist, the developer should be careful to have the roads designed in accordance with accepted practice to avoid both extra costs in upgrading under-designed facilities and liabilities resulting from operations on such roads.

Geometric requirements for road design include dimensional features such as alignment, sight distances, and clearances. Values for geometric standards range in accordance with the design speed usually given in 10 m.p.h. increments. Sight distance requirements include stopping sight distance, passing sight distance, and corner sight distance. The developer should assure that geometric design requirements are considered in preliminary planning to avoid revisions later by reviewing authorities.

Standard drawings, usually available in most agencies, include configurations and details for typical street sections, commercial entrances and private driveways, cul-de-sacs and turnarounds, intersections, interchanges, medians, bridge and bridge approaches, guard rail, signalization, signing, lighting, and others. Standard construction details are usually available for drainage items, curbs and gutters, erosion control features, sidewalks and paved approaches, pavement joints, traffic control equipment, lighting, signing equipment, and numerous others. Care should be exercised in the use of standards to insure that adequate designs are properly used. Standards are subject to constant revisions and may become obsolete rapidly.

In general, sidewalks along streets in industrial areas are not functional except under unusual circumstances. Far more appropriate is a system of pedestrian pathways linking buildings to each other, to parking areas, and to transit stops. Where plans include significant amounts of open space, and residential and commercial uses are located within convenient walking distance, this system of walkways will be even more justifiable and desirable. Sidewalks in an industrial area are an unnecessary expense both for first cost and for maintenance.

In addition, consideration should be given to the value of bicycle paths using similar criteria to that suggested for sidewalks and walkway systems. Bicycle paths in street rights-of-way are being recommended in many areas, but the hazard of heavy truck traffic intermingled with bicycle traffic needs careful evaluation. As ULI Industrial Council member Lowell Eastman points out, "In industrial areas bicycle paths when appropriate should be taken off the streets whenever possible and placed in the green areas or in the yard setbacks next to collector streets or major thoroughfares." He further points out that the introduction of bicycle paths

4-9 Pedestrian pathways are used to integrate recreational and employment areas in Erin Mills Southern Business Park.

requires the provision of bicycle curb cut ramps at street crossings.

Other Design Considerations

Conformance with design policies, geometric criteria, and appropriate standards does not insure that adequate design plans will ensue.

Safety ranks with economy in importance in roadway design. For example, the removal of roadside hazards and the utilization of recovery zones along major streets are being implemented more frequently.

Proper *drainage* is essential for operational performance and for protection of the roadway.

Earthwork grading design is essential for establishing economical gradelines consistent with good design practice. *Erosion control measures* during construction, and afterward, may be required.

Maximum earth slopes for roadways in cut or fill sections range from 1½:1 to 4:1 depending upon the height of fill or generally speaking, slopes steeper than 3:1 should be avoided since they cannot be maintained easily with mowing equipment.

Aesthetic factors cannot be overlooked in good design practice. The alignment should complement existing physical features and blend into the terrain. Visual or physical obstruction near the roadway should be minimal. Beautification is an important design consideration, particularly within the development. Landscape design should be incorporated into the street plans.

Design Innovations—Porous Pavements

A pavement research program which involves porous pavement construction, appears to have application to industrial development. Porous pavement consists of a relatively thin course of gap-graded asphalt over a deep layer of permeable crushed stone. Gap-graded asphalt has only stone aggregate, no fines. Thus stormwater penetrates the pavement percolating through the crushed stone base and into the ground. Of course, the underlying soil must be reasonably permeable.

Advantages of the system include:

- Elimination of curbs, drains, and storm sewers;
- Surface flooding may be averted in some areas;
- Groundwater conditions in water deficient areas may be improved; and
- Pollution from overflowing combined storm and sanitary sewers can be reduced.

Researchers report that normal dust and oil dripping will not clog the surface and weeds will not sprout through the base. Also, with free draining soil there is no possibility of frost heave; ice will

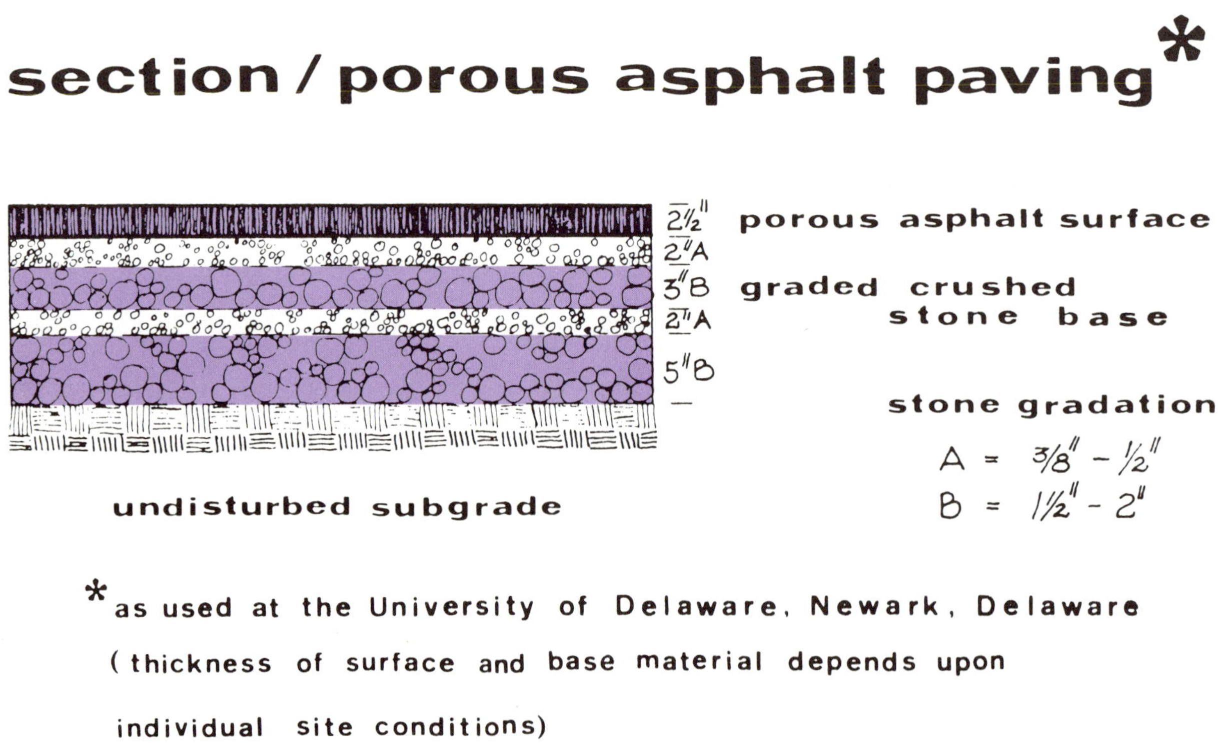

4-10 Design innovation—porous paving.

clog the pavement only in very severe weather.[11]

Porous pavement currently holds its greatest promise for developers in terms of cost savings on public roads within the development. Greatest cost savings would occur on the higher level design thoroughfares such as four-lane roadways. Because of their normally minimal pavement design, parking lot pavements may offer no cost savings as porous pavement facilities, but may provide storm drainage benefits and cost savings which should be evaluated.

Street Lights and Alternatives

The subdivision ordinance of the locality usually will establish the requirements, or a general guide, for lighting within the industrial development—primarily as a matter of public safety and convenience. Evenly spaced streetlights are needed in areas of high pedestrian activity but are not necessary where most movements are by private automobile, or by various modes of public transportation. Accordingly, municipalities may allow deviation from the standard street lighting procedure which requires lighting fixtures at certain spacing. Where sidewalks, pedestrianways, and bicycle paths are provided, the need for appropriate well-spaced lighting should be considered. Normally, even these facilities would have little or no nighttime use, and therefore night lighting would appear unnecessary.

Because sizable industrial developments are frequently remote from city services, municipal street lighting may not be available. Therefore, considerable latitude in lighting may exist. Alternatives to street lighting include highlighting of building facades, parking lot lighting, lighting of street intersections only, and combinations of these.

The developer may be required to pay for the installation of the required street lighting system, and may be responsible to the municipality or other authority for the system's operation and maintenance. Such expense may be recovered by inclusion in the assessment levied on property owners. The developer may be required to submit for approval his method of assuring that the lighting system operation is properly funded. Such assurance may be by maintenance agreement, trust indenture, or other instrument naming the party responsible for the assessment and collection of monies and the method of collection of said monies. Because regulations vary, it is important when planning the development, to be familiar with local requirements for lighting public rights-of-way.

4-11 Street lighting occupies a landscaped median at the South San Francisco Industrial Park.

Grading and Drainage

Grading

Industrial developments are usually sited in flat terrain to minimize grading costs. Consequently, the general philosophy for grading will be established by the requirements for drainage. In order to minimize the volume of earth to be moved, preliminary flowline and channel sections should be determined for stormwater runoff. Hydraulic gradients for the design storm along these channels should be determined. Where a closed sewer system is planned, corresponding elevations should be established. With design stormwater surface elevations determined, the adequacy of the elevation of the existing topography across the site can be determined. Areas with unsuitable low elevations can be brought up to minimum practicable elevations by filling with earth excavated at least in part from proposed stormwater channels or stormwater retention basins. Where practicable, grades should be established for roads which allow lot frontage to slope toward the street. The street gutter will then become a collector for stormwater runoff directing flows to the roadway inlet system. Grades for the

[11] "When It Rains, It Pours Through The Pavement," *Engineering News Record* (New York: McGraw-Hill, October 11, 1973), p. 38.

remainder of the site should generally direct runoff to side and rear lot lines in open channel drainage systems. Such practice generally produces the most economical site plan.

While the costs of excavating and grading for industrial buildings, parking areas, and other facilities properly should be considered as building rather than as site development costs, the developer, in his planning, should consider the problems of drainage, balancing of cut and fills, and disposition of excavated materials resulting from building operations.

In evaluating the suitability of the proposed grading plan, it should be recognized that the lowest area of a lot will usually be the truck loading dock area or depressed rail track serving an industrial building or warehouse. Where feasible loading areas when excavated will be sufficiently above storm drainage facilities to allow adequate drainage of pavements. Otherwise an internal system of gutters and trench drains, together with ejector sump pumps, will be required to provide adequate drainage. In other words, a good drainage plan is one that will assure economical development of individual sites for purchasers as well as economical site development costs for the developer.[12]

Grading cost estimates should be developed including costs for clearing roots, and for building foundation excavations, as well as for stripping vegetation and topsoil. The developer's engineer should refer to soils reports during preparation of the land plan and grading plans so areas of deep organic or otherwise unsuitable or difficult soils —such as highly plastic clays—may be anticipated and properly evaluated in terms of removal and/or treatment to achieve adequate density in fills. In cases where areas may be too costly to improve due to poor soil and drainage conditions, the land may be planned for open space or stormwater retention.

Excavated materials considered unsuitable as foundation soils frequently may be used as earth berms. These can become attractive landscaped site features with a functional value as screens for parking lots and building maintenance areas.

In preparing grading cost estimates, it should be recognized that site grading in highly plastic clays could be considerably more expensive than in low plasticity clays or silts due primarily to more time spent in dewatering excavations and drying operations prior to compaction. Treatment of highly plastic clays with hydrated lime is common practice; and, while this method generally requires less time, it may be more costly than spreading and drying. Relative cost of the two procedures depends upon the type of soil and severity of the problem, the season of the year, and the presence of cover.

Storm Drainage

One of the principal costs of industrial development is construction of storm drainage facilities. For this reason, a detailed analysis of storm drainage alternatives should be made during the planning process. Is it more economical to excavate open channels and sacrifice the land required for such a system? Or, should an underground piping system be utilized thus freeing the land for parking or other uses? An economic comparison of alternatives should be made for each project. Perhaps the value of the land will not support the cost of an underground system and a decision will be made to excavate open channels in the first phase of the project. The initial decision should be reevaluated as the area develops and as land values increase.

Procedures and criteria applicable to the design of storm drainage facilities are available from the agency having jurisdiction. Studies have determined that upwards of 90 percent of municipalities and agencies currently use the Rational Method of storm drainage design.

Rational Method

The Rational Method was introduced in 1889 by Emil Kuichling. It is a synthetic method for describing in an approximate way what happens after a rainfall or storm.

This method has gained popularity over the years primarily due to its ease of application. It is not indicative of actual storm runoff conditions, and therefore, is considered a conservative design approach because it produces a peak runoff at each point of design along the channel or pipe system.

[12] An example of a grading plan which proved uneconomical is given by Marshall Bennett, ULI Industrial Council chairman: "In Bennett & Kahnweiler's Montbello Park in Denver, the market strategy called for sequential development of four 125-acre sections. Site grading was done initially on the first two sections and development proceeded with no problems. When the last two sections were to be improved, it was necessary to go around the developed areas with heavy equipment, requiring longer hauls. This resulted in additional grading costs of $250,000 or $1,000 per acre which could have been avoided if the entire site had been graded initially. These kinds of problems needed to be considered in the initial planning of grading and grading operations."

This method is currently being challenged by many engineers who claim it is a basic cause of overdesign.[13]

While the Rational Method continues to be of value in designing smaller storm drainage systems, it does not lend itself to vigorous analysis of systems where storage of stormwater is contemplated. For projects where ponding is employed, other methods—such as the unit hydrograph method, and for small watersheds the triangular hydrograph method—provide a more realistic design approach.

Simulation Models

When the scale of an industrial development is large, significant opportunities may exist for more innovative drainage analysis modeling with potential cost savings. The availability of the digital computer has made it possible to compute or simulate runoff hydrographs from programs based on mathematical models which reflect hydrologic characteristics of precipitation and hydraulic characteristics of both the natural and artificial drainage systems of the site. With adequate information available on infiltration, soil moisture, overland flow, surface retention, groundwater, channel flow, etc., any storm precipitation can be routed through the site and the storm drain system, including detention ponds, to produce the hydrograph of runoff from the site. As for unit hydrographs, measured

[13] *Residential Stormwater Management: Objectives, Principles and Design Considerations* (Washington, D.C.: Urban Land Institute, National Association of Home Builders, and American Society of Civil Engineers, 1975).

4-12 Excess surface water drains into a storm water retention pond at Busch Corporate Center. This pond is located adjacent to Interstate 71 and a sculpture has been placed in it.

records of runoff are desirable from several good storms on the site. With such data, the model can be calibrated for the site in that condition, presumably undeveloped. Then, the model can be adjusted to various alternate stormwater management plans. Subroutines can be added to the main program to compute costs of constructing, operating, and maintaining alternate plans.[14]

Most of the urban runoff simulation models currently available do not estimate quantities of pollutants discharged with the stormwater. The Environmental Protection Agency, in Washington, D.C., is assisting in the development of such models.

This sophisticated modeling approach is most valid when applied to an entire watershed area. The extent to which the site is a major portion of a watershed, and the extent to which watershed plans are responsive to this approach, will be the governing factors in the future use of simulation models.

Detention Ponds

The demand for more economical methods of handling stormwater runoff, and the trend in planning to provide a more attractive work environment through use of lakes and ponds, has led to increased use of water bodies in modern industrial parks. A number of developments have utilized basins for stormwater retention in attractive open space settings. As pointed out by ULI Industrial Council member, John French, many communities are aware of these opportunities and now are requiring maximum consideration of on-site detention and retention techniques.

In areas where drainage problems exist, the developer can minimize the area of land devoted to lake or channel ponding by establishing a set of selective ponding priorities. A specific area may be established as a permanent lake with perimeter areas set at selectively higher levels. For example, the normal lake surface would be the lowest elevation, followed by landscaped areas, followed by parking areas and private drives, and in turn by public streets. Building floors would be at the highest elevations on the site. The philosophy of allocating space for ponding of stormwater is well developed in engineering literature.[15]

Utilities, Layout, and Design

Water Supply and Distribution

Industrial developments require water for domestic uses, manufacturing processes, and fire protection. This supply may be provided either by public or private utility. At the master planning stage, the engineer, in cooperation with the water utility, will determine the capability of the existing off-site system to supply the proposed development.

The water usage of an industrial development depends upon the nature of the business within the development. The amount and type of industrial manufacturing and warehousing within the development will determine the water requirement. Previous studies have shown that water usage varies widely, as indicated by the following table.

4-13

TYPICAL WATER USAGE FOR INDUSTRIAL DEVELOPMENTS

Location	Type of User	Average Water Usage gal./day/acre
St. Louis County*	Manufacturing plant	6,000
	Mainly warehousing	200
	Mixed area including offices, warehouses and processing plants	1,325
Kansas City**	Industrial park	3,620

* St. Louis County, Missouri—water consumption records.

** *Water and Sewage Works Magazine* (Chicago: Scranton Publishing Co., July 1967).

From the data above, it is evident that the type of industry anticipated to occupy the park will have a significant effect upon water needs. In a development with mixed uses, a daily water usage of 1,500 gallons per acre frequently has been found to be adequate. Where manufacturing and office uses predominate, a higher usage rate should be

[14] John C. Schaake, Jr., Brendan M. Harley, and Guy Leclerc, *Evaluation and Control of Urban Runoff* (New York: American Society of Civil Engineers, 1973)—a description of the MIT Catchment Model; Constantine N. Papadakis and Herbert C. Preul, "The University of Cincinnati Urban Runoff Model," *Journal* (New York: American Society of Civil Engineers, Hydraulics Division, 1972); *Studies in the Application of Digital Simulation to Urban Hydrology* (Palo Alto: Hydrocomp International, Inc., 1971).

[15] William J. Bauer, "Economics of Urban Drainage Design," *Journal* (New York: American Society of Civil Engineers, Hydraulics Division, 1962).

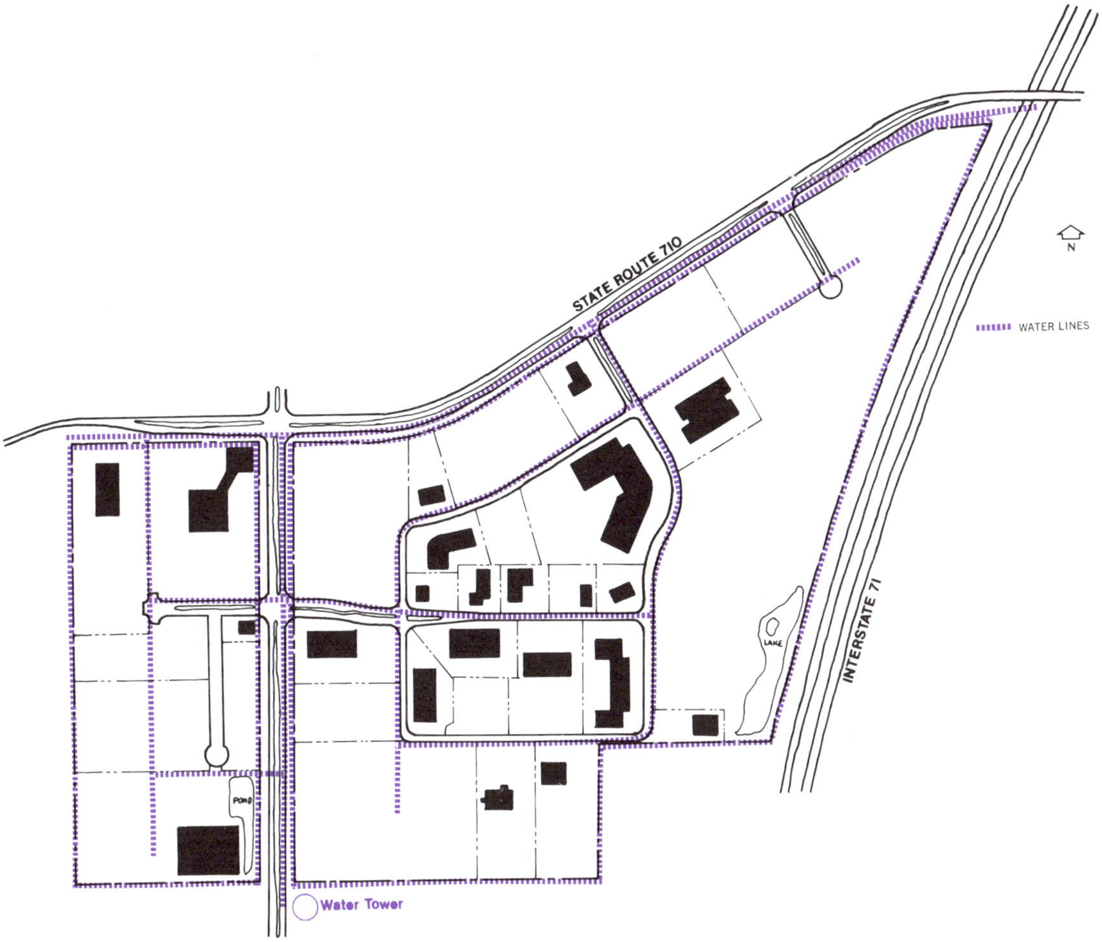

4-14 Water System Plan, Busch Corporate Center.

used for design of the water distribution system. Where warehousing predominates, a usage of 500 gallons per acre may be adequate.

The water system must be designed to meet all the water demands imposed on the system. Peak demand factors which create the variations from average daily and seasonal water usage requirements of the development will vary with the water requirements of industrial occupants.

An important water capacity determinant is the fire protection requirement. For industrial developments of a few hundred acres, the fire protection requirement will normally determine the size of the pipes comprising the system. Pipe sizes 8-inches and larger are recommended. The state fire inspection bureau should be contacted to determine the minimum fire flow rate requirement and fire fighting duration.

In laying out the system, the water mains should be looped to provide continuous service to users. In the event of main break or fire fighting activity, looping provides reliability and is desired by insurance underwriters.

Insurer requirements are related to the class of fire protection available within the fire protection district. Common requirements vary from 1,000 to 5,000 gallons per minute for a minimum duration of two to five hours. A minimum residual pressure of 20 psi is generally required in the system; however, sizing the water distribution system within the industrial development for 60 psi or more may allow service to some multi-story buildings without the need for fire pumps within the building. Also, better fire protection for highly combustible substances in warehouses will be provided. Such features could improve marketability of the land.

In order to obtain fire protection insurance at reasonable rates, the insurance underwriter usually will require high risk industries to have readily available large flow capacities for fire fighting purposes. The required flows may exceed the capacity of the water distribution system and may thus require construction of separate storage facilities by the particular high risk industry, at its own expense. If a developer expects to attract a group of high risk industries, he should consider the economic advantages of offering, in his marketing program, a mechanism for joint construction of separate storage facilities to this industry subgroup.

Flow tests are made at hydrants within the system from time to time, in cooperation with the fire department, to substantiate the fire protection classification of the local area for insurance purposes. These tests will provide information on flows and related pressures. With both the test information and the estimated water usage within the development, the engineer may properly design the water distribution system within the industrial development.

Occasionally, a development may require water in excess of that which can be supplied by the local utility. In such a case, the developer may choose to provide his own source of water. The choice of developing a water supply to meet the requirements of a development must be predicated upon producing a potable water which meets state and federal standards.

The production of such potable water is a function of the raw water characteristic and the desired finished water characteristic. Selection of type of treatment and design of needed facilities must be determined by a competent engineer.

Where potable water is in short supply, or expensive, a separate source and system for irrigation water should be explored. This is particularly true if the storm drainage system has been designed with on-site detention ponds.

Sanitary Sewers

A complete system of sanitary sewers must be provided to serve each lot in the industrial development. Sewers are normally designed on the same flow criteria as domestic water consumption. While there may be an occasional high volume water consumer such as a bottling plant, it is generally suitable to design for about 1,500 gallons per acre per day (gpapd) for industrial developments when the occupants are not known in advance and the development is to consist of a substantial variety of users. If the development is limited to dry industries—such as warehousing—500 gpapd may be adequate; however, reviewing agencies may require higher design flows. The design flows for each planned or zoned use other than industrial, such as commercial and office, should be estimated. These estimated flows may be converted to design flows for sewer design by application of a peaking factor.

Care should be used in designing the sewer sys-

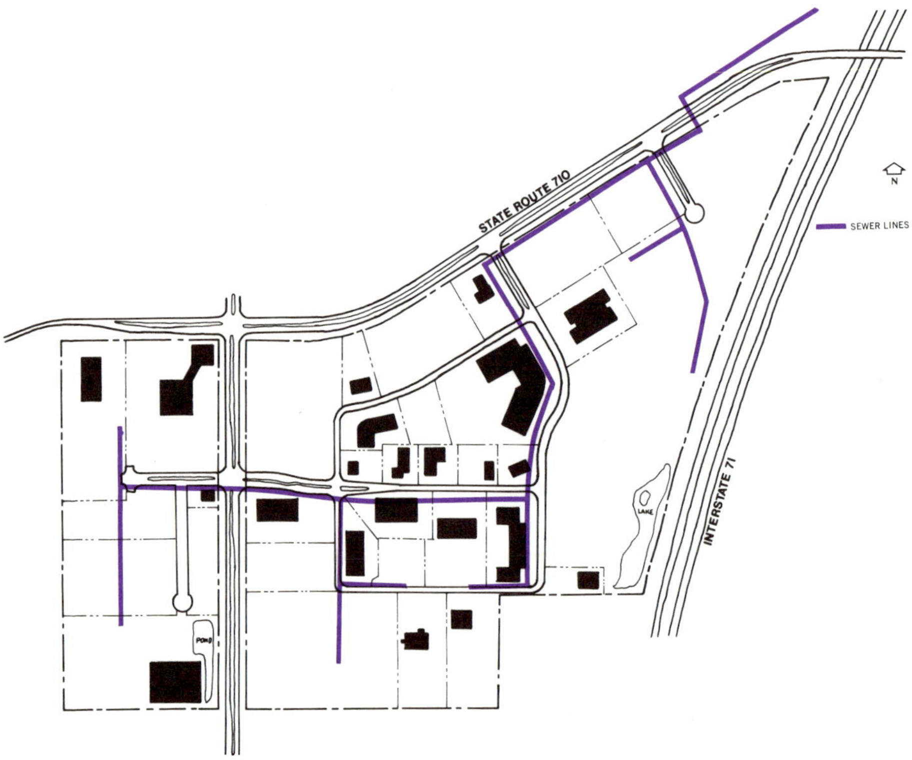

4-15 Sewer System Plan, Busch Corporate Center.

tem in order to assure self-cleansing velocities in the lines. State and local criteria will specify minimum velocities to be maintained for the design flow—usually 2.0 to 2.5 feet per second. Such criteria are specified to minimize maintenance of the system and to avoid septic conditions. Conservation of energy within the sewer system is important also. This is best accomplished by using the total available energy to produce movement of the water in a near-horizontal direction rather than wasting energy in drops, falls, chutes, or cascades. Conservation of energy is particularly important in flat terrain where sewage must be moved along by pumping. Energy conservation in such terrain will maximize the distance sewage will flow by gravity and will minimize the number of sewage pumping stations required in a project.

Sewer design and construction standards may be obtained from the state or local agency having jurisdiction. In recent years, the federal Environmental Protection Agency (EPA) through the state EPA offices, has assumed much of the authority for regional planning of sewerage systems. The developer, through his engineer, should be thoroughly familiar with current regulations and practices of the authorities to assure that adequate capacity is available in the off-site trunk sewer serving the property. It is also important to know whether adequate sewage treatment facilities are available in the watershed and whether they will be available through completion of the development. A thorough investigation of this matter should be made, and documented, for adequate assessment of the potential costs of sewage treatment. State and local authorities are usually cooperative when problems in existing systems occur. However, regulations and procedures have become more complex in recent years. The developer should be prepared for occasional lengthy delays in obtaining sewerage proposal approvals when existing downstream sewerage and treatment facilities are near capacity.

Sewage pumping deserves consideration in any conceptual sewerage plan. The federal government's strict controls on wastewater discharges has led to adoption of a regional concept of sewage treatment by municipalities and to the discouragement of small individual plants at the lower end of individual watersheds. Such policy, along with the additional cost of more effective treatment facilities required in individual watersheds, has led to more frequent use of sewage pumping stations for the discharge of wastewater from individual watersheds toward a single interceptor and point of treatment. The developer, through his engineer, will be aware of the criteria for sewage lift station design and particularly of the requirements for emergency provisions for handling sewage during power outage or malfunction of the lift station.

Sewage lift stations, particularly those of small size, are available from a number of manufacturers. The stations are comprised either of metal or of fiberglas tanks complete with sewage pumps, discharge piping, valves, and electrical controls. While these items are pre-engineered and manufactured for ease of installation and operation, state regulations may require provisions for emergency operation in the event of malfunction or loss of electrical power.

Authorities have required telemetric monitoring of major sewerage systems when the project includes several sewage pumping stations. Sophisticated systems include print-out equipment giving the specific nature and location of malfunction at a central, manned facility, such as a guardhouse or maintenance building.

Usually, the on-site sewerage system will be constructed by the developer at his own expense and will be inspected during construction by the local sewer authority. The authority may accept dedication upon completion of all or part of the sewerage system and may assume operation and maintenance of the system. A private sewer company, rather than a local authority, may be franchised to serve the area and may have a different policy for developer participation in cost of the sewer system. The developer should be familiar with the practice in his area. A detailed discussion of the design of sewer systems may be found in many texts.[16]

Sewage Treatment

Treatment of wastewater discharges from the development will usually be done by the municipal agency, or private sewer company, franchised to serve the area.

Sewerage treatment plants are designed to treat sewerage of domestic quality as defined in applicable regulations. To the extent that most industrial wastes are similar to residential sewerage, no problems should arise provided there is adequate

[16] See *Design and Construction of Sanitary and Storm Sewers*, Manual of Engineering Practice no. 37 (New York: American Society of Civil Engineers, 1969).

4-16 A flocculator-clarifier cell at Bethlehem Steel Corporation's Burns Harbor, Indiana secondary waste treatment plant. The system, designed in conjunction with the Indiana Department of Health and the State Stream Pollution Control Board, also includes two treatment lagoons, each with a surface area of 720,000 square feet. Treated effluent drains into Lake Michigan.

capacity. However, waste which is unsuitable for the local treatment plant, will have to be pretreated at some other location—usually on the property of the facility generating the wastewater—prior to release to the sewer system for treatment at the municipal or private plant.

Until recent years, adequate treatment was usually available using sewage lagoons; however, current practice in many states is to require mechanical treatment. In any case, when treatment must be carried out by the developer, on-site lagoons are not practical due to the large amount of land required.

Mechanical treatment plants are available in a number of sizes. If the developer finds it necessary to construct his own sewage treatment plant, it may be economical to construct the facility in increments. The size and number of increments should be determined by the engineer in consultation with the developer. The most economical solution will be coordinated with the loading produced by occupants of the development. Since occupants are usually not known at the time the treatment plant is designed, the plant should be sized for estimated daily sewage flows.

ULI Industrial Council member Ben Dyer advises,

> "The size and type of on-site, pre-treatment plant will be determined by the amount and characteristics of the waste generated by the industry.
>
> "Until recent years, some developers have found it feasible to construct a sewage treatment plant to serve their project. With the advent of much more restrictive environmental controls, this option is becoming less desirable. Current federal regulations require that all existing treatment plants be upgraded to meet very high effluent standards. High capital outlay, high operating cost, and possible required upgrading as new technology becomes available, make it generally uneconomical for a developer to provide his own treatment plant.
>
> "It is imperative that the developer determine which sewage treatment requirements are in force or are likely to be imposed in his area, during the construction phase of his project. These requirements could be in the form of flow restrictions, standards of quality which necessitate pre-treatment, or capital contributions to the cost of treatment plant expansion."

Utilities Layout

Utilities such as natural gas, electricity, and telephone may be installed within the industrial development by the utility company at no expense to the developer. The trend in recent years toward underground installation of electric and telephone lines is due to the demand by industrial developers for a more attractive environment for their commercial and office occupants. The electric utility frequently will require a surcharge to be paid by the developer for underground services. During the planning of the project, the developer should interview each utility to determine its capacity to serve the project and its extension policies.

Utilities are most efficiently dealt with by placement in a common trench. However, recent interviews with local utility representatives indicate a preference, by all parties, for the other utilities to be located on the opposite side of the street from a buried electric cable. This is understandable due to safety precautions necessary in working near energized electric cable. A potential difficulty is the extensive easement width sometimes required by the electric utility to allow room for future parallel cables. Heat generated by buried cables usually precludes placement of electrical cables side by side. Thus considerable easement width may be required as future expansion of service occurs along the street or other utility corridor.

The procedure recommended in establishing utility locations is to place sanitary sewers on one side of the street along with direct burial electrical cables and to place water, gas, and telephone systems on the opposite side of the street. Regulations generally require a minimum separation of 10 feet between water mains and sewers which makes practical their location on opposite sides of the street. ULI Industrial Council member John French notes, "The location of water and sewer behind the curb, rather than under the pavement, which is common practice, is more suitable to industrial development when the location of future water and sewer service connections are not known. This location allows connections using boring and jacking techniques which eliminate cutting and patching of street pavement." The pairing of electrical lines with sanitary sewers is practical for three primary reasons: 1) sanitary sewers generally precede electrical cable placement and are considerably deeper in the ground so that conflicts are minimized; 2) water main, telephone conduit, and gas main construction can proceed without conflict with electrical work; and, 3) the electrical utility company may need to construct additional cables adjoining existing cables as demand builds. Such future cable placement is more practical when easements are shared with the deeper sanitary sewer than when shared with several shallow utility lines. Utility planning requires a substantial coordinating effort with the various

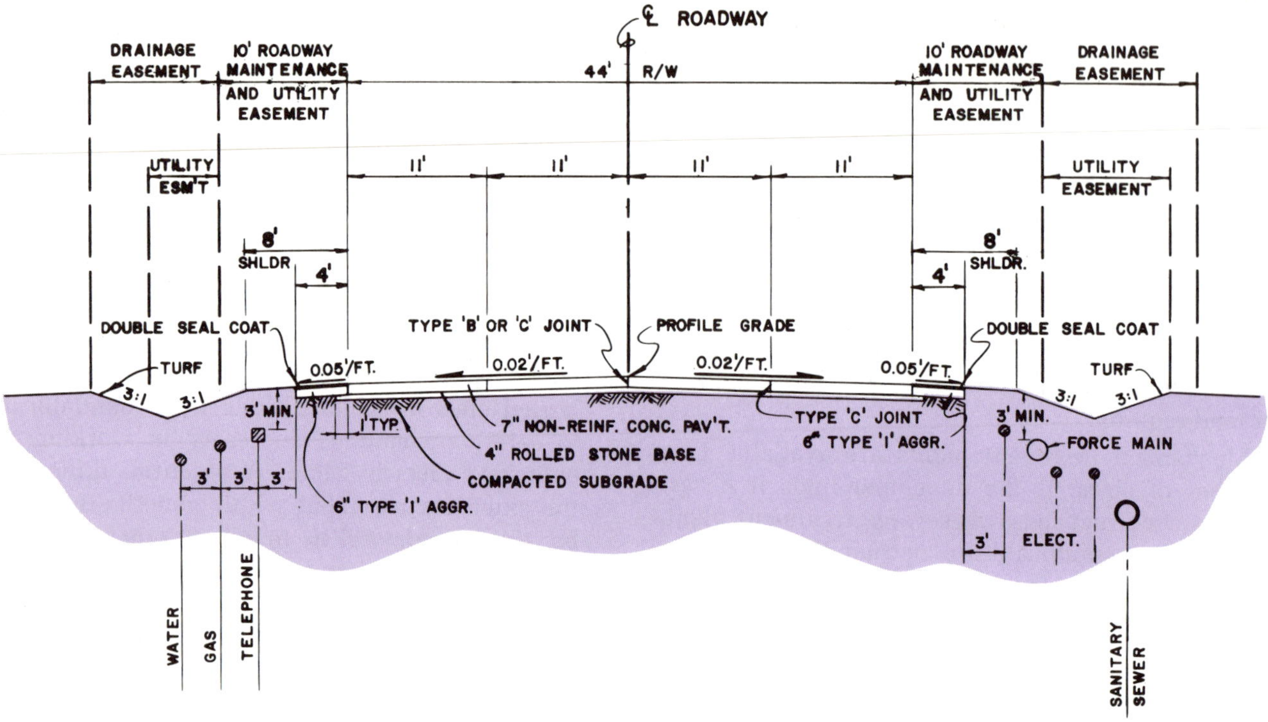

UTILITY AND PAVEMENT RELATIONSHIP IN AN INDUSTRIAL PARK IN ST. LOUIS COUNTY, MISSOURI

4-17

utilities so the facilities can be suitably located and so services will be available when needed.

Primary easements for utilities usually are located along street rights-of-way. Easement widths along side and rear lot lines may be established to provide for additional utility line extensions.

Relocation of Existing Utilities

Occasionally, a tract of land to be developed will have a major utility pipeline such as a gas main, a water main, an oil pipeline, or a power transmission line, traversing the property. The abandonment or relocation of such line is usually not permissible by the owner or, if permissible, could be accomplished only at extreme cost. Consequently, it is important for the developer to recognize that this utility line may affect both the configuration of the master plan, and the net salable acreage. The amount of reduction in net salable acreage will depend primarily upon the skill of the master planner. For example, a master plan was developed recently for a tract of land in Madison County, Illinois. The tract is approximately square; however, two gas mains owned by Mississippi River Transmission Corporation traverse the property from the southwest to the northeast crossing the central portion of the site on a diagonal. One main is 14 inches in diameter and the other 24 inches in diameter. According to available records, the two mains are parallel and 20 to 25 feet apart. In order to provide access to these lines by the utility and to minimize reduction in net salable acreage, the major roadway through the park was established along the alignment of the pipelines. Thus, this single feature established the configuration of the roadway system and for the most part the master plan.

If the existing utility traverses a minor part of the tract, it may be possible to obtain permission from the utility to allow certain uses within the pipeline easement which will allow the land to be sold without a reduction in price. For example, the owner of the pipeline may permit the easement to be paved for parking—provided access to the pipeline is maintained at all times for its full extent across the property. It is possible that such requirements will not be considered a significant restriction by the purchaser of the lot traversed by the pipeline.

Where utility relocation is desirable, or necessary, the developer must contact the utility owner. Usually, the owner will engineer and construct the relocation but will require the developer to pay all costs. If utility relocation is contemplated by the developer, he should contact the owner and obtain in writing both the relocation policy of the owner and cost estimate of the relocation. Such a step is essential for the cost of this work to be included along with the cost of other improvements in the analysis of financial feasibility of the project.

Design Considerations

Site Planning—Some Physical Design Guidelines

Planning specific building sites in industrial developments requires a number of considerations. These considerations include setback requirements, truck loading and maneuvering depths, vehicular parking needs, building coverage, and rail service requirements.

Building setbacks from the fronting street will vary between setting the building on the property line and therefore having a zero front yard, to a setback of 50 feet or more. If vehicular parking is placed in front of the buildings, and allowed on one side only, 40 to 45 feet should be provided for the driveway and parking stalls. If parking is allowed on both sides of the driveway, 60 to 65 feet should be provided. Side and rear building setbacks usually will be less than front setbacks. They typically range from no setback required to 15 feet. Truck and rail loading needs may control building setbacks along side and rear lot lines.

Truck loading and maneuvering depths from edge of dock to edge of maneuvering area are variable depending upon anticipated traffic. Successful

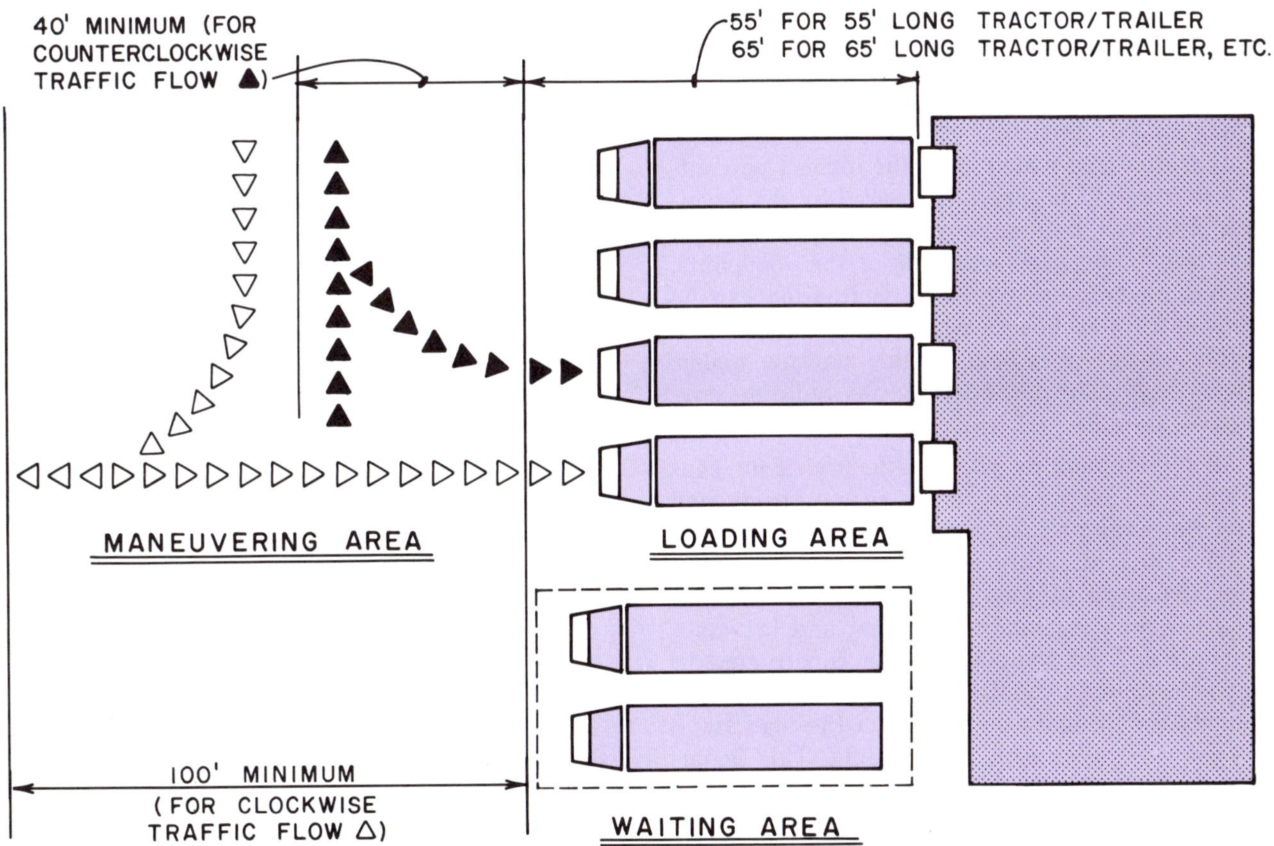

TRUCK LOADING AND MANEUVERING CONFIGURATION

4-18

warehousing operation can occur with as little as 85 feet of truck docking and maneuvering depths whereas the recommended depth for trucking terminals for larger trucks (45-feet trailer length) is as much as 129 feet.

Vehicular parking needs will depend on building use. Warehousing generally will have fewer occupants which enables auto parking to be accommodated near the front end of the truck loading area. Higher densities will require devoting as much as one side of the building to vehicular parking. Seventy-five feet should be allowed for parking and landscaping if one aisle, and parking stalls on either side, are considered sufficient. The ratio of parking spaces required to building will relate to employee densities and to employee commuting habits. Keeping the area of pavement to the minimum required to accommodate parking needs will have significant effect on storms drainage design. The developer, users, and public agencies will be well advised to design parking standards with this objective in mind and to provide minimum parking with initial site development with provision for additional expansion as empirical studies indicate.[17]

To provide maximum return on investment, building coverage which gives the highest percentage site coverage is generally desired by the developer. However, building coverage may be limited by zoning, or by parking, needs of the occupants. Building coverage above 50 percent often can be achieved for warehouses whereas offices and light manufacturing, with their sizable parking needs, may be in the 30 percent range of building coverage.

Rail service requires about 15 feet from the centerline of the spur track to the rear property line when the drill or lead track is centered on the property line. If, however, the rail drill track is in its own right of way, then inside building rail service is most economical of land and improvement costs if the building can be constructed to the property line. An allowance of 150 feet should be made for bringing the rail from the lead track to a point parallel with the building. This is important in planning the distance from the building to the side property line because rail spur geometrics will require an easement on the property next door if the building must be set near the side property line.

Building dimensions are variable; however, most structures fall within the square to two-to-one ratios of length to width.

Platting Techniques

A primary objective in preparing a preliminary plan for the industrial development is to provide maximum flexibility. Because the needs of prospective users cannot be known in advance, the layout of streets and rail leads must be done to create a plan which provides lots of various depths.[18] Lots 200 to 300 feet in depth are popular. Large lots may range from 500 to 700 feet in depth. Establishing lots for such depths will permit later introduction of a short cul-de-sac street to break these deeper lots into two medium-depth parcels.[19]

Cul-de-sac streets should end in a paved turnaround 100 feet in diameter. This diameter will accommodate larger trucks, including 45-foot trailers, and will allow a 180-degree turn without backing.

A further objective of lot layout should be to minimize the number of at-grade rail crossings of major roads within the development. In areas with high density uses, such vehicular traffic interruptions are particularly annoying; and, automatic crossing protection is often required at the developer's expense.

Submittal of a preliminary plan to the community reviewing agency will establish a pattern for the orderly submittal of record plats. While practices vary, it is desirable to make a minimum initial commitment when filing a record plat of roads and parcels. This minimum commitment will retain flexibility and will provide control over the amount of funds paid in filing fees, when these fees are based on area recorded.

---

[17] *Special Traffic Generator Study—Industrial Generations*, Report no. 2 (Dover, Delaware: State of Delaware, Department of Highways and Transportation, 1973). This report prepared detailed analyses of 22 industrial users, the parking ratios ranged from 0.21 spaces per 1,000 square feet of floor area to 20 spaces per 1,000 square feet of floor area. The firm with the lower ratio manufactured chemical products; the firm with a higher ratio was a clothing manufacturer. Of the 22 industrial firms studied, 11 had ratios of less than one space per 1,000 square feet of floor area; six had ratios of less than two spaces per 1,000 square feet of floor area; and only five had over two spaces per 1,000 square feet of floor space and with the exception of the one with 20 spaces per 1,000 square feet of floor area, all of these were below a four-to-one ratio.

[18] For one developer's experience in industrial site layout, see Paul P. Shepherd, "Planning and Designing an Industrial Park," *Developer's Diary* (Arlington, Virginia: National Association of Industrial Parks, March 1972).

[19] ULI Industrial Council Chairman Marshall Bennett, contrary to the views of some Council members, feels that cul-de-sacs should be avoided in industrial development if at all possible.

Rail Service

When rail service to industrial developments is contemplated, the developer should contact the railroad company's industrial development department as early as possible to determine which provisions are necessary for service. As ULI Industrial Council member O. G. Linde points out, such things as reciprocal switching limits, frequency of switching service, car supply, and general rate considerations, can be very important. He suggests, "A project might experience slow development simply because the development is located outside reciprocal switching limits, thereby making it unattractive for industries to locate at a local station on one railroad when a significant share of their rail traffic originates or is destined on other railroads."

Assuming the desirability of rail service has been determined and the mix of land to rail served versus nonrail has been determined by market studies, several engineering and design factors must be considered—including topography, soil conditions, drainage, existing improvements, right-of-way access, building setback requirements, operational requirements, elevation and alignment of existing tracks, desire for in-plant rail service, and most suitable point(s) of intersection with existing tracks. As ULI Industrial Council member Otto Pongrace points out, "The location of rail access and the direction of that access from the main line must be discussed with the serving railroad since they may object to the point and direction of service that is most desirable for the developer."

While procedures vary, the cost of the lead track through the development is generally borne by the developer including the switch and spur track up to the property line of the individual tenant; the tenant generally pays for the remaining spur length. The railroad may participate in the costs of providing lead tracks to serve the several

4-19 An interior view of a rail-served building at Columbia, Maryland's General Electric Appliance Park.

TYPICAL TRACK SPACING REQUIREMENTS

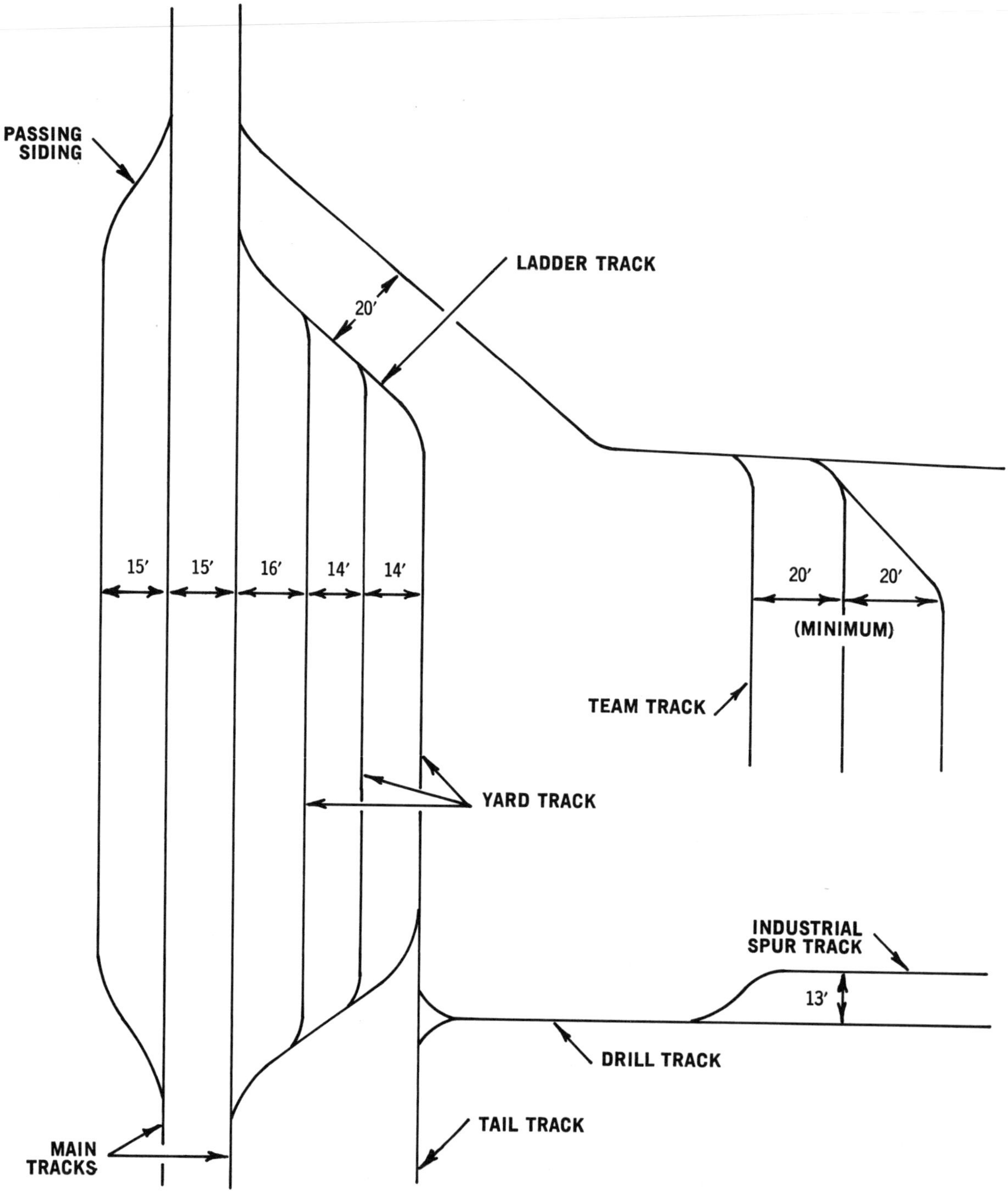

4-20

industrial sites within an industrial development through refunds to the developer based on car loadings. Switching tracks, sidings, and yards for storage may be necessary if required by the railroad. This additional trackage is of considerable value to the railroad since it facilitates operations. The developer should negotiate the cost of such additional trackage with the railroad.

Historically, the community has not been involved in the provision of rail service. It has been the function of the developer to work with the railroads and share the costs based on mutual benefits. This function is in contrast to other on-site and off-site improvement such as streets and utilities. Frequently, communities have lent public powers such as improvement district financing in order to provide these facilities to further development of an employment and industrial tax base for the community. ULI Industrial Council member O. G. Linde suggests this policy may change and points out a case in California:

> "The Southern Pacific Industrial Development Company negotiated with a community to include under the public assessment district procedure the cost of constructing industrial rail trackage in an industrial park. After construction, the trackage will be deeded to the Southern Pacific for ownership and maintenance and under a separate agreement the Southern Pacific will refund the cost of the trackage to the city based on $10.00 per car with the refund contract not to exceed 20 years."

All rail service must conform to the requirements of the railroad company which will be operating over the facility. Also, rail service is subject to public service commission requirements. Many states have rules and regulations prescribed by lawful authority for clearances to or under adjacent buildings, structures, or physical obstruction of any kind.

Approval of railroad plans and construction is at the discretion of the railroad company. Close contact with railroad officials is desirable throughout all phases of planning, design, and construction. Agreements must be made between the railroad company and owner for operation and maintenance of private tracks.

The railroad company may furnish guidelines and standards for design. Every effort should be made to obtain the required information prior to any detailed layout.

For design purposes, the following information should be obtained from the railroad:

- maximum horizontal curvature and minimum tangent distances allowed for the type of layout,
- maximum grades allowed for transfer and storage,
- vertical curve requirements and maximum rate of change of grade,
- standard vertical and horizontal clearances—for lateral clearances, between centerlines of track and to fixed objects,
- turnout numbers to be used from the existing track and within the proposed layout,
- lead distances for the turnouts to be used,
- weight of rail for existing track and for proposed track,
- typical sections for roadbed width, slopes, ballast, ties, and rail configurations, and
- technical specifications.

Track design standards, such as standard rail section, turnouts, guard rails, frogs, plank crossings, signals, and others, are available from the railroad company.

Other design considerations include drainage, earthwork, slope stability analysis, crossings, rights-of-way, and special conditions required for the project.

Whether done by the railroad or by the developer, the construction of all tracks and appurtenances should conform to the best construction practices as prescribed by the Manual of the American Railway Engineering Association (AREA).[20]

As a rule-of-thumb, when a development is rail-served: 1) the rail lead track and nearby street elevations will generally approximate one another; 2) the minimum radius of curvature of the track will be between 350 and 400 feet; 3) maximum permissible gradients along spur tracks will be about 1½ to 2 percent; and, 4) the dock height should be set from 3.5 to 4.0 feet above top of rail of the spur track.

In deciding whether to provide rail service, and what portion of a development should be rail served, it is important to consider the amount of land in an industrial development which would be consumed by rail service. Admittedly, some of this land would be in required yard setbacks but this area could be used for parking, truck loading, or in some instances, structures—if not required for rail.

---

[20] For additional information of significance to the developer, see Robert E. Boley, *Industrial Districts: Principles in Practice*, Technical Bulletin no. 44 (Washington, D.C.: Urban Land Institute, 1962), pp. 16-18.

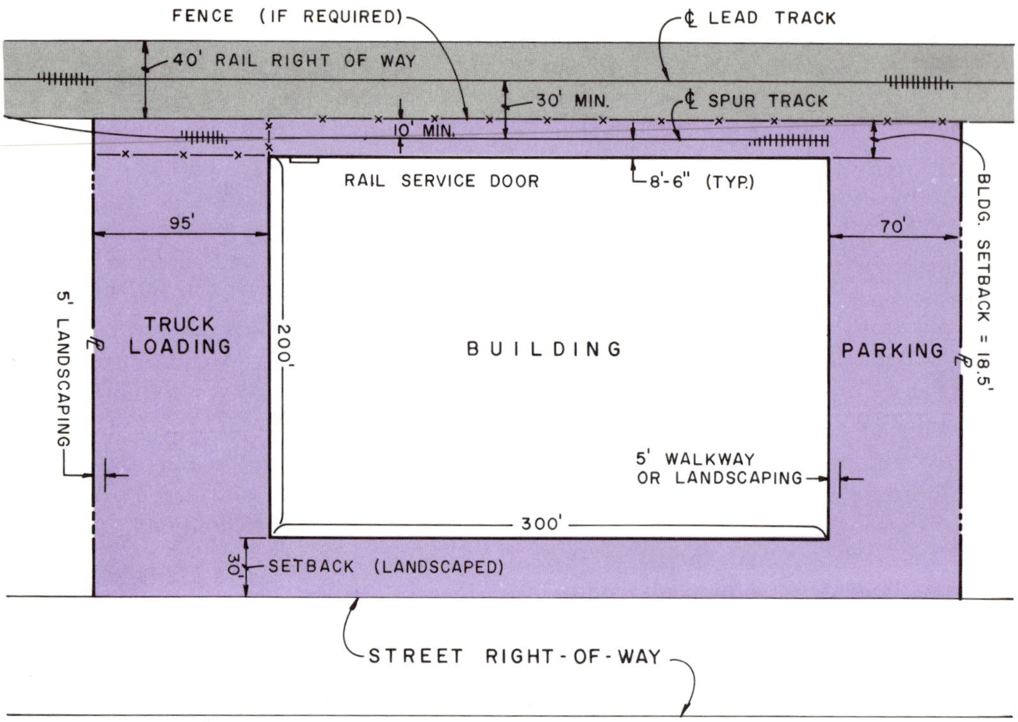

4-21 **RAIL SERVED BUILDING LAYOUT**

BUILDING COVERAGE = 52% OF SITE

RAIL IN RIGHT-OF-WAY, BLDG. SETBACK, SPUR ALONGSIDE BLDG.

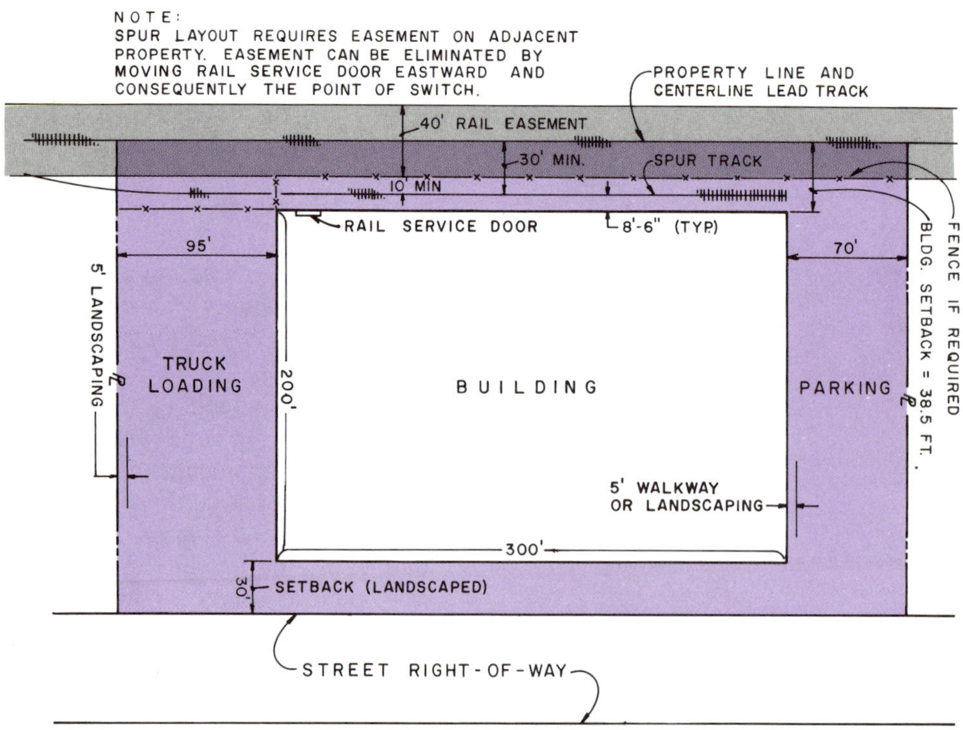

4-22 **RAIL SERVED BUILDING LAYOUT**

BUILDING COVERAGE = 48% OF SITE

RAIL IN EASEMENT, BLDG. SETBACK, SPUR ALONGSIDE BLDG.

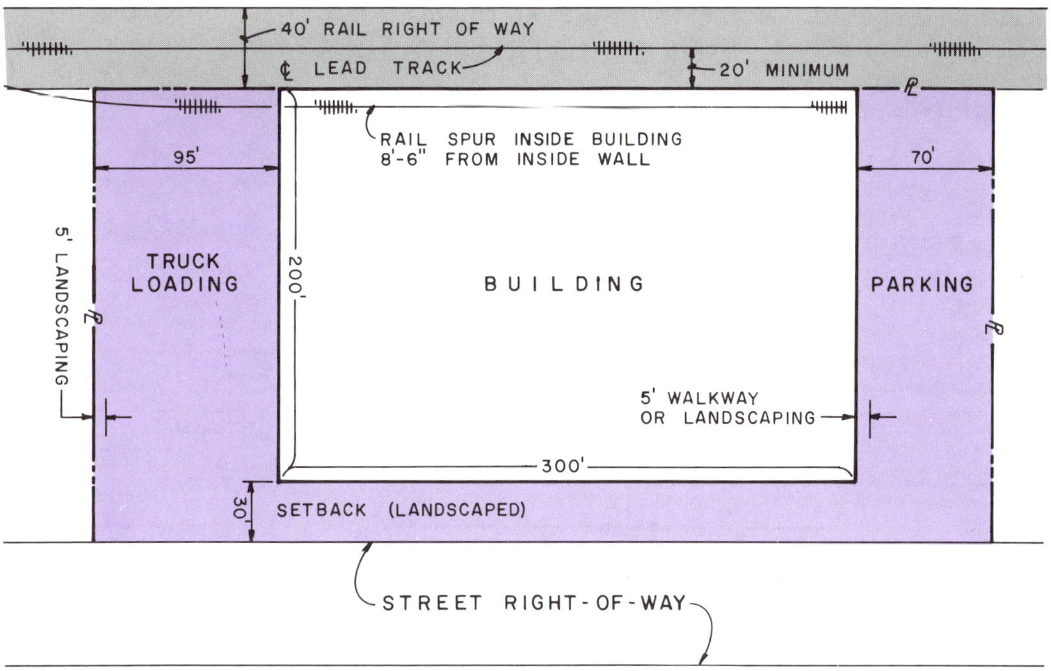

RAIL SERVED BUILDING LAYOUT

BUILDING COVERAGE = 56% OF SITE

4-. RAIL IN RIGHT-OF-WAY, BLDG. ON PROPERTY LINE, SPUR INSIDE BLDG.

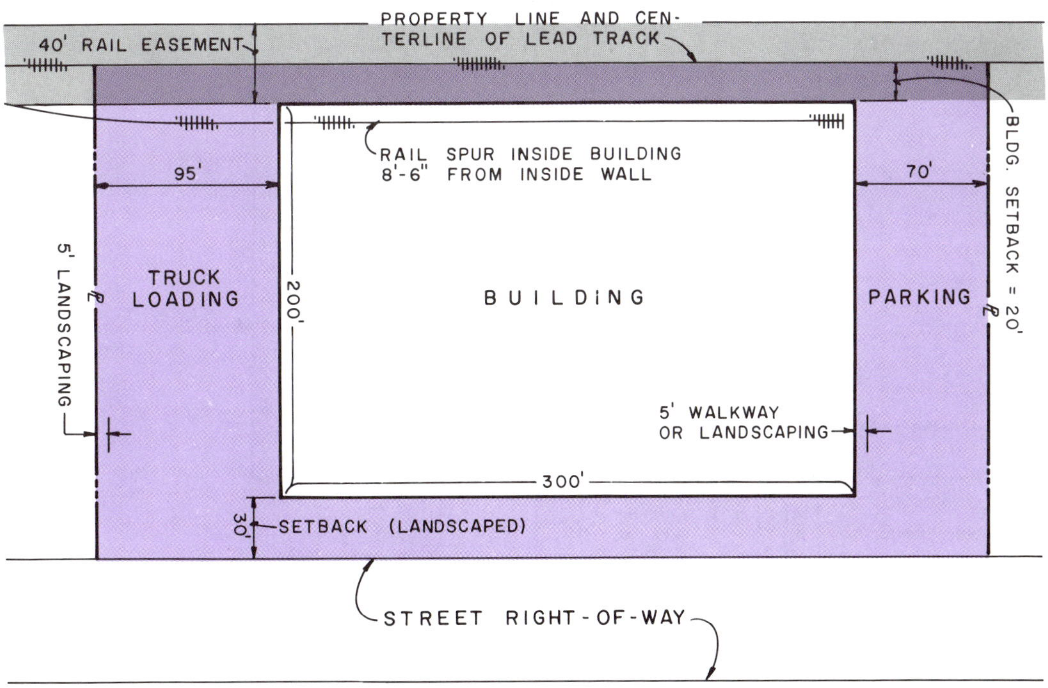

RAIL SERVED BUILDING LAYOUT

BUILDING COVERAGE = 52% OF SITE

4-24 RAIL IN EASEMENT, BLDG. SETBACK, SPUR INSIDE BLDG.

TYPICAL PLANS FOR INDUSTRIAL TRACKS

Typical Roadbed and Ballast Section

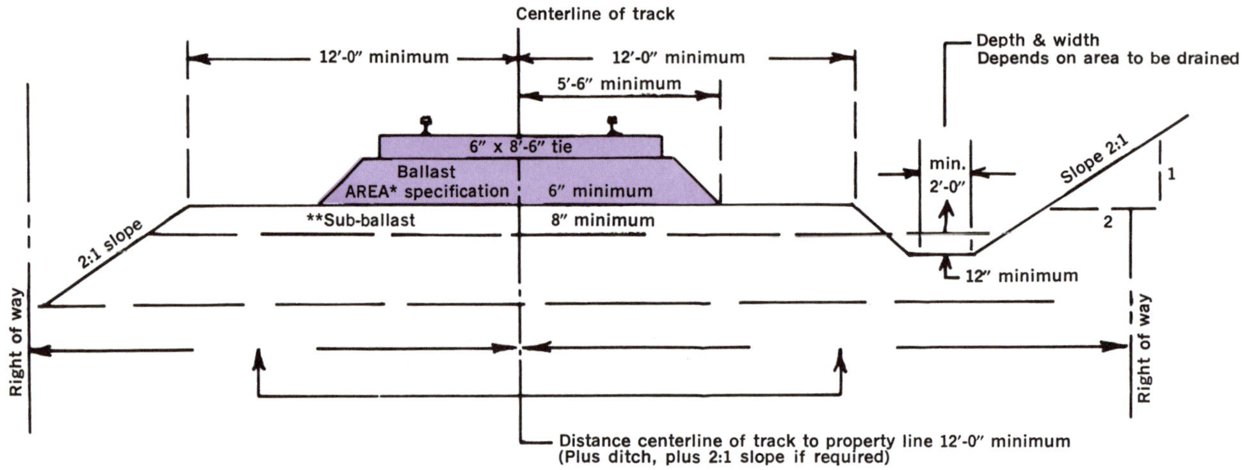

*American Railway Engineering Association

**If natural ground does not provide good drainage, 8″ of porous material must be installed. Note: No draining to be diverted to railroad ditches without consent of railroad and agreement with railroad.

4-25

TYPICAL CLEARANCE SECTION INSIDE BUILDING

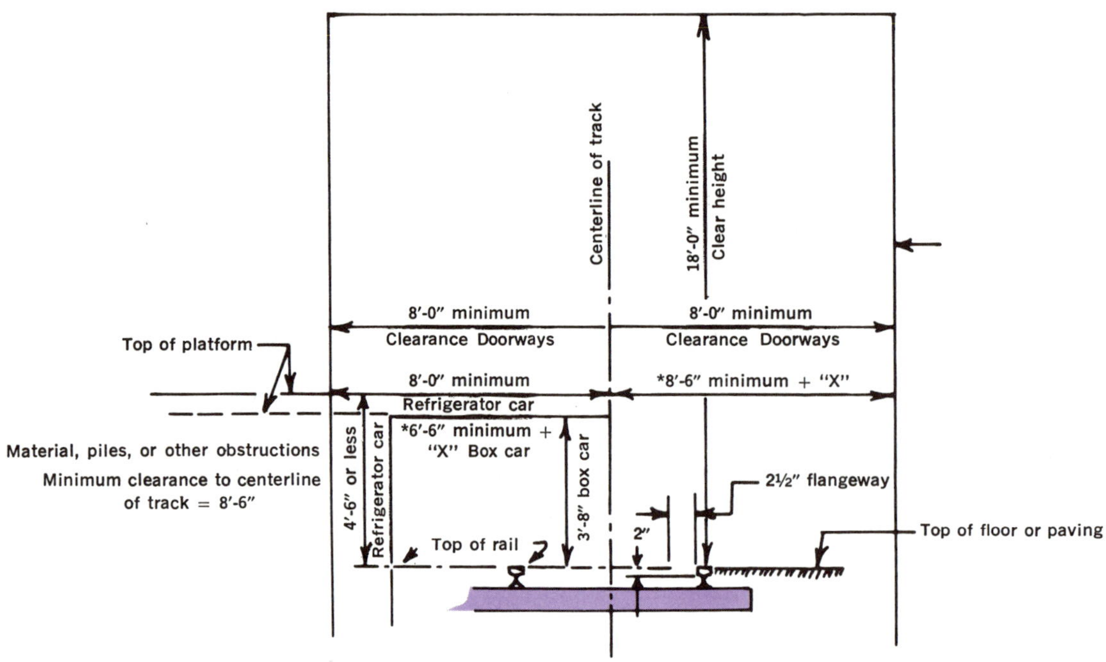

*Minimum distance can be reversed, as long as 8′-6″ minimum + "X" is maintained on one side. Note: "X" = Add 1″ per degree of curvature until car is totally on tangent track.

4-26

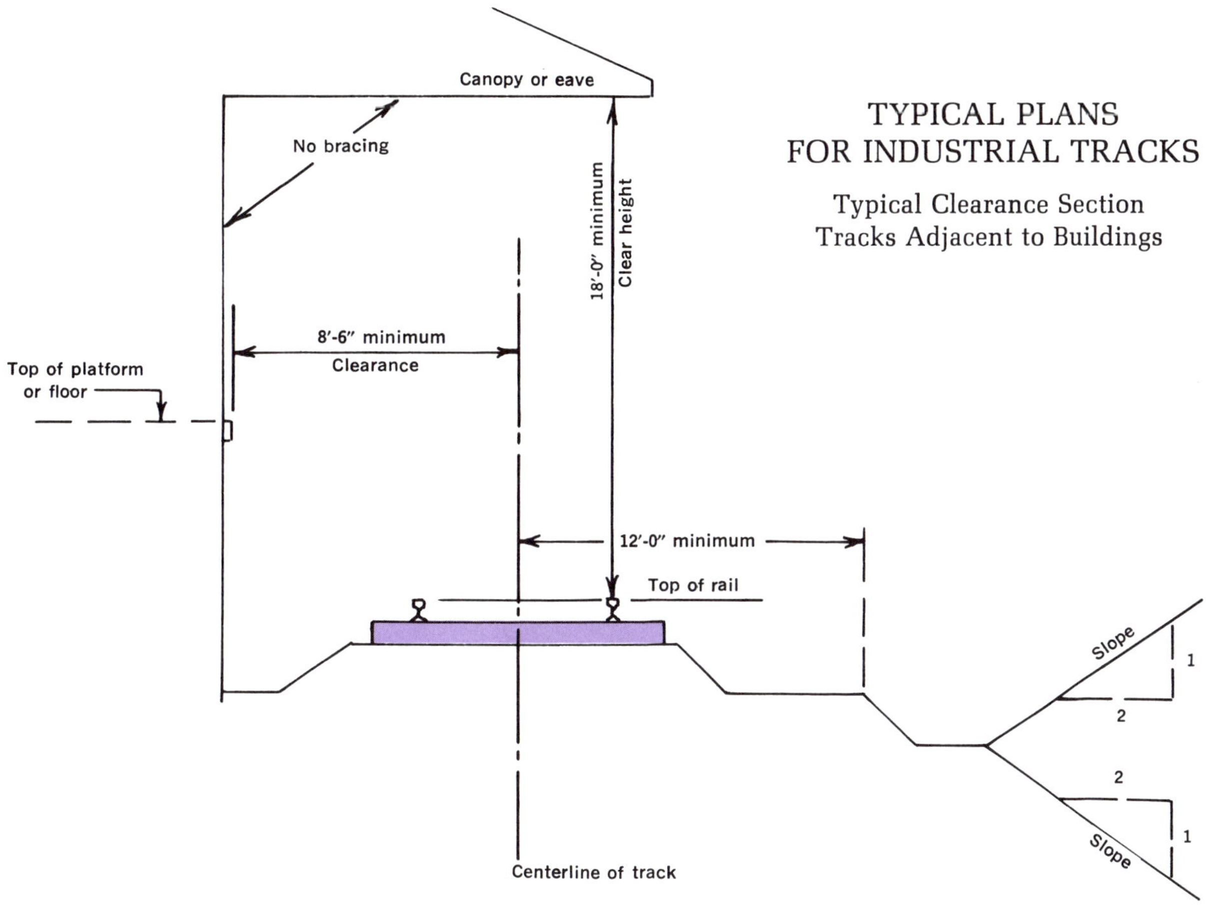

TYPICAL PLANS FOR INDUSTRIAL TRACKS

Typical Clearance Section
Tracks Adjacent to Buildings

4-27

Engineering Considerations in Lowlands Development

Industrial development in urbanized lowlands, protected from flooding by levees or floodwalls, is an area of special concern. Many of the nation's industrial districts have been developed over the years in this physiographic setting. Such areas have special problems with where improvements are to be constructed and with the land prepared for building construction. These areas characteristically consist of flat-lying alluvial soils of variable texture ranging from coarse sands and gravels to silts and clays. In areas which are former river or creek channels or sloughs, clays of high plasticity may cover many acres, and may create special limitations due to potential settlement or shear failure under structural loading. In addition, problems due to groundwater and stormwater runoff from tributary hillside watersheds, are possible. These problems can be recognized and delineated in a thorough site study by a qualified engineering consultant. Development of protected lowland areas has been successfully accomplished in many areas of the United States due to thorough analysis and skillful site engineering.

As part of the consideration given to purchasing lowlands for development, the degree of flood protection needed should be determined. Recent federal legislation, under the National Flood Insurance Act of 1968, effectively prohibits banks or other agencies from financing or otherwise assisting in the development of lands in flood-prone areas since insurance of loans cannot be obtained. Applicable regulations should be checked out by the developer prior to commitment.

Specific engineering design features applicable to lowlands development include:

- use of water tight joints for sewer construction,
- use of subdrainage beneath street pavements to drain off accumulating water and thus improve pavement performance,
- use of open channels and roadside swales to minimize drainage costs,
- subdrainage systems around rooms below grade

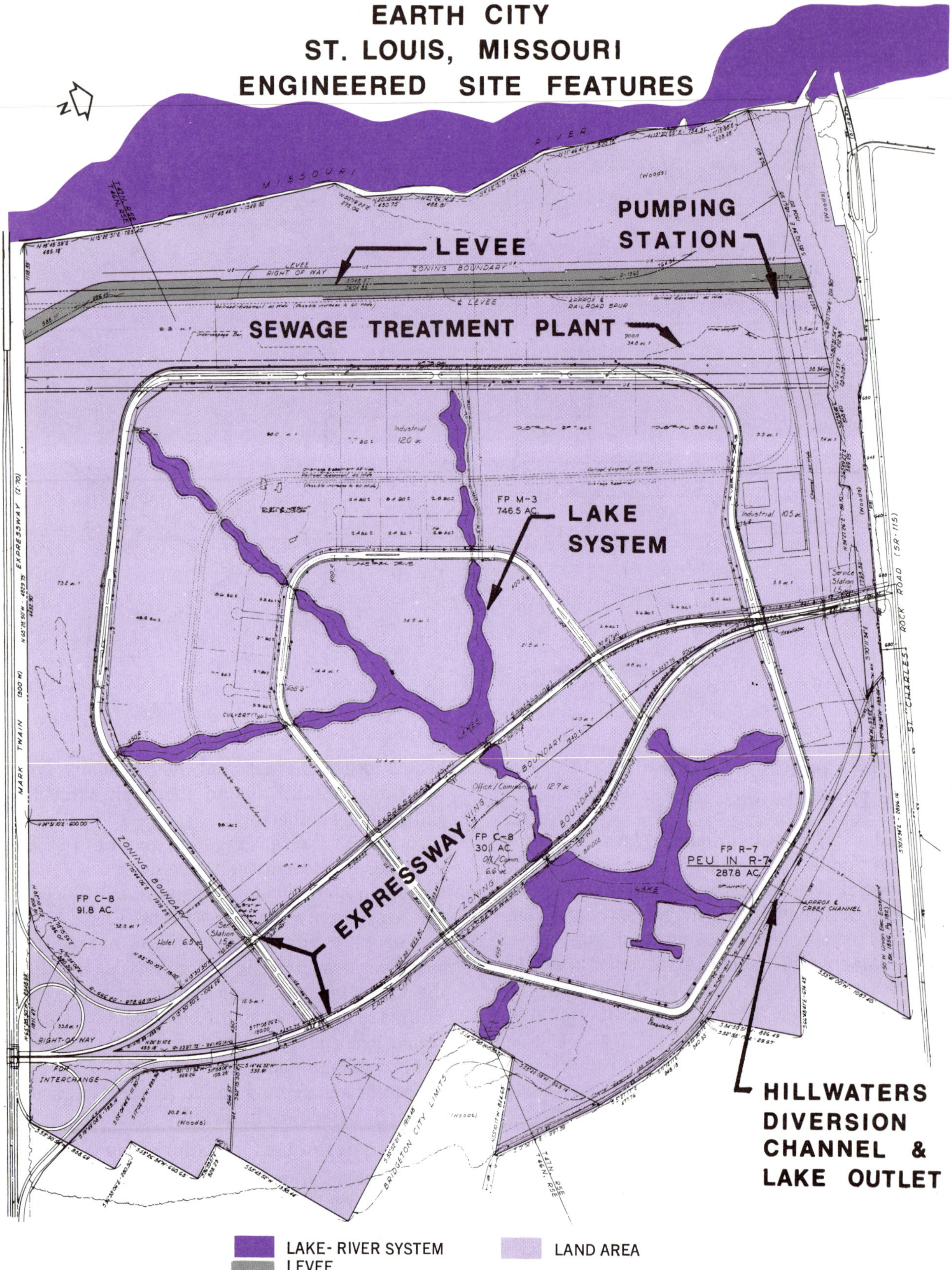

4-28 Site plan—Earth City, St. Louis County, Missouri. Development in the flood plain of the Missouri River necessitated a system of drainage lakes and a diversion channel. There is a levee between Earth City and the River.

including use of sump pumps to relieve hydrostatic pressures,

- slope protection along drainage channels with berms as applicable to assure stability during bleedout of groundwater through slopes,
- use of stormwater retention basins or lakes to accommodate stormwater runoff from hillside areas,
- use of smooth sewer pipe to reduce friction and thus minimize the number of sewage lift stations required in the development, and
- anchored provisions for underground storage tanks to prevent flotation.

One of the most important considerations in developing lowland areas is how to deal with relatively high, and fluctuating, groundwater levels. When excavations are made in granular soils (primarily in fine sands) below the water table, the bottom and sides of the excavation will become unstable due to the movement of groundwater into the excavation. Placement of sewers and foundations under these conditions may be extremely difficult.

Since instability is due to movement of soil particles along with the groundwater, some success has been achieved by placement of a blanket of relatively coarse sand over the finer foundation sand prior to removal of accumulated groundwater from the excavation. The coarse blanket material serves as a filter allowing groundwater to bleed through while holding the fine sand particles in position. Such a procedure can provide adequate stability against light hydrostatic pressures up to a few feet. Stability for more severe conditions can be accomplished by use of a dewatering system. Dewatering systems in common use consist of well points or deep wells. Both employ the principle of pumping down the groundwater to a level below the bottom of the planned excavation and maintaining this condition throughout the duration of the construction work—twenty-four hours per day, seven days per week. Obviously, such procedures are extremely costly.

Because groundwater levels fluctuate primarily in response to river levels, and other influences—such as seasonal rainfall, and local withdrawal from wells by industry and municipalities—the developer should refer to available river hydrographs and well records.

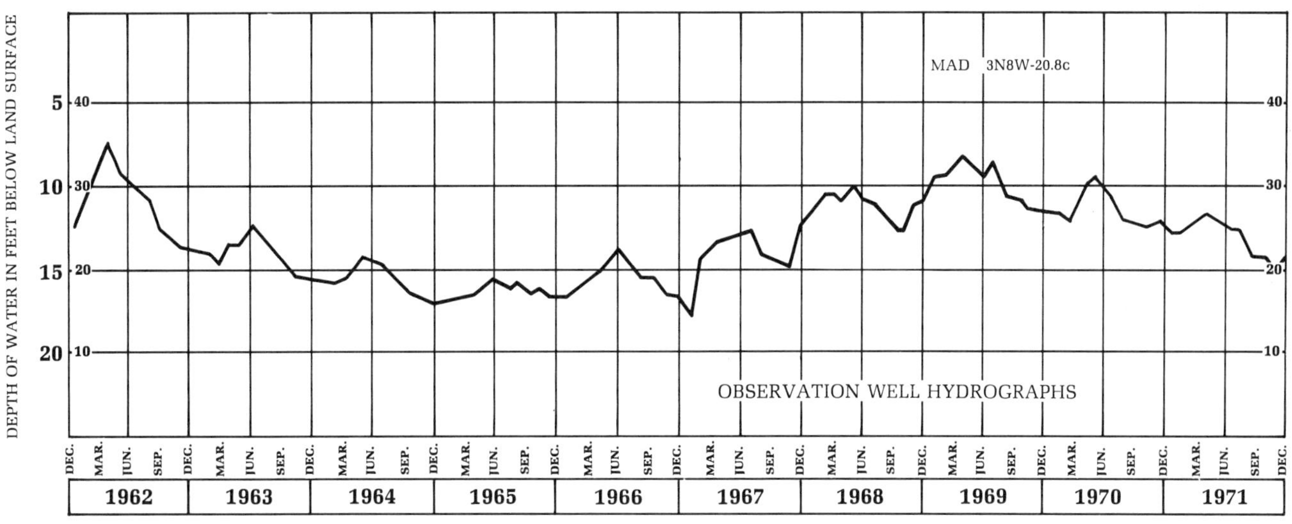

Source:
PGA—Engineers
St. Louis, Mo.

4-29 Seasonal Fluctuation in Groundwater Level—Madison County, Ill.

Historical information on local rivers may be obtained from the U.S. Army Corps of Engineers' district office. Groundwater data may be obtained from water level records in observation wells maintained by state agencies or United States Geological Survey Offices in the area. For example, the Illinois State Water Survey in Urbana, Illinois, has excellent records of groundwater levels at selected locations in the American Bottoms along the Mississippi River in Madison County and St. Clair County, Illinois.

During a 1973 study of potential development of a large tract of land in American Bottoms available records were invaluable. The ten-year record (1962 to 1972) indicated that groundwater depths had seasonal fluctuations ranging up to eight feet.

Groundwater depths in the observation wells during this period ranged from a minimum depth of 7.6 feet to a maximum depth of 29 feet. Average depth for the ten-year period was 17 feet below ground surface. It could also be seen that groundwater levels generally receded in the late spring, summer, and early fall when discharge from the groundwater reservoir by evapo-transpiration and groundwater runoff to streams is greater than recharge from precipitation and induced infiltration from the Mississippi River and nearby water courses. Water levels begin to recover in early winter when conditions are favorable for the infiltration of rainfall to the water table.

Based upon groundwater observations, it was concluded that groundwater levels were at sufficient depth below the ground surface to permit development without substantial expense due to groundwater conditions. Late fall and early winter were determined to be the best time for sewer construction and excavation for stormwater retention basins.

Architectural Character, Signage, Graphics

The overall appearance of an industrial park, in large measure, is determined by the control the developer exercises over the architectural character of individual buildings, signage, and graphics. Normally, these controls take the form of predetermined guidelines established by the developer and understood both by potential tenants and by the developer.

Graphics and signage are used for purposes of identification, communication of information, and vehicular control. These are devised by the architect/designer and applied according to a uniform format to street signs, interior and exterior directional signs, building numbers, and tenant names. The primary goal is to communicate and to identify visually, from a vehicular and pedestrian perspective, a design solution which accomplishes functional purposes and also constitutes a unifying and attractive element within the development. Elements used to create the desired effects include type face, color, and materials. A logo should be created which will serve as a graphic representation of the industrial development project. It is particularly important to establish a suitably landscaped identification sign at the entrances which highlights the graphics theme applied throughout the development. This may take various forms ranging from a simple sign to an elaborate pylon or sculpture.

4-30 Graphics—Freeway Commerce Center, Tempe, Arizona.

The developer has considerable flexibility in the area of graphics although street signs, building numbers, and the form of signs to be used on dedicated streets must be approved by municipal authorities. The institutionalized signage of industrial users will have to be accommodated within this design program but total acquiescence to the user's preconception of their signage needs may damage the very quality environment which attracted them to the development.

Once the basic graphics framework has been established, architectural controls are used to insure proper placement of buildings within the industrial park; to set the tone for the entire development; and to protect property values by insuring high quality appearance, operation, and maintenance. These controls, which supplement governmental land use controls, are delineated in protective covenants devised by the developer. Design guidelines incorporated in these covenants apply to such factors as exterior appearance of buildings, landscaping, site coverage, lighting, signs, and placement of buildings. Certain things not in conformity with the design plan, such as rooftops signs, and flashing lights, will be specifically prohibited. Normally, a review process is established by the developer to enforce these regulations. Plans of park tenants are submitted to a review board which evaluates them for compliance with the regulations.

4-31 Signage—Busch Corporate Center, Columbus, Ohio.

Perimeter Sites

Perimeter sites with high visibility usually command higher sales prices than interior locations. Yet, many new industrial subdivisions continue to be planned and developed with buildings oriented toward internal streets with the rear of the structures facing freeways or thoroughfares.

In preparing the development plan for the industrial park, it should be remembered that lots which are adjacent to major highways are fronting on these streets and highways regardless of building orientation or site access points. In some instances, a lot may be bounded by a freeway, a major highway, and also be a corner lot within the park. Thus, some sites within the park will have no backyard or area in which architectural standards for the building, landscape treatment, and maintenance, can be relaxed.

Protective covenants provide protection to the community and to future occupants of the industrial park. The covenants should state specifically the requirements governing development, use, and maintenance of lots which have several front exposures.

Creating Value on Interior Sites

Industrial park developers should consider ways to create prestige sites in the interior of the parks. Some have provided plazas, sitting parks, gardens, recreation facilities, or other open space areas as a focal point around which higher value interior sites can be planned. This design technique has been successfully used in the Irvine Industrial Complex, where a series of freestanding buildings on a loop street front onto a landscaped central open space.

Another means of creating unique interior sites is in the design of the internal road system. Many developers, however, are committed to an unrelieved gridiron pattern which does not create vistas. Modest modifications of gridiron streets can retain essentially rectilinear sites and can achieve the impression of varied streetscapes and vistas.

A zoning incentive—setback requirements which are decreased when public open space is provided in the development—is one way of providing land for shared open space at no expense.

Unkempt Speculative Buildings

Not infrequently an industrial developer's marketing policy permits the sale of lots to speculative builders who begin to erect multi-tenant shell buildings, to be finished later to the specifications of future clients.

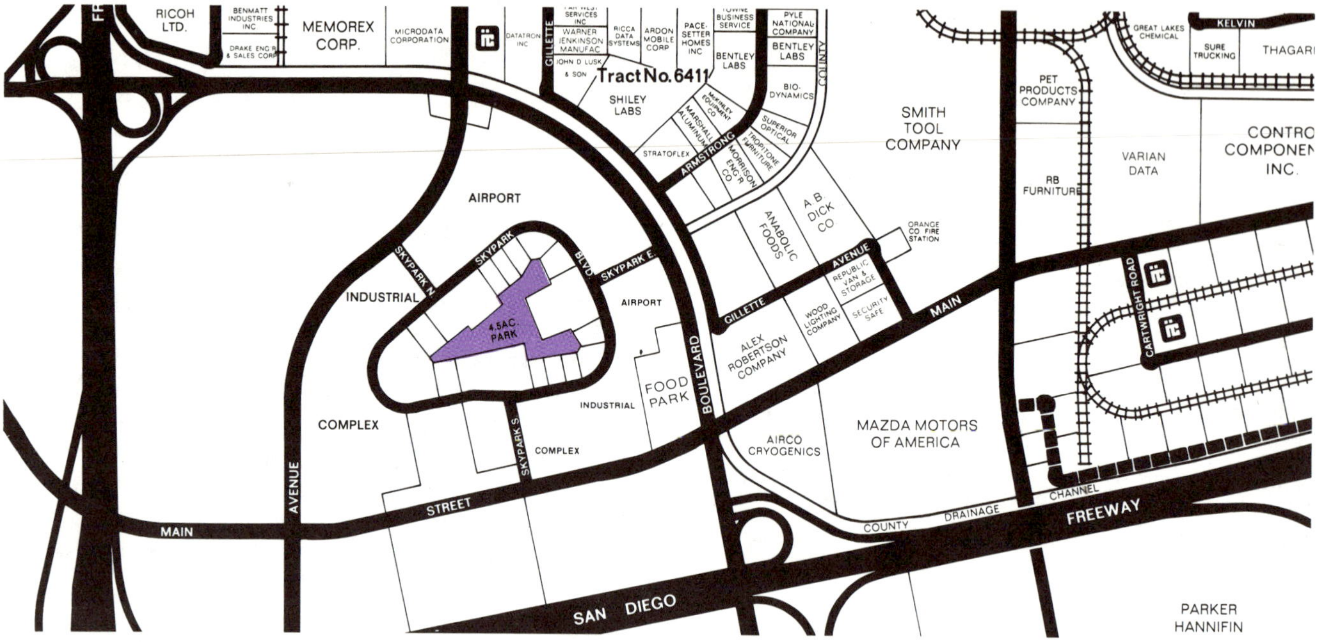

4-32 A small park was established within a loop street at Airport Industrial Complex–Irvine.

In some instances the shell building, with an unfinished exterior and devoid of interior features, remains empty for many months. If the environs of such a building do not conform to occupied structures, if the rubble of construction is not removed, and if the land is not restored by landscaping, the structure will give the appearance of long neglect and vacancy. In an otherwise successfully growing industrial development, the appearance of an unfinished shell building might give the appearance of poor management. To the present occupants, the structure becomes a nuisance and a potential threat to property value. To the potential client for an industrial site, the empty shell with its air of neglect says, "Be cautious—the neglect may spread."

When the industrial developer sells or leases land to a speculative builder, a performance bond should be required to assure within a specified time period:

- completion with final finish of all exterior walls, except possibly one designated for future extension, and full installation of all windows and doors;
- paving of all driveways, walks, parking lots, and truck loading areas;
- removal of all construction debris;
- complete landscape restoration of all remaining land including tree planting to specified size and type at all designated locations; planting of flowering shrubs, other perennials and sod in the front yard, and sowing of grass seed or other ground cover on remaining unpaved open land; and, provision of a permanent maintenance system including underground watering facility.

Off-Street Parking[21]

Off-street parking in a planned industrial development is important because:

- It promotes efficient intermingling of auto and truck traffic;
- It avoids the use of either paved lanes or unpaved rights-of-way which is an expensive use of land, not intended for parking;
- It adds to the appearance of the park, and makes the development aesthetically pleasing.

Off-street parking regulations are some of the most difficult standards of all industrial park restrictions to enforce. This problem usually arises when buildings constructed for one industrial use are subleased or sold to another. For example, in the very carefully planned New England Industrial Center, an original owner/occupant—a department store warehouse—required 50 parking spaces. Later, the warehouse was subleased by the owner to an electronics firm with 500 employees. Even

[21] Dorothy A. Muncy, *Principles of Off-Street Parking and Loading for Industry* (Columbus, Ohio: National Industrial Zoning Committee, 1962).

4-33 Off-street parking is essential in industrial park development. In this instance, the parking area was divided into smaller bays. Berms and landscaping are effective methods of screening these areas from the view of passers-by.

though the new tenant leased additional parking spaces several hundred yards away, employees continued to park on the street. Meters were installed by the community but this only reinforced on-street parking resulting in violations of the time limits and the waste of community police time.[22]

A provision for an occupancy permit, to be required where any change of occupant or activity on the site takes place, may be included in the community zoning code. If a new occupant will bring about a change in traffic patterns and additional parking (either surface or in structures) is the appropriate solution, this would have to be provided prior to the municipality granting the new occupancy permit. The occupancy permit inhibits subleasing or sale of an existing facility to an industry which does not accommodate the building or the site. Also, it protects the long-range financial interests of the developer and all the other occupants of the industrial park. It frees the developer from having to assume the role of policeman, and saves community police staff for more important public safety tasks.

Design of off-street parking should be given the same attention as other aspects of site development controls. Where parking will be permitted in front yards, appropriate screening requirements should be established which might include landscaping, low walls, earth berms, or depressed surface elevation. Equal concern should be given to relationships between adjoining parcels. Where large parking lots are anticipated, landscaping requirements within the parking areas should be considered to relieve the appearance of a sea of asphalt.

The location of landscaping in relation to parking areas requires careful attention. Car overhang dimensions should be considered in the placement of plant material to prevent damage from adjacent car parking.

---

[22] *Industrial Parks: Their Growth and Impact on the Industrial Real Estate Market* (Washington, D.C.: Society of Industrial Realtors, 1971), p. 19.

Relationship to Adjacent Residential Development

Planning the industrial development so it is a compatible neighbor to adjacent commercial and residential areas is a matter of critical importance. This can be accomplished in a number of different ways. Perhaps most common is to utilize an arterial street or freeway as a buffer zone. The effectiveness of this approach can be enhanced by creating a landscaped median in the right-of-way to enlarge the separation between homes, businesses and industrial activity. In Elk Grove Village, Illinois a power line easement occupies the median of the arterial street separating the industrial and residential portions of the community.

Back-up treatment involves orienting industrial and residential uses away from arterial streets toward interior circulation elements in conjunction with appropriate landscaping and screen-fencing along thoroughfare frontages.

A more subtle approach is to abut the uses directly. When this is done, landscaping, berming, creation of deep lots and other site planning techniques are needed to visually separate the uses. In addition, industrial park occupants placed in peripheral locations must have appropriate user characteristics and structural designs. They should generate low traffic volumes and exhibit no outward evidence of the industrial processes carried on within. Low profile building designs are preferable and structures should be given four-sided architectural treatment.

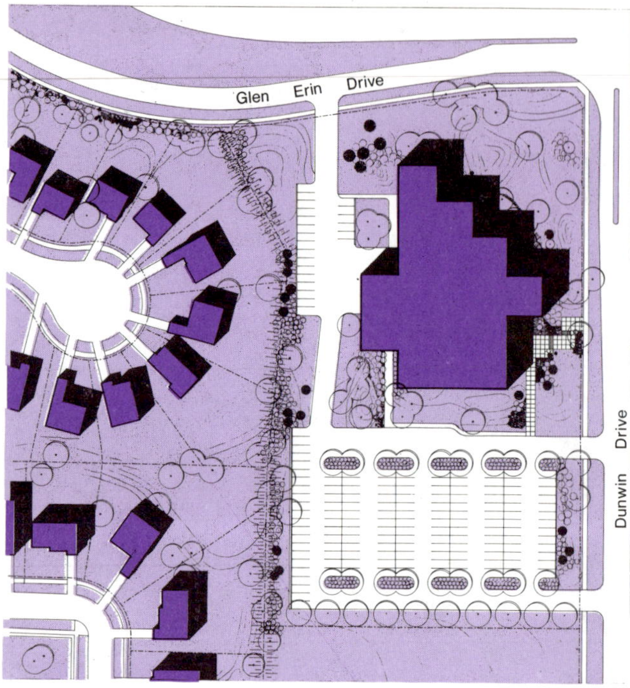

4-35 Site plan shows use of long lots as a way of minimizing industrial/residential conflict. The photograph below looks north toward Glen Erin Drive.

4-34 Separation of residential and industrial areas is graphically depicted in this photograph taken at Erin Mills, Ontario. Note wood fencing, chain link fencing, landscaping, and placement of parking adjacent to the property line.

Gas Station Locations

Auto repair shops and gas stations are useful additions to the principal industrial occupants. Proper site selection, however, is critical. Prominent and accessible sites at the entrance to the industrial development should be avoided, because these are likely to be the locations selected by oil companies, if given a choice.

The following example, drawn from an industrial park less than four years old, proves the point. The developer owned several thousand acres along an interstate highway. Part was being developed for houses, and part for an industrial park. The entrance to the industrial park was at the first grade intersection several hundred feet from the freeway. The land uses which the developer put at this intersection were: one corner occupied by an expensive redwood stockade-type fence which screened an attractive new residential development; on each of the remaining three corners was a gas station. In the fall of 1972, one of the two gas stations which formed the entrance to this multi-million dollar industrial park, closed, and left behind a site too small to be readily converted into a dignified use for half an entrance to the industrial park.

A more suitable solution to gas station location is illustrated by Stanford Industrial Park. A single gas station was permitted on one corner of the major crossroads in the center of this park. It is easily accessible to all industrial occupants of the park, in a prominent location, but not at the entrance. This gas station performs an essential function in an inconspicuous manner.

Construction Planning

Sequence of Constructing Improvements

Ideally, the sequence of constructing improvements will be that which results in the least cost to the developer. Sequence of construction will vary in different parts of the country depending on climate, time of year, and availability of materials. Normally, sewers are placed first. However, good practice suggests that topsoil removal and stockpiling, along with rough grading of cuts and fills within the street rights-of-way, should proceed before the placement of any improvements.

Sanitary and storm sewer construction should be scheduled in advance of other improvements due to their greater depth. Constructing sewers in ad-

4-36 Construction at an early stage at Busch Corporate Center. Initial development included streets and utilities with some buildings going in simultaneously. A high water table and poor soil drainage characteristics hampered construction.

vance of other improvements is particularly advisable in mixed developments with curvilinear streets because of potential conflicts with straight sewer alignments alongside curved streets. In developments consisting primarily of warehousing, the streets usually have a straight alignment. If sewers can be placed far enough from pavement to avoid damage to streets, sewer work can follow pavement placement. Sewer work done under open field conditions when groundwater is not a problem usually can be accomplished with minimum shoring of the excavation. Occupational Safety and Health Act requirements should be referred to when planning this work as well as all other construction.

When groundwater is encountered, particularly in lowland soils comprised of silt or fine sand, dewatering of the trench may be required. Such conditions should be determined and evaluated by the engineer during preliminary planning of the project. Dewatering can add as much as $20 to $25 per lineal foot to the cost of trenching for sanitary and storm sewers and can be an important additional cost factor.

When sanitary or storm sewers extend beneath areas to be covered by pavement for street construction, trench backfill should be either granular material such as sand or crushed limestone and screenings flooded with water, or compacted with vibratory equipment as placement proceeds, or should be compacted earth backfill so that a firm foundation will be provided for streets and trench subsidence beneath pavements may be avoided. Such provisions usually are included as requirements in the construction standards of the agency having jurisdiction.

After sanitary sewers, storm sewers, and rough grading are completed, streets should be constructed. Street construction may be followed by other utilities such as gas, electric, telephone, and water. Utility companies may, however, construct their facilities before or after street construction since they are placed fairly shallow and close together. Where the street is in place prior to utility construction, the pavement becomes a convenient reference for horizontal and vertical control of utility work as well as a convenient platform to facilitate construction.

When the street pavement is in place prior to the utilities, boring beneath pavements is required at street intersections. In this instance, pipe casings to accommodate water mains should be set prior to paving streets. The ends of these casings should be staked and this information should be provided to the water utility. Water mains can then be pushed through the casing instead of the more costly method of boring under the street for water main installation.

Phase Construction

Construction of improvements should be phased in accordance with anticipated rate of marketing the land. The most appropriate location within the development for marketing the first parcels of land should be thoroughly studied. Marketing considerations may dictate that early construction phases be located near an existing major off-site roadway. Such location would result in the shortest initial roadway construction but perhaps a substantial length of storm and sanitary sewer and water main construction. Considerations will vary with each project depending upon a number of physical and marketing factors. It is the purpose of this discussion to alert the developer to early involvement of the engineer in determination of the appropriate phasing of improvement construction.

Construction phasing should be related to the project's economic model. Such relationships are not part of the engineering portion of the handbook but are covered under development strategy.

Erosion Control

Erosion control may be accomplished by seeding or by placement of sod, ground cover, or by planting of trees or other vegetation. Seeding and planting seasons vary throughout the country so that care should be taken in selecting the appropriate time for each location. Seeding of large areas frequently is done quite economically by power spraying from a truck mounted system. Seeds, fertilizer, and mulch are mixed with a liquid asphalt and the mixture sprayed onto earth slopes. The asphalt tends to hold the mass in place until germination. Seeding of temporary, fast-growing grasses often is most desirable when final grading cannot be done until a later date.

Rock riprap may be dumped and spread by construction equipment as a means of erosion control for sloping ground or along drainage channels. Prior to placement of riprap along drainage channels, or other surfaces susceptible to occasional inundation, one or more layers of granular material should be placed on the natural soil slope. Blanket material particle sizes should be well graded. Succeeding blanket courses should have increased particle sizes as the thickness is built up so that a filter will be developed which prevents foundation soil particles from washing through the riprap. Such washing eventually would result in collapse of the erosion protection and progressive deterioration of the land. Filters may be designed by methods well established in engineering practice. Such engineered erosion protection can be placed rather simply using standard construction equipment and have been found to be quite effective. On slopes of extensive length, benching is commonly done. The long slope is interrupted by a flat surface or berm. This surface may be sloped back toward the upper slope to intercept stormwater runoff. The bench may be seeded, sodded, or paved to direct runoff longitudinally along the slope to some point of discharge.

Sedimentation due to erosion has been with us for a long time. However, in recent years the problem has elicited greater concern, particularly where large areas are graded. This sediment, consisting of dirt and debris, is carried downhill and deposited on streets or adjoining property, or in a natural water course. The federal government has produced a number of publications which give helpful suggestions for controlling erosion.[23]

Land development erosion control has been implemented in many areas of the country in recent years. This form of environmental protection has been strongly encouraged by the federal government because of damage to soil and water re-

[23] *Controlling Erosion on Construction Sites,* Soil Conservation Service Bulletin no. 34 (Washington, D.C.: U.S. Department of Agriculture, 1970); *Control of Erosion and Sediment Deposition from Construction of Highways and Land Development* (Washington, D.C.: U.S. Environmental Protection Agency, Office of Water Programs, 1971).

4-37 Crocker Industrial Park, Brisbane, California, was built on steeply-sloping terrain between two mountainous areas. A substantial amount of terracing was needed to ready this property for industrial use.

sources of developing communities.[24]

Sediment can be controlled during construction by installing weirs in down-slope drainage swales. A weir may be constructed of nothing more than a few planks stacked on edge and supported in place in the water course. A desilting basin may be created by placing an earth fill across the water course with a pipe projecting through the fill. At the upstream end the pipe is provided with an elbow and a short length of perforated vertical pipe. The earth fill may be provided with paving to protect the fill during passage of stormwater over the surface. Silt is trapped for the most part in the area behind the earth fill. Sediment control devices should be left in place until the period of significant sediment potential is passed.

Control of erosion and sedimentation is good construction practice. It is also good public relations and may even avoid unnecessary risk of liability.

Estimating Development Costs

A vital input to determination of project feasibility is the development cost estimate. The master plan for the project will be worked out with the objective of maximizing marketing potential while minimizing the cost of site development. Items which normally will be included in the development cost estimate are:

- water distribution system,
- sanitary sewerage system—including pretreatment,
- stormwater drainage system—including detention facilities,
- street network,
- earthwork—including clearing and grubbing,
- landscaping and seeding,
- rail lead track system,
- street lighting,
- entries, other special features, and signage,
- utility relocations,
- underground electrical and telephone service (in lieu of overhead),
- utility connection fees,
- maintenance costs of facilities and systems prior to acceptance by authorities, and
- engineering and planning fees.

Development costs can vary widely depending upon factors such as the physical character of the site, anticipated lot size and availability of off-site municipal services.

Development costs will generally range from $5,000 to as much as $20,000 per net salable acre. The higher costs are particularly applicable to developments of one or two hundred acres with small lots, extensive grading, and an underground storm drainage system. Large lots usually can justify drainage easements necessary for open channel drainage without discounting the space occupied by the easement in the purchase price of the lot. However, small lots—one and two acres in size—on which such easements and open channels are located, generally will not bring full price. Also, small lot development will increase lengths of street and utilities per unit of area developed resulting in increased development cost per acre. Net salable acreage also will be lower for small lot developments.

In areas where the developer must pay for installation of the water distribution system, costs may be partly recoverable through refunds based upon volume used by customers. If sewage treatment is available through a private company rather than a municipality, a substantial connection charge to the developer is likely. Rail lead track construction costs may be recoverable in part as a function of car loadings by agreement with the railroad.

For accuracy, development cost estimates should be prepared carefully using cost experience of the developer and the engineer. Local contractors should be contacted during estimate preparation to verify prices. The contractor can provide accurate up-to-date cost data. Further cost input, including current and near-future trends, can be obtained from publications such as *Engineering News-Record*.

Net salable acreage should be calculated for the project so that development costs per net salable acre can be determined. Net salable acreage is determined by deducting from the gross acreage of the site, the proposed dedicated rights-of-way and areas devoted to sizeable drainage channels or otherwise unusable. Such deductions usually will be 8 to 12 percent of gross acreage. A somewhat higher percentage may occur when stormwater retention basins are necessary.

A further breakdown of development costs should be made for each phase of the development. This will be necessary to properly evaluate costs in the financial planning of the project. The cost per acre

[24] See Mel D. Powell, William C. Winter, and William P. Bodwitch, *Community Action Guidebook for Soil Erosion and Sediment Control* (Washington, D.C.: National Association of Counties, 1970).

for developing the area included in the first phase of a project may be found to be substantially above the average cost per acre for the entire project. This frequently occurs due to the need for constructing complete storm and sanitary sewerage and outlet facilities which will ultimately serve the total project during the first phase.

A minimum 10 percent contingency factor should be applied to any costs developed during the early investigations and should be maintained for each item of work until the work is under contract. Additional escalation factors should be used when phased construction is planned to stretch out for more than two years.

Obviously, true costs will not be known until the development work is completed. However, forecasting costs as accurately as possible, and anticipating potential extraordinary costs, is vital to the developer. Cost estimating should begin early in the development analysis and should continue through to project completion. This is discussed in further detail in the development strategy section of the Handbook.

Case Studies

This section contains two case studies. Special engineering requirements for each project are analyzed in detail. The purpose of this section is to show how engineering and design considerations were incorporated into the master planning and into the actual development of the project. Both projects offered different challenges in terms of engineering and design of the site to minimize cost and maximize attractiveness in the market place.

Busch Corporate Center

Busch Corporate Center is a 153-acre business and industrial park located adjacent to the Anheuser-Busch Brewery in the city of Columbus, Ohio. The park—a development of Busch Properties Inc.—includes light manufacturing, warehousing, office, and commercial uses.

The site has little topographic relief. High ground occupies the central part of the site which slopes toward ditches along Interstate Highway 71 to the

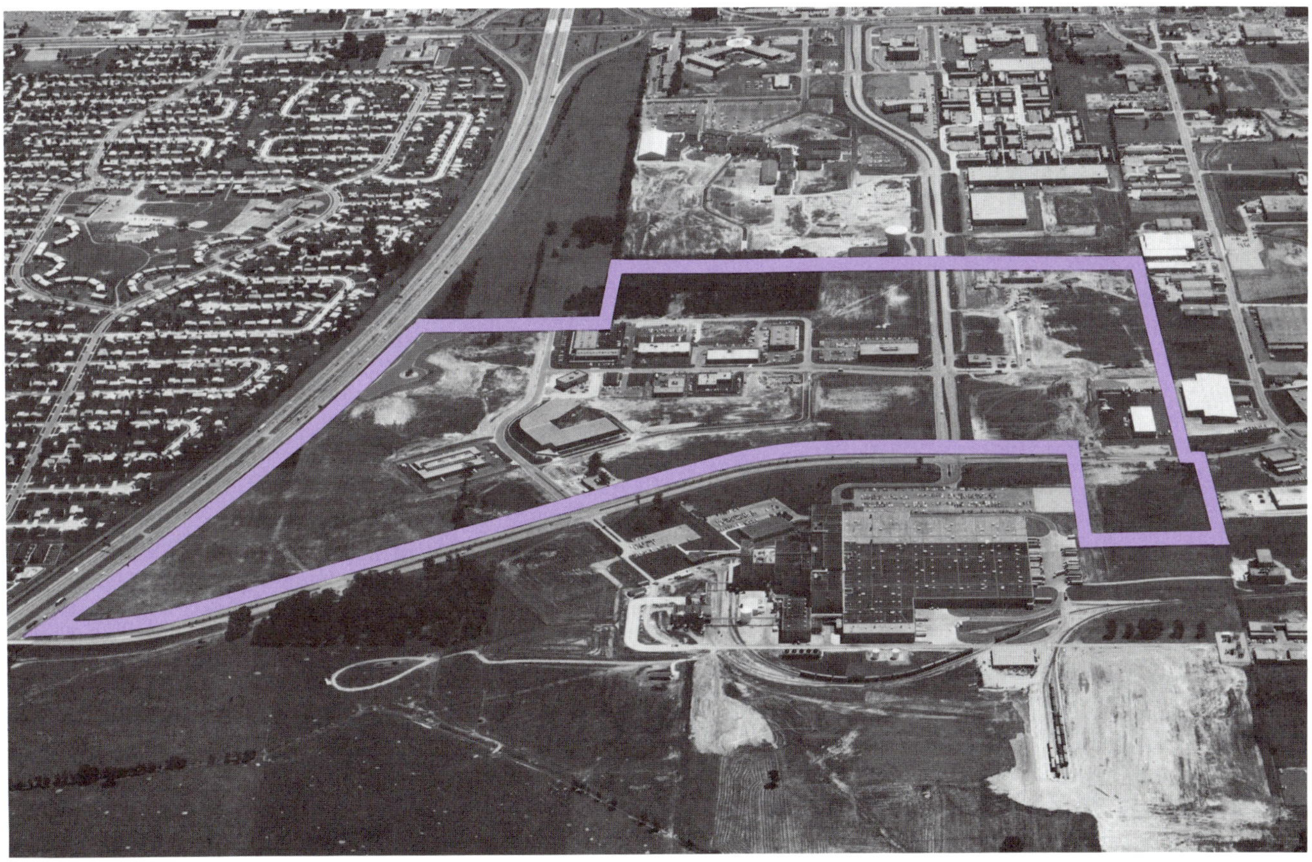

4-38 Busch Corporate Center—Columbus, Ohio. Anheuser-Busch Inc.'s Columbus brewery is located in the foreground. The site has ready access to Interstate 71, but substantial on-site arterial street improvements were required. State Route 710 was rebuilt as a four-lane arterial with a median.

east and along Busch Boulevard and Schrock Road within the site. Both Busch Boulevard and Schrock Road have been improved as State Route 710. The land west of Busch Boulevard slopes gently to the southwest.

The area is well located for vehicular access. State Route 710, which penetrates the site, is a four-lane road with existing openings in the median and left turn lanes at three locations.

Busch Corporate Center was conceived in 1970. A master plan was prepared the same year. As of 1974, more than 60 acres of the property were sold or leased and 21 buildings had been constructed.

Coordination with Local Agencies

As an early step in planning and engineering studies, local agencies and utility companies were contacted for basic data and design criteria. Contacts were made with the Ohio State Highway Department; city of Columbus Department of Public Services—water, streets, and sewerage and drainage, divisions; Ohio Inspection Bureau (fire protection); Columbus and Southern Ohio Electric Company; Ohio Bell Telephone Company; and, Columbia Gas of Ohio, Inc.

Primary engineering considerations included stormwater drainage, sanitary sewers, water distribution, roads, grading, building foundation conditions, electrical, telephone and gas service, development lighting, and pipeline relocation.

Stormwater Drainage

Stormwater drainage for the project was accomplished by a system of enclosed storm sewers and a retention pond. The pond overflow structure discharges to the storm sewer system. Inlets along the roadways and within parking and loading areas connect to the storm sewers.

Special attention was given to storm drainage engineering because preliminary estimates indicated that more than 25 percent of the development costs might be devoted to storm drainage. Consequently, this area offered the greatest promise for cost saving efforts.

The design of a stormwater drainage system for the site was affected by several basic considerations: generally flat topography; availability of more than one outlet for stormwater; and, limited capacity of the existing sewer in Busch Boulevard.

Flat topography is most economically drained by taking advantage of the temporary ponding capabilities of the site. Under the stormwater drainage plan for Busch Corporate Center, a low-lying marshy area about two acres in size at the southeast corner of the development near Interstate 71, was reserved for ponding. Creation of a lake in this location was also a popular idea with both the land planners and the developer because of its pleasant appearance and the exposure such a feature has when viewed from Interstate Highway 71.

The diversion of stormwater to the lake rather than to a storm sewer eliminated the need to extend a storm sewer through this part of the site and reduced the size of the trunk sewer downstream of the lake overflow structure. The net saving on development cost due to use of a lake instead of a storm sewer is estimated to amount to $30,000 or about $200 per salable acre. While the area occupied by the lake represents a two-acre reduction in salable ground, the value of the land is more than offset by savings in sewer costs and by enhancement of land values in the vicinity of the lake.

Two main outlets for stormwater drainage are available—a tributary of Alum Creek to the north to which 102 acres discharge, and an existing storm sewer in Busch Boulevard to which 41 acres discharge. Storm sewers were sized using the Rational Method and the two-year frequency storm design values of the city of Columbus Sewerage and Drainage Division. Sewers designed in this manner met the design requirements of the city. During storms of less frequent occurrence, it is expected that the capacity of the storm sewer system will be exceeded and that excess stormwater will pond temporarily in depressed areas. These low areas will drain, however, when the runoff rate reduces. In order to test the proposed storm drainage plan, a 10-year frequency storm was routed through the system. The amount of ponding in the lowest areas of the park was determined using a triangular hydrograph. It was found that ponding would occur in these areas to a depth less than 12 inches for a period of approximately 30 minutes. The sewer system could have been designed to eliminate such ponding; however, such a system would be substantially in excess of the standards required by the city, would have induced faster runoff, and would have required an increase in pipe size all along the line at an estimated cost of $40,000.

It should be noted that metropolitan areas with flat terrain generally design for more frequent storms, say 2 to 5 years than areas with steeper terrain where 15 to 20 year frequencies are often used. This differential in design standards is due to the substantial increase in size of storm sewers, and therefore, cost in flat terrain. For example, in relatively moderate terrain, a 36-inch diameter storm sewer at a slope of one foot in 100 feet (1 percent grade) will carry 67 cubic feet per second; while this same discharge in flat terrain at 0.3 percent grade would require a 48-inch diameter pipe. Consequently, metropolitan areas with flat terrain permit ponding of stormwater for short periods of time as a practical matter.

Innovative design and cooperation with local authorities was necessary in the handling of stormwater from the 35-acre tract west of Busch Boulevard. Because of the limited capacity of the storm sewer in Busch Boulevard, it was determined that stormwater runoff from the 35-acre area would have to be carried through the sewer system serving the 118-acre area east of Busch Boulevard which drains to Alum Creek. Such connection would have resulted in larger trunk sewer sizes through the 118-acre area at considerable extra cost. Such increased pipe sizes were avoided by devising a plan to drain the 35-acre tract to a small retention pond and by providing an overflow sewer from the pond to the sewer system east of Busch Boulevard. The advantage of this plan was that no increase in trunk sewer size would be required for the 118-acre area.

The Lake Design

The lake along Highway 71 and its spillway to the main storm sewer system was sized using a 10-year frequency design storm. Natural low grades along the highway are such that lake overflows from major storms will discharge toward a ditch along the highway much as occurred prior to land development.

The lake has been excavated to a depth of approximately eight feet. This depth will support a fish population and discourage plant growth. The lake has sufficient tributary watershed to maintain a satisfactory pool elevation even during periods

4-39 Looking west over Interstate 71 to the storm water drainage retention pond.

of prolonged dry weather. The lake has a water surface area of approximately three acres and a watershed of approximately 34 acres, producing a watershed/lake area rate of eleven-to-one. With such a watershed/lake area ratio, it is estimated that a drop in the normal lake level of more than one foot due to evaporation during a drought period would be extremely rare. According to U.S. Weather Bureau data, the mean annual lake evaporation for the Columbus area amounts to approximately 33 inches per year with 78 percent of the total, or 26 inches, occurring in the May-October period. This would amount to an average evaporation rate of 4.3 inches per month. A drop of one foot in the lake surface due to evaporation would require a three-month period during which no stormwater runoff occurred. Such a circumstance would be extremely rare.

Also, lake design should consider possible seepage losses. Soil borings in the vicinity of the proposed lake indicated that significant leakage from the lake was not likely. The soils are generally clay-silt mixtures which would provide an excellent basin for the lake. It was determined that no special treatment of the lake sides and bottom would be required. Dumped stone was recommended around the shoreline to produce an attractive appearance and to protect against shoreline erosion. The lake was excavated in 1972 and is functioning as designed, providing an attractive setting for occupants of the development.

Sanitary Sewers

A complete system of sanitary sewers was designed to serve all lots within the park. Sewers extend along streets as do other utilities within the park. Such practice provides maximum flexibility in planning lot size and lot line locations as marketing prospects develop, yet allows construction of improvements to proceed so that improved lots will be available when needed.

The area east of Busch Boulevard drains by gravity to the northeast to an existing 21-inch diameter sanitary sewer; the area west of Busch Boulevard discharges by gravity to an existing 18-inch sanitary sewer flowing south in Busch Boulevard.

The sanitary sewer system was designed in accordance with the criteria of the city of Columbus Sewerage and Drainage Division. These design criteria included a substantial flow allowance for infiltration. Such allowance is based upon observations made by the city. The sewerage system is designed to adequately serve a development with a population density of 50 people per acre plus infiltration allowance.

Water Distribution

Water service to the development is supplied by the city of Columbus Water Division. The Anheuser-Busch Brewery is served by a 24-inch main along Busch Boulevard and a 36-inch main along Schrock Road. These mains also serve Busch Corporate Center. Adjacent to the development is a 2,000,000-gallon elevated water storage tank. The large mains and storage facilities in close proximity to the development provide the site with excellent service.

A complete system of water mains was designed for the development. These mains extend along the roads and are cast iron pipe, eight and ten inches in diameter. The water distribution system also is designed to serve a development with a density of 50 people per acre and in addition provides fire protection in accordance with the requirements of the Ohio Inspection Bureau. Fire hydrants are spaced at intervals of 300 feet along all roadways in accordance with local criteria. Water main construction has progressed in phases, along with streets and sewers. Busch Properties, Inc. has constructed the system and dedicated mains and easements to the city.

Streets

Vehicular access to the development is from a single entry on Busch Boulevard, and from two entries along Schrock Road. Street widths for the three dual entry roads to the park, and for the interior collection road and minor roads, are based

4-40 Collector street, 41-foot right-of-way, 25-foot pavement.

upon city of Columbus standards for industrial subdivision streets and upon a traffic study by a private consultant. Standards are given below. The protective covenants for Busch Corporate Center provide for off-street parking. Such requirements preclude additional lanes on public streets for parking purposes, and have resulted in corresponding cost reductions for street construction. This prohibition of on-street parking also greatly enhances the flow of traffic, provides better safety, and creates a much better appearance.

- Pavement width:
  —Collector streets—41 feet. Two 25-foot pavements are constructed where the road is divided by a median. A 12-foot left-turn lane is constructed at major intersection where traffic warrants.
  —Minor streets—28 feet.
- Right-of-way:
  —Ninety feet for four-lane road with left-turn lane and medians.
  —Eighty feet for four-lane road with median and no left-turn lane.
  —Sixty feet for four-lane road without median.
  —Fifty feet for two-lane (minor) roads.
- Pavement materials and thickness:
  Seven-inch portland cement concrete base with 1¼-inch asphaltic concrete leveling course and one 1¼-inch asphaltic concrete surface course (9½-inch total thickness);
- Gradient and alignment requirements per city standards; and,
- Utility easements extend beyond the road right-of-way as required.

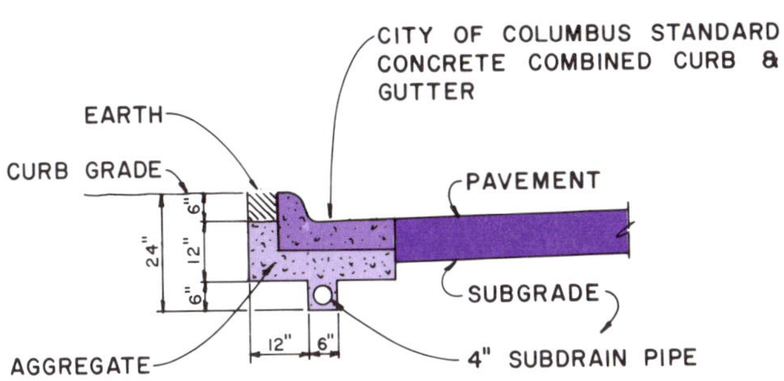

CITY OF COLUMBUS, OHIO
PROVISION FOR PAVEMENT SUBDRAINAGE
TO IMPROVE SUBGRADE STABILITY

4-41

Rail Service

Busch Corporate Center is essentially a truck-served development. While there was demand for rail-served facilities in the general area, the development did not lend itself to economical layout of rail trackage. The features which precluded economic rail service were:

- the need to cross Schrock Road at grade; and
- the limited size (35 acres) of the potential rail-served area south of Schrock Road and west of Busch Boulevard. A further detracting element of this land was the added cost of single loading for such a railroad lead track. The rail would have had to be located along the western boundary of the 35-acre parcel. With the track in this location, only the property to the east of the lead track could be rail-served.

In 1974, the demand for a mixture of truck-served and rail-served lots caused Busch Properties, Inc. to reevaluate nearby land as potential rail-served property.

Soil and Foundation Conditions

Soil conditions on the site were evaluated on a preliminary basis during the planning of the project. Six borings were made at significant topographic locations on the site. In addition, a number of borings were taken at the site of three proposed buildings planned by the developer along the south side of the main collector road (Kingsmill Parkway). The total boring program provided significant data on soil conditions likely to be encountered throughout the project.

Soils comprising the site are of glacial origin. The upper soils of the area typically exhibit significantly lower strength than the deeper glacial deposits; consequently, when wall and column loads will exceed 1,000 pounds per square foot, it was recommended for three of the earliest buildings on the site that several feet of existing soil be excavated beneath the building site then replaced in layers with each layer compacted to high density. It was estimated that the compacted soils can achieve a bearing capacity of approximately 4,000 pounds per square foot through proper compactive effort and moisture control. Achievement of such high strength often allows use of spread footings instead of more costly drilled piers or piling. Structural loads which require greater bearing capacities than 4,000 pounds per square foot can be carried by drilled-in piers. Allowable soil bearing capacities at depths 18 to 25 feet beneath a proposed office building near Busch Boulevard will support pier loads up to 10,000 pounds per square foot. Piles, which are normally considered as an alternate foundation element to piers, were determined not to be economical at this location due to the relatively shallow depths to adequate bearing stratum.

In accordance with good practice, the developer has been advised that soils investigations be made for each proposed structure within the park in conjunction with building foundation design.

Area Grading

Rough grading for the site was coordinated with the stormwater drainage plan for the project. Because of the relative flatness of the original ground, minor rough grading was required. Street grades were generally established at minimum permissible gradients, to facilitate removal of stormwater from the surface and into the stormwater inlet system. Street grades were established at elevations slightly below the parking area within the development except for buildings which have truck-loading docks.

Earth materials excavated to create the lake were utilized for filling on the site in the vicinity of the lake and for fill beneath building floor slabs.

Topsoil, and near surface highly organic soils, were stripped from areas to be paved. The upper six inches of all subgrade were uniformly compacted to not less than 100 percent of ASTM Specifications D 698 maximum density at optimum moisture.

Grading work was planned in accordance with seasonal climatic conditions prevalent in the Columbus area. The most favorable atmospheric and soil moisture conditions generally occur between July 1 and October 1. Quite favorable conditions may occur in May, June, and October, and possibly extending into November. During March, April, and December, soil moisture is usually excessively high and prevailing atmospheric conditions are poor for reduction of soil moisture. In December, January, and February, the unfavorable conditions are compounded by freezing temperatures.

The quantity of groundwater in excavations during construction was anticipated during the preliminary soils investigation; and, it was expected to be adequately handled with a small volume pump. This anticipated condition was generally confirmed in the field; however, excavation for the depressed portion of one building became quite costly due to movement of large quantities of water into the excavation.

If perched water table conditions exist, the developer suggests to potential tenants that new buildings be established slightly above the street grade. Such practice will allow positive drainage toward the street and will take advantage of the stabilizing effect of subdrainage piping which extends along both sides of the street, intercepting subsurface water. Subdrainage piping placed beneath the perimeter of parking and loading area pavements, and connected to storm drains, also was considered effective by the developer.

4-42 Architectural character.

Electrical Service

The Columbus and Southern Ohio Electric Company supplies electric power to the area. All lines are constructed underground using direct burial cable. The utility company installs all utility lines at no cost to the developer. High voltage pad-mounted switchgear are provided for sectionalizing the distribution system, thus providing alternate feed from two directions to all major load areas.

All switchgear is screened from view by earth berms, landscaping, or other suitable means.

Telephone service for Busch Corporate Center is furnished by the Ohio Bell Telephone Company. All lines are buried in the ground at no cost to the developer.

Gas Service

Busch Corporate Center is within the area supplied by Columbus Gas Company of Ohio, Inc. There is an existing eight-inch main in Busch Boulevard and existing ten-inch main in Schrock Road serving Anheuser-Busch Brewery.

While gas mains are convenient to the property, a serious inadequacy exists in the availability of natural gas. When the project was in the planning stage in 1970, gas was available on an interruptible basis. In case of a gas shortage, highest priority was given to residential customers and industry had to use supplementary fuel sources. New gas service to industry became unavailable in 1973. Subsequently, new construction has turned to electricity as the primary energy source.

A study of the project's energy requirements was made during the planning stages. This study concluded that buildings should be designed for all-electric services in view of the anticipated curtailment of natural gas. As an energy conservation measure, buildings constructed for lease by Busch Properties, Inc. utilize thermopane for all glass surfaces and reduce the roof insulation U-factor to 0.04. This latter value was achieved through use of four-inch thick roof insulation.

Development Lighting

During the planning stage of the development, it was determined the property was beyond the city street lighting department's service area. There were no city ordinances requiring minimum street lighting levels. It would have been possible to lease a dusk-to-dawn mercury vapor type lighting system from the utility company. It was decided, however, that this type of lighting would not be compatible with the overall lighting and landscape concept proposed for the park.

The absence of requirements, and the limitations of conventional systems, provided a unique opportunity to install a system totally integrated with the development plan. As developed, the lighting and landscape plan provides for a series of handsome focal points to define significant features. Street intersections have received special lighting and landscaping treatment. Buildings are designed with supplementary interior as well as exterior illumination. Since there are no sidewalks in the development, the level of lighting normally provided by street lights at intervals along the curb was not necessary. In addition to reduction in cost compared to a conventional street lighting program, the lighting concept does provide an acceptable level of illumination and enhances the appearance of the park.

Conclusions

One of the major factors in the success of the Busch Corporate Center marketing program has been the protective covenants. This document lends credibility to the engineering and master planning concepts. Such covenants assure purchasers that these concepts will be preserved throughout the life of the project.

Busch Corporate Center received a comprehensive engineering design analysis during the master planning stage. By such analysis, accurate overall development costs were determined so that economic feasibility could be properly evaluated prior to commitment to the project by Busch Properties, Inc. The project has been highly successful, due in no small part to thorough planning and engineering analysis.

In the words of Walter E. Diggs, Jr., president of Busch Properties, Inc., "A carefully prepared engineering analysis impacts every element of a development such as Busch Corporate Center. We will continue to utilize similar engineering and planning techniques in our future development projects. We also accept the importance of the developer's role in such planning processes and will continue to be actively involved in the development of the design data."

As can be seen from this case study, all the factors previously discussed under the *Handbook's* Engineering and Design section were considered in the design and development of Busch Corporate Center. Even so, it was difficult to predict some costs, such as foundation problems resulting from water infiltration. Therefore, even the best engineering analysis and cost estimates should allow for contingencies.

Freeway Commerce Center

General Description

Freeway Commerce Center is a 157-acre commercial and business community located within the Phoenix metropolitan area on Interstate Highway 10 in Tempe, Arizona. The project is being developed by Visioneering Properties for the State Farm Insurance Company. Uses within the park include office, warehouse, light industrial, and commercial facilities. Lots range upward from one acre in size. The development features paved streets, curbs, landscaped parkways, lighted intersections, and underground utilities.

It would not serve any useful purpose to repeat in detail the general analysis of this project since the process was similar to, and as comprehensive, as that described for Busch Corporate Center. The design solution however was different, particularly with respect to the management of stormwater runoff.

Stormwater Runoff—An Innovative Design Solution

Although Tempe is situated in the arid southwest, and experiences an average annual rainfall of approximately eight inches, storms with high intensity rainfall of short duration frequently do occur. This condition is illustrated in the table below which compares short duration storms for Tempe, Arizona with St. Louis, Missori.

While Tempe has only 21 percent of the average annual rainfall of the St. Louis area, the 30-minute rainfall in Tempe is 55 to 67 percent of the 30-minute rainfall in St. Louis.

COMPARISON OF RAINFALL OF SHORT DURATION

Frequency	Duration	Tempe, Arizona	St. Louis, Missouri
1 year	30 minutes	0.6 inch	1.1 inches
2 years	30 minutes	0.8 inch	1.3 inches
5 years	30 minutes	1.0 inch	1.6 inches
10 years	30 minutes	1.2 inches	1.8 inches

SOURCE: U.S. Weather Bureau Technical Paper no. 40.

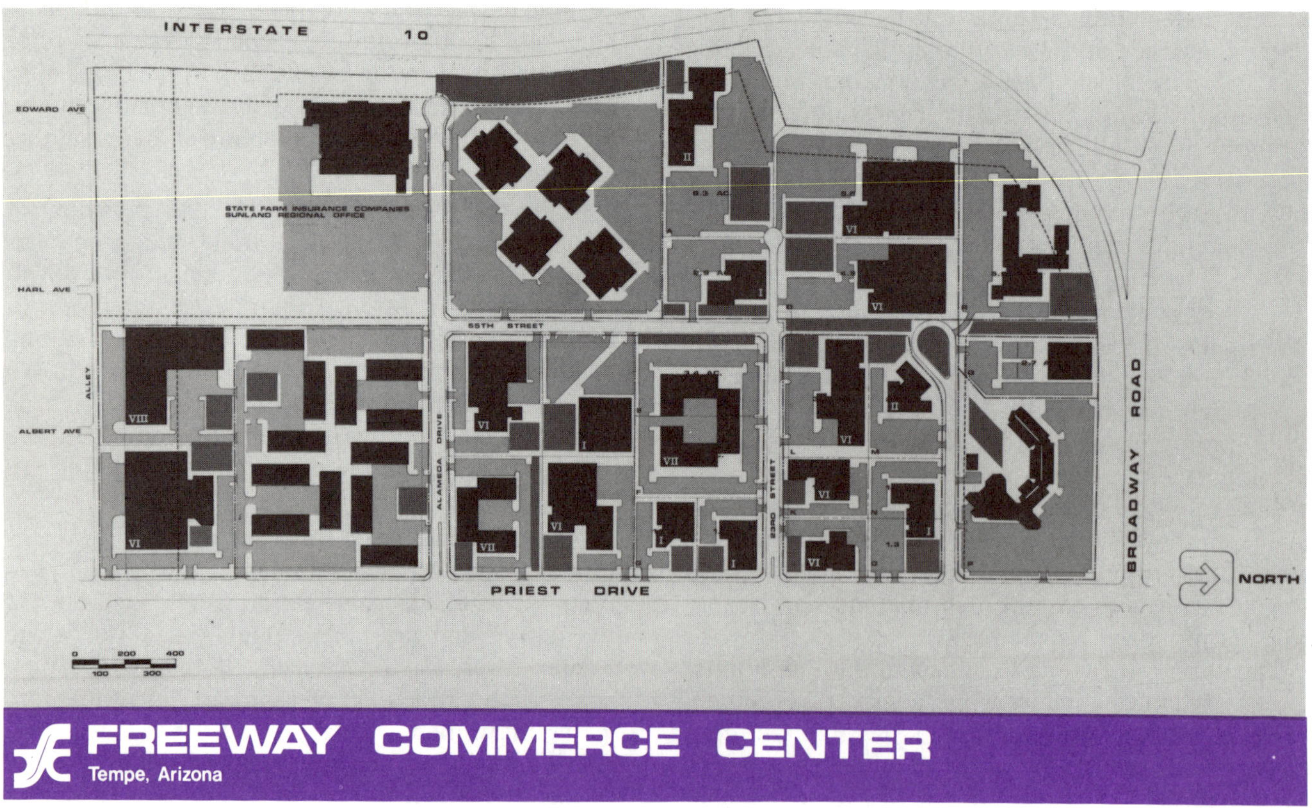

4-43 Freeway Commerce Center site plan.

In an analysis of the stormwater situation, and the topography of the site, the following features were noted:

- flooding potential due to inadequate downstream (off-site) drainage facilities; and,
- an existing 100-foot-wide drainage easement established to accommodate a future stormwater drainage channel.

Protection from flooding could have been accomplished by hauling in large amounts of earth fill to raise building and parking lot elevations. However, such a remedy was determined to be extremely costly.

In order to solve the problem, temporary retention basins were introduced into the project design. The zoning ordinance and protective covenants require a certain amount of open space in each parcel of developed land. This open space was partially developed by excavating and landscaping a series of drainage basins. Drainage from parking lots, other tributary areas, and other overland runoff, is piped or drained to each basin where it is temporarily retained. Some water stored in the basins discharges by normal evapo-transpiration, but the principal manner of discharge is into the ground through dry wells. These wells penetrate to a depth of 15 feet or more, discharging water into a layer of river gravel 15 to 20 feet below the ground surface. The dry wells are sized and distributed within the basins to discharge stored water into the gravel within 36 hours after the storm. This maximum ponding duration was established by the city as a condition for approval of the retention basin plan. The 36-hour ponding period was specified by the city in order to minimize mosquito nuisance, to avoid stagnation of the water, and to preclude the creation of otherwise unsightly conditions related to extended submergence of these areas.

There is no underground piping system for stormwater within the development. Consequently, there is no outlet for the basins other than the dry wells. The basins are designed to be adequate for storms up to two-year frequency. Runoff from storms in excess of the two-year frequency, but less than the 100-year frequency, will be contained in parking lots adjacent to the basins. Streets and

4-44 One leg of the retention pond which stores storm water from the streets of the Freeway Commerce Center. One catch basin/drywell in the splash box at the far end takes nuisance water before it reaches the pond.

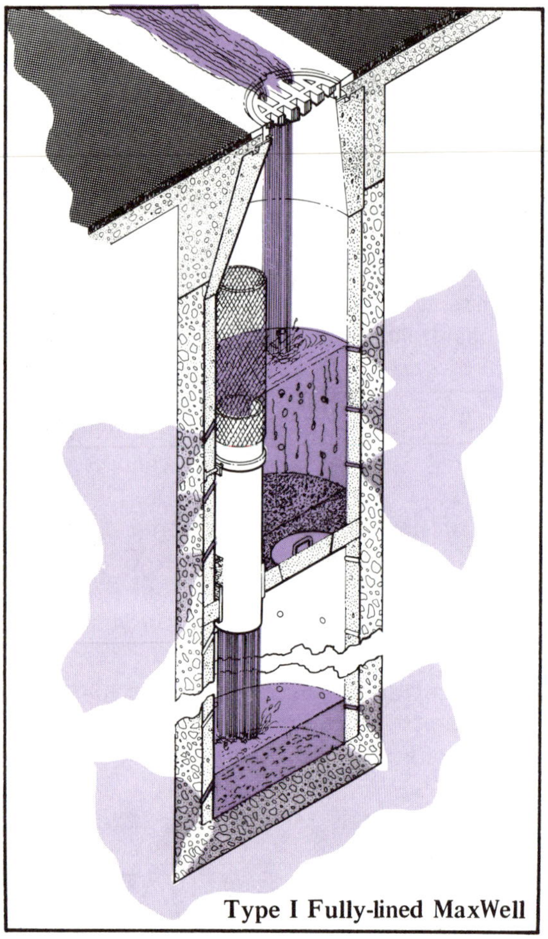

4-45 Max Well, a means of accommodating storm drainage on-site by permitting maximum percolation. Runoff drains into the well and from there into the water table.

building floors are established above the 100-year frequency ponding elevations.

Environmental advantages other than flood control are gained from the basins as well. The basins blend with the landscape, create terrain on an otherwise flat site, and provide a park-like setting for the development. Plants in the basins were selected for adaptability to the 36-hour flooding period. These areas provide a comfortable environment for employees during leisure periods of the business day. A basin located on a motel site has been planned to accommodate a tennis court. The basins also provided fill material for building sites, earth berms, and parking lot grading, thereby eliminating the need for any importation of fill. Use of the basins resulted in the vacation of an existing 100-foot-wide easement over part of the property, and restored this area for development and to the tax rolls. Finally, the use of the basin concept helps to recharge groundwater levels by the periodic introduction of stormwater into the aquifer.

In order to assure that retention basins are used on properties sold by the developer and thus contribute to the success of the drainage plan, the specific requirement for storage of runoff has been included in documents recorded with the property. Many economic benefits are produced by the use of retention basins including substantial reduction in cost of storm drainage piping; however, occasional maintenance of the basins will be required —particularly keeping the dry wells open and functioning.

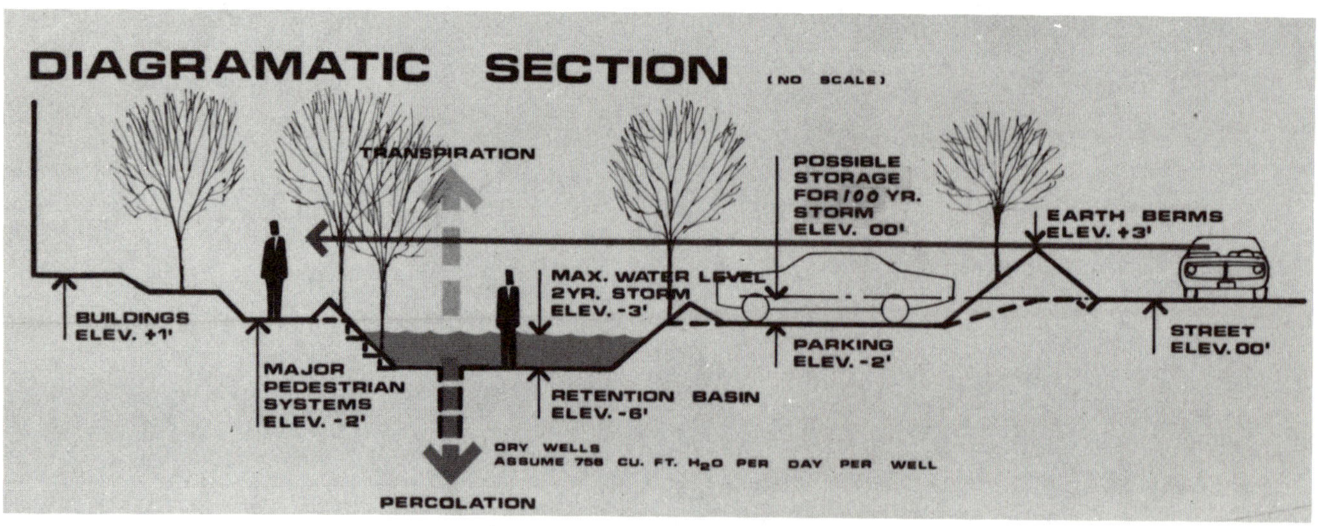

4-46

4-47 Patio restaurant at the Fiesta Inn in Freeway Commerce Center is below grade. It depends entirely on drywells like the one in the foreground for drainage.

Conclusions

Much effort and time is required to develop new concepts and innovative ideas such as this one. Once identified, energy and funds must be expended to demonstrate the soundness of new concepts to local agencies in order to win approval for their use. Once implemented, however, these concepts can provide substantial cost savings to the developer and create a better environment for all.

5.
Development Strategy

Introduction

In all real estate development, money makes the difference between a concept and the actual implementation of that concept. Traditionally, real estate developers commit very little of their own money to the long term or permanent financing of their own projects. They do provide the start-up or "front-end" money which is then replaced by financing as the project develops. As true entrepreneurs, they are idea men and not necessarily cash contributors. Many entrepreneurs contribute the idea of property, time, effort, and, finally and hopefully, know-how rather than cash. Consequently, financing is required in practically all real estate projects. This section basically deals with the industrial developer's investment requirements to satisfy both his sources of financing and himself that the project under consideration is, in fact, feasible.

The planning and engineering efforts described in earlier sections are started just prior to the formulation of the financing program. This work is undertaken by outside consultants if the developer does not have this capability on his own staff. The documents prepared during the planning and engineering phases initially are used internally by the developer to evaluate alternative development concepts. Then, along with several other elements relative to the mortgage submission proposal or application, these documents are externalized to seek the necessary development capital. It is at this point that the developer and his staff prepare the packaging of the concept in order to procure financing for implementation of the concept.

A profitable industrial development may require the multidisciplinary inputs of various specialists, including real estate economists, appraisers, planners, engineers, marketing specialists, brokers, accountants, tax advisors, legal advisors, and others. The manner in which these combined efforts are coordinated by the developer will determine the degree of financial success achieved by any project. However, it is important to remember that determining feasibility of an industrial development requires the skill of many disciplines rather than any single individual. Each discipline introduces a specific set of criteria and a framework which helps to define the overall parameters within which the project may be undertaken.

5-1

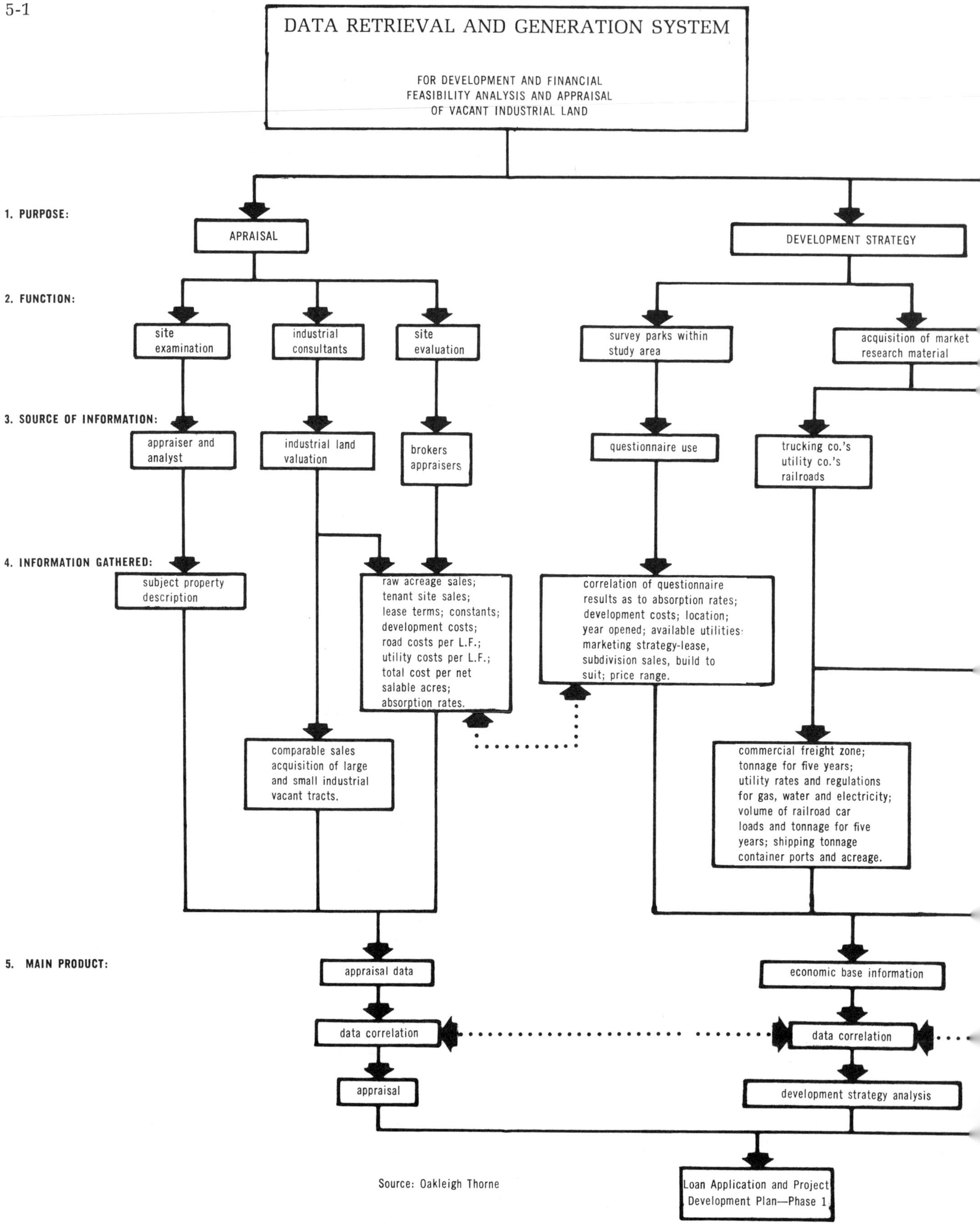

Source: Oakleigh Thorne

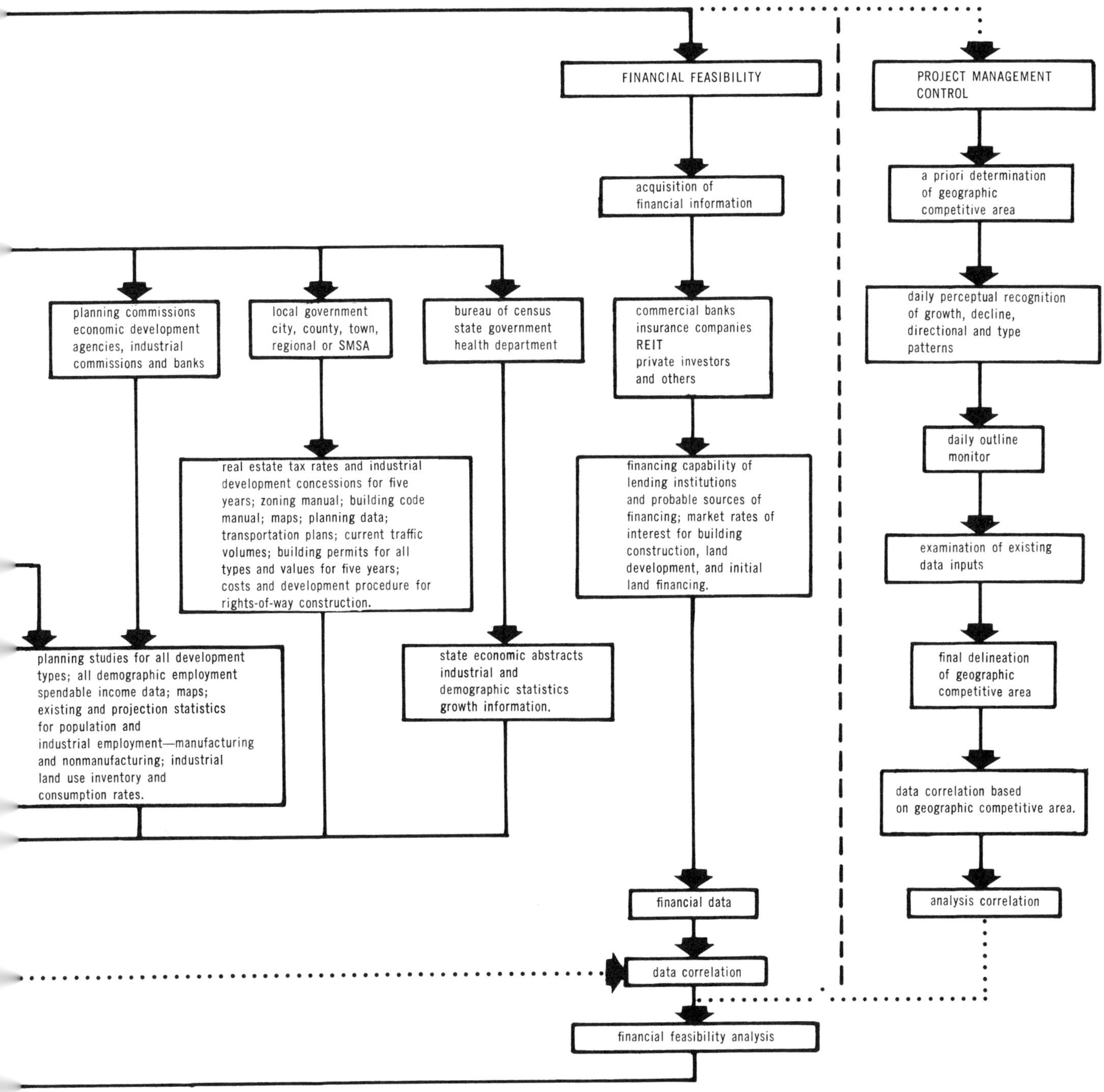

FINANCIAL FEASIBILITY
PROJECT MANAGEMENT CONTROL
acquisition of financial information
a priori determination of geographic competitive area
planning commissions economic development agencies, industrial commissions and banks
local government city, county, town, regional or SMSA
bureau of census state government health department
commercial banks insurance companies REIT private investors and others
daily perceptual recognition of growth, decline, directional and type patterns
daily outline monitor
real estate tax rates and industrial development concessions for five years; zoning manual; building code manual; maps; planning data; transportation plans; current traffic volumes; building permits for all types and values for five years; costs and development procedure for rights-of-way construction.
financing capability of lending institutions and probable sources of financing; market rates of interest for building construction, land development, and initial land financing.
examination of existing data inputs
planning studies for all development types; all demographic employment spendable income data; maps; existing and projection statistics for population and industrial employment—manufacturing and nonmanufacturing; industrial land use inventory and consumption rates.
state economic abstracts industrial and demographic statistics growth information.
final delineation of geographic competitive area
data correlation based on geographic competitive area.
financial data
analysis correlation
data correlation
financial feasibility analysis

Detailed Market Analysis

The demand for industrial space generally conforms to two representative types. These are: derived demand, and replacement demand. Derived demand represents the greater portion of the demand for new space and land. Derived demand arises from new firms wishing to locate in the area, or the expansion of existing firms already in the area, and is oriented almost solely to growth. Replacement demand, on the other hand, is demonstrated by industrial firms wishing to dispose of their obsolete facilities and move to a new location which offers a planned environment featuring architectural design quality, lower operating costs, more parking and structural economies, truck maneuverability, adjacent transportation, and similar advantages. The most obvious example of replacement demand is the shift during the 1950s and 1960s of many center city, vertically designed industrial plants, situated on crowded sites in areas of high traffic congestion, to spacious suburban locations. Although replacement demand, in some areas, can represent a substantial portion of total demand, the industrial developer more often looks to derived demand for his major marketing potential.

The data and generation retrieval system illustrated in Figure 5-1 outlines in detail the information which is incorporated in a market analysis study. The actual work may be prepared either by the ownership group or by outside advisors. Much of the data involved will be needed for inclusion in the loan submission package. If an outside appraisal is not required, the one requirement of all loan submission applications will be the comparable sales of adjacent land, and some kind of conclusion as to the value of the property under consideration. However, the lending institution may require an outside appraisal of the land, both in its existing condition and as it is planned for development. The retrieval system illustrates the need both for economic base information and for a detailed analysis of the competition either through physical inspection or through the administration of a mail questionnaire. This information, gathered from the community, is in turn incorporated into the financial feasibility study or into the financial projection, both of which are discussed later. The purpose of the detailed analysis described in the following pages is to determine the economic base of the community relative to current economic trends, patterns, potentialities, and deficiencies. Analysis of these factors establishes the potential for industrial growth and development of the macro or metropolitan area. The economic base information also will be used to identify the types of industrial users or investors which constitute the most likely sources of demand for space within the project. Using these various factors, a determination of the project's possible share of that future demand for industrial land can be estimated. An analysis of the competition provides other important elements in the marketing strategy such as competitive sale and leasing policies, range of products offered, absorption trends, extent of the architectural and engineering services offered to prospective tenants, necessary and required utilities, parcel availability, financing arrangements, and development costs.

Analysis of the Competition

Analysis of competitive districts and parks will determine the remaining unsold inventory which has to be considered as competitive with the site being evaluated for industrial development. In obtaining information from competitive development, the following information is critical: the year open for occupancy, gross acres at the time of opening, zoning, any unusual physical conditions, occupancy level at the date of the survey (acres sold and the percent occupied), estimated annual average acreage absorption (number of acres sold, divided by the number of years open for sales), development costs per acre (either actual or estimated), opening price per square foot, current price per square foot, major highway access availability, rail service, and available utilities.

The price data from the survey can be used to test the reasonableness of the initial price structure proposed for the project as well as the potential for incremental increases during the development period. The total demand (total of all acreage sold) should be calculated on an annual basis for all districts and parks surveyed. This survey also should determine the percentage of absorption related to project size, annual absorption patterns, and the impact of new projects on overall absorption patterns. The remaining unsold inventory in competitive projects should be examined physically to determine if it will be competition for the project under consideration. The likelihood of competition from well-located properties also should be considered even though no development plans have been announced. Some of the land remaining in some of the competitive projects may not be suitable for development from the standpoint of location and/or topography and can be discounted. ULI Industrial Council member William Dillard points out: "A careful analysis of currently available industrial lots may indicate a void in the marketplace and therefore can point to an opportunity."

A good analysis should include an evaluation of the indirect or behind-the-scenes activity which led up to the demand evidenced by the sales history of competitive projects.

Inclusion of a summary of competitive developments in the loan submission application is a good way to establish credibility of potential markets.

Economic Base Study

The approach to the economic base study outlined here is a formal one describing the detail needed to depict the weaknesses and strengths of any community. It will concentrate on one segment of an overall economic base study; namely, employment and its changes over time. Growth and its relationship to import and export of goods and materials will be examined. Published government census data provides the most abundant source of information for the data needed in this analysis. However, some primary field work will be required as well to update or check the published data.

Frequently, an industrial developer can rely on studies which have been done by a public or quasi-public agency such as a regional planning organization or an area economics development group. However, where these studies are not available or are dated, the following paragraphs describe how such a base study is undertaken.

Employment Analysis

The purpose of an employment analysis is to determine what changes in employment have occurred in the last ten years in the metropolitan area which can affect demand for industrial land. Once the historic data has been assembled, it is possible to project either a continuation of those trends in the future, or some alteration based upon reasoned analysis. Figure 5-2, which illustrates a typical metropolitan employment analysis, tabulates two decennial years of local employment in the metropolitan area by employment category as published in standard government census data. Employment distribution is shown for 1960 and 1970; the percentage and numerical changes between the two terminal years provide an indication of increase or decrease by employment category. A fundamental consideration in metropolitan economic analysis is the distribution of employment into its basic and nonbasic sectors. The basic economy consists of export activities which bring outside money into the economy. Nonbasic economic activity consists of the provision of goods and services which meet entirely local needs.

The determination of basic/nonbasic industries is undertaken by two different methods: (1) Locational Quotient (LQ); and (2) Shift-Share Analysis. These two methods in combination help identify which industries in the community are exporting (basic) and which do not export (nonbasic). In the sample, the comparative numerical and percentage distribution calculations for the metropolitan economic base area are compared to the United States using a Locational Quotient.

Locational Quotient

The Locational Quotient (LQ) indicates the degree to which a given industrial group is of greater or lesser importance locally rather than nationally. The LQ may be calculated by dividing the percentage of local employment in a given industry by the percentage of national employment in the same industry. Uniform productivity within each industry group throughout the nation is assumed in this technique. While the technique is not foolproof, it gives a strong indication of industries which provide the underlying strength of an economy.

An LQ of 1.00 for an industry group in a community indicates the industry group has the same proportion of employment locally as nationally. This ratio implies that the area just takes care of its own needs for whatever goods or service the industry group provides. An LQ of less than 1.00 implies the area must import a portion of its needs for the type of goods or service involved; and, an LQ of larger than 1.00 indicates the area exports this category of goods or services.

Judgment, however, must be exercised in interpreting the LQ. For example, LQs of .96 (stone, clay, and glass products) and 1.06 (apparel and textile mill products) indicate that the area is probably self-supporting for all practical purposes and that the industries are nonbasic. An LQ of 1.64 (instruments and related products) indicates substantial exports and marks the industry as overwhelmingly basic. In the other extreme, an LQ of .39 (lumber and wood products) implies imports of goods and services and labels the industry as nonbasic.

METROPOLITAN (SMSA) ECONOMIC BASE ANALYSIS (000s

	Employment Analysis						
	1960 Annual Employ-ment	Percent Distribution	1970 Annual Employ-ment	Percent Distribution	Percent Change 1960-1970	Locational Quotien 1960	1970
Total Employment	542,346	100.00	705,933	100.00	30.16	—	—
Mining	6,122	1.13	5,231	0.74	(14.55)	.08	.09
Contract Construction	28,859	5.32	33,815	4.79	17.17	.82	.92
Manufacturing	167,960	30.97	193,490	27.41	15.20	1.21	1.16
Durable Goods	94,594	17.44	111,958	15.86	18.36	1.26	1.19
Ordinance and Accessories	2,200	0.41	2,420	0.34	10.00	1.10	1.04
Lumber and Wood Products	4,627	0.85	3,612	0.51	(21.94)	.72	.39
Furniture and Fixtures	3,830	0.71	4,657	0.66	21.59	.82	.90
Stone, Clay and Glass Products	6,044	1.11	6,413	0.91	6.11	.81	.96
Primary Metal Industries	12,313	2.27	13,165	1.87	6.92	.98	1.01
Fabricated Metal Products	11,350	2.09	13,800	1.96	21.59	1.59	1.28
Machinery, Except Electrical	14,796	2.73	19,823	2.81	33.98	1.14	1.15
Electrical Equipment and Supplies	14,720	2.71	19,178	2.72	30.28	1.61	1.52
Transportation Equipment	15,697	2.89	17,992	2.55	14.62	.90	.62
Instruments and Related Products	5,115	0.94	6,631	0.94	29.64	1.47	1.64
Miscellaneous Manufacturing Industries	3,902	0.72	4,267	0.60	9.35	.79	.83
Nondurable Goods	73,366	13.53	81,532	11.55	11.13	1.35	1.33
Food and Kindred Products	17,900	3.30	17,803	2.52	(0.54)	1.07	1.03
Tobacco Manufacturers	744	0.14	392	0.06	(47.31)	1.47	.85
Textile Mill Products	9,248	1.71	9,566	1.36	3.44	1.21	1.06
Apparel and Other Textile Products	12,336	2.28	13,659	1.94	10.73	1.54	1.58
Paper and Allied Products	6,015	1.11	7,060	1.00	17.37	1.26	1.33
Printing and Publishing	9,119	1.68	11,203	1.59	22.85	1.64	1.45
Chemicals and Allied Products	8,287	1.53	10,495	1.49	26.64	1.68	1.62
Petroleum and Coal Products	2,278	0.42	2,077	0.29	(8.82)	3.31	3.45
Rubber and Plastic Products	3,798	0.70	5,800	0.82	52.71	1.07	1.13
Leather and Leather Products	3,641	0.67	3,477	0.49	(4.50)	.52	.43
Transportation and Public Utilities	40,046	7.38	44,093	6.25	10.11	.94	1.01
Wholesale and Retail Trade	113,884	21.00	149,036	21.11	30.87	.64	1.06
Wholesale Trade	30,004	5.53	38,012	5.39	26.69	.97	.93
Retail Trade	83,880	15.47	111,024	15.73	32.36	.99	1.08
Finance, Insurance and Real Estate	26,698	4.92	38,791	5.50	45.30	1.18	1.35
Services	75,243	13.87	116,121	16.45	54.33	1.03	1.05
Government	83,534	15.40	125,356	17.76	50.07	.82	.95

Note:
Parentheses ( ) denote loss.

† The complete effectiveness of this method depends upon the use of industrial questio naires for a more detailed understanding of the numerical changes in each employme category.

Shift-Share Analysis †			Numerical Change
N.G.S.*	I.M.*	R.S.*	1960-1970
—	—	—	163,587
(452)	(373)	(66)	(891)
7,588	(3,976)	1,344	4,956
21,760	1,638	2,132	25,530
22,761	(18,416)	13,019	17,364
1,052	(639)	(193)	220
(924)	(73)	18	(1,015)
648	(372)	551	827
423	(39)	(15)	369
1,085	34	(267)	852
3,628	(2,899)	1,721	2,450
4,461	(584)	1,150	5,027
5,747	(681)	(608)	4,458
3,659	(1,655)	291	2,295
2,617	266	(1,367)	1,516
251	75	39	365
6,897	273	1,542	8,166
(31)	(18)	(48)	(97)
(243)	(102)	(7)	(352)
551	(127)	(106)	318
1,478	(50)	(105)	1,323
933	(24)	88	1,045
3,489	(1,005)	(400)	2,084
2,525	(1,845)	1,528	2,208
(877)	(570)	1,246	(201)
2,184	(1,012)	830	2,002
(53)	(85)	(26)	(164)
3,267	(117)	897	4,047
25,490	8,421	1,241	35,152
10,120	(5,441)	3,329	8,008
19,478	6,283	1,383	27,144
11,577	(3,039)	3,555	12,093
39,326	1,461	91	40,878
45,070	(3,275)	27	41,822

National Growth Share
Industrial Mix
Regional Share

SHIFT-SHARE EXAMPLE EQUATIONS

	1960	1970
U.S. Total Employment Growth (000s)	59,692	74,055
Manufacturing—Durable Goods (U.S.)	10,406	10,884
Metropolitan SMSA Durable Goods	94,594	111,958

National Growth Share

$$\text{N.G.S.} = Ei\left(\frac{US^*}{US} - 1\right)$$

Where: Ei = Regional employment in the ith (durable goods) at the beginning of the period.

US* = Aggregate national employment at the end of the period.

US = Aggregate national employment at the beginning of the period.

Hence:

$$\text{N.G.S.} = 94.594\left(\frac{74{,}055}{59{,}692} - 1\right)$$

$$\text{N.G.S.} = 94.594\ (.24062)$$

$$\text{N.G.S.} = (+)\ 22{,}761$$

Industry Mix

$$\text{I.M.} = Ei\left(\frac{USi^*}{USi} - \frac{US^*}{US}\right)$$

Where: USI = National employment in the ith (durable goods) industry at the beginning of the period.

USi* = National employment in the ith (durable goods) industry at the end of the period.

Hence:

$$\text{I.M.} = 94.594\left(\frac{10{,}884}{10{,}406} - \frac{74{,}055}{59{,}692}\right)$$

$$\text{I.M.} = 94.594\ (.04594 - .24062)$$

$$\text{I.M.} = 94.594\ (-.19468)$$

$$\text{I.M.} = (-)\ 18{,}416$$

Regional Share

$$\text{R.S.} = Ei\left(\frac{Ei^*}{Ei} - \frac{USi^*}{USi}\right)$$

Where: Ei* = Regional employment in the ith (durable goods) industry at the end of the period.

Hence:

$$\text{R.S.} = 94.594\left(\frac{111.958}{94.594} - \frac{10{,}884}{10{,}406}\right)$$

$$\text{R.S.} = 94.594\ (1.18356 - 1.04594)$$

$$\text{R.S.} = 94.594\ (+.13762)$$

$$\text{R.S.} = (+)\ 13{,}019$$

Therefore: Aggregate change in ith industry is:

$$\triangle i = \text{N.G.S.} + \text{I.M.} + \text{R.S.}$$

$$\triangle i = 22{,}761 + (-18{,}416) + 13{,}019$$

$$\triangle i = 17{,}364$$

Corollary: $111{,}598 - 94{,}594 = 17{,}364$

Shift-Share Analysis[1]

Shift-Share Analysis, in general terms, defines a region by relating it quantitatively and, to some extent, qualitatively to the national norms. This method characterizes the economic structure of a region in relative terms and highlights the extent to which various aspects of growth are uniform across the nation. These employment elements in the metropolitan area identify and describe key economic relationships and aid in depicting which segments of the economy are growing or declining.

The first factor, the National Growth Share (NGS), measures the needed employment change in an industry group in relation to national employment growth. The second factor, the Industrial Mix (IM), represents what would have happened to the employment structure of an industry group if it were to gravitate toward the overall national pattern of industrial mix. If the NGS is larger than the IM, the particular metropolitan area component is a growth component and is contributing to area growth. Conversely, when it is smaller, the particular component is a slow growth component and has a negative effect on the employment of the area. The third factor, the Regional Share (RS), measures the competitive position of the metropolitan area or component in the area to the rest of the nation. A positive difference signifies a shift in the particular employment component into the area, while a negative difference indicates a shift out of the area.

The objective of this analysis is to identify the changes which have occurred over the last 10 years and to anticipate how they will affect the sale of industrial land for the next 10 years. The economic base analysis (employment characteristics, appropriate analysis for Shift-Share, and LQ) can be interpolated into the projected demand for industrial land via density ratios.

Projected Demand for Industrial Land

A common method for depicting *aggregate* demand for industrial land is via employment projections in relation to employment densities. When the competition is surveyed as part of the program for the market analysis, a classification of industry and employee density relationships should be established. These employment densities will vary from area to area and from time to time as technology forces changes upon the distribution and manufacturing processes. Figure 5-3 is a classification of employment densities per acre, divided into broad categories. Industries with a low ratio of employees per acre in excess of 1,000 square feet per employee are termed *labor extensive.* Industries having a high concentration of employees per acre, below 200 square feet per employee, are termed *labor intensive.* The middle group is referred to as *intermediate extensive.* When there is a low ratio of employment per acre, it implies that the plant must have a fairly large area per employee, and that it is probably a *capital intensive industry.* Capital intensive means there is a high ratio of capital investment or equipment per employee. Warehousing and distribution activities are labor extensive but are not necessarily capital intensive. Intensive labor industries include instrument manufacturing, electronics, and other operations using highly skilled scientific and professional personnel. Because of high plant and equipment costs, these industries are also capital intensive.

Numerous classifications exist between labor intensive and labor extensive, but for purposes of this analysis and illustration, only three different categories have been used. However, substantial differences in the employment density relationships for specific industrial categories can vary significantly from the straight curve line shown in Figure 5-3. Therefore, current employees per gross acre should be established by actual survey and a relationship should be established, both for employees per gross acre and for employees per square foot of building area, when possible. Density relationships tend to change over time. They are shown in Figure 5-4 as they existed in 1970 for a particular community with a projected density for 1980. The increase in employment between 1970 and 1980, divided by the density equivalent for the industry group, equals the amount of land that is required to meet the new employment needs.

In the example shown, there is a projected future demand for an additional 5,696 industrial acres between 1970 and 1980. The survey of the competitive supply will indicate the acreage which is available to meet that demand. For a new project to be viable, the probable 10-year demand should significantly exceed the existing supply. However, it always must be kept in mind that much land zoned for industrial use will be occupied by non-industrial users. This is the result of frequent situations in which industrial zoning

[1] For more detailed discussion of shift-share analysis see Edgar M. Hoover, *An Introduction to Regional Economics* (New York: Alfred A. Knopf, 1971), p. 292f.

5-3

STRUCTURAL AND LAND EMPLOYMENT DENSITIES BY INDUSTRIAL ACTIVITY GROUPS

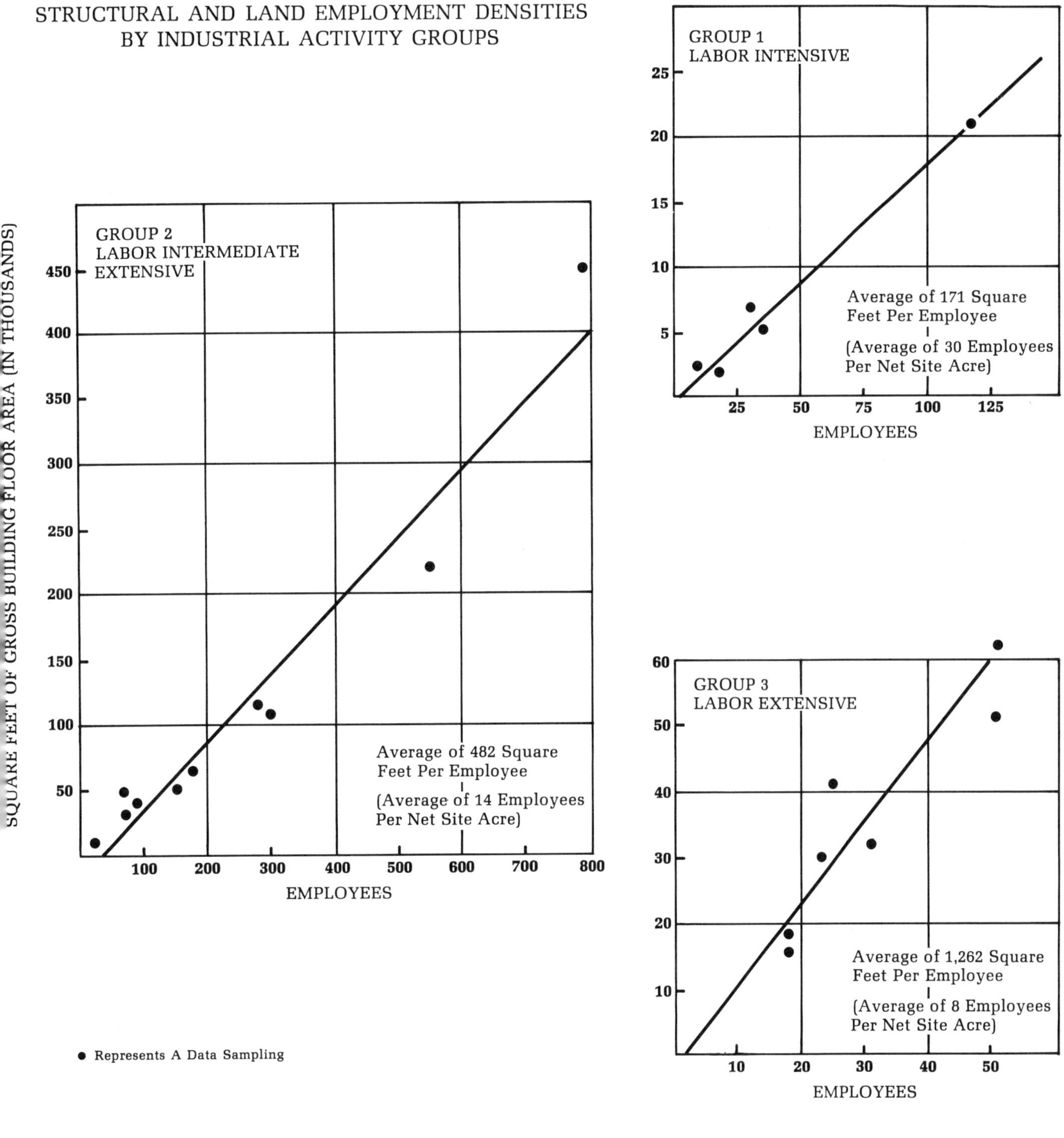

5-4 INDUSTRIAL EMPLOYEE DENSITIES AND PROJECTED DEMAND FOR INDUSTRIAL LAND METROPOLITAN AREA (000)

Industry Use Group	Type of Major Industry Group	1970 Annual Employment	Employees Per Gross Acre—1970	1980 Annual Employment	Employees Per Gross Acre—1980	Decade Employment Increase	Additional Demand for Industrial Land	Percent Distribution
1 — Intensive		68,663	26*	87,889	24	19,226	801	14.06
	Electrical Equipment and Supplies	19,178						
	Transportation Equipment	17,992						
	Instruments and Related Products	6,631						
	Apparel and Other Textile Products	13,659						
	Printing and Publishing	11,203						
2 — Intermediate Extensive		112,468	12*	137,211	10	24,743	2,474	43.43
	Ordinance and Accessories	2,420						
	Lumber and Wood Products	3,612						
	Furniture and Fixtures	4,657						
	Primary Metal Industries	13,165						
	Fabricated Metal Products	13,800						
	Machinery, Except Electrical	19,823						
	Miscellaneous Manufacturing Industries	4,267						
	Food and Kindred Products	17,803						
	Textile Mill Products	9,566						
	Paper and Allied Products	7,060						
	Chemicals and Allied Products	10,495						
	Rubber and Plastic Products	5,800						
3 — Extensive		161,395	6*	180,762	8	19,367	2,421	42.50
	Stone, Clay and Glass Products	6,413						
	Tobacco Products	392						
	Petroleum and Coal Products	2,077						
	Leather and Leather Products	3,477						
	Wholesale Trade	149,036						
	Totals	342,526		405,862		63,336	5,696	100.00

* The net acre density has been adjusted to a gross acre density different from that indicated in the graph.

allows many non-industrial uses as a matter of right. Also, there is a long history of overzoning for industry with the expectation of high attrition into other zones over time.

ULI Industrial Council member Thomas C. Wolff, Jr. notes, "growth patterns and corridors change and large areas originally planned for industrial use may become undesirable and are not therefore really a part of the 'existing supply'."

The additional industrial acreage required in the example derives from employment growth. From the average of 500 acres per year required, a site share for the proposed project must be allocated. This step in the market analysis process takes a great deal of interpretive skill; and, the judgments made will depend to a great extent upon the experience and reputation of the developer or ownership group. Given similar sites, one development group may be able to outpull another because of reputation and track record.

Legitimate percentage relationships with regard to a site's share of the total acres, demanded on an average annual basis, cannot be derived mechanically from the analysis. For example, in some small communities, 100 percent of the demand may be going to one site. In some metropolitan locations, on the other hand, the demand evidenced at a particular site may be as small as 10 or 12 percent of the projected metropolitan area needs. At the very least, the market analysis can highlight the growth sectors in the local economy. Examination of the absorption history will provide some insights to the competition's success in meeting past and future demand. The analysis of industrial land absorption should distinguish between land sold to other developers and land sold and/or buildings leased to users. Projection of the trends will pinpoint selective industrial employment segments which the project might capture from the overall market demand.

Conclusions

The results of a market analysis are a determination of both overall demand for industrial land and an estimation of the share of that demand which might be attributed to a specific location. This must be compared with the historical evidence of demand as determined by a survey of the competitive facilities and through evaluation of information from a multitude of other sources, including the research departments of commercial banks, transportation companies, utility companies, and industrial real estate brokers. ULI Industrial Council member Robert R. Walker, Jr. notes, "The most important information to be sought from the various sources mentioned is the actual absorption rate of land for the uses which the developer plans to attract." The metropolitan area should be examined for a period between three to five years to identify trends or cyclical patterns in the economy. Then, this information must be compared with the projected demand in order to establish the credibility of the projections. ULI Industrial Council member Thomas C. Wolff, Jr. suggests, however, that, "market projections based on historical facts will not reflect the unique ability of a developer to identify a way to attract a group of related economic activities such as aerospace firms, and research and development groups."

Another way to test the strength of a market is through development of a pilot project. A number of companies with excellent financial strength have entered a metropolitan area and acquired a number of small sites in industrial districts and parks and built speculative buildings. The characteristics of the rent-up or sellout period are examined and judgment decisions then can be made concerning the relative strengths or weaknesses of the economy and whether or not major capital should be committed to acquire a large tract of land for industrial development. This test marketing procedure is one of the ways to estimate the potential for success in a given area. Success or failure of a test marketing procedure should certainly take precedence over results of a market analysis; however, the two methods can be used most effectively in combination. Paul Shepherd of Cabot, Cabot and Forbes notes, however, "a market test such as that described might be misleading. It might prove a market for speculative buildings but this could not be transferred to land sales or build-to-suit potential."

Financing

Sources and Types of Financing

The financing function for any real estate project requires an understanding of the following three basic concepts and activities:

- What type of financing is being sought?
  - Land acquisition cost
  - Land development costs
  - Building-construction
  - Building-permanent financing
  - Equity or joint ventures
- What are the sources of the type of financing being sought?
  It is important to be aware both of statutory lending limitations of various institutional sources and of lending preferences of particular ones within a general classification. Thus, it is generally fruitless to apply to a commercial bank for long-term financing, and it is useful to know that the ABC Insurance Company is primarily a credit-oriented lender which only makes loans on buildings leased to top credits for long terms.
- What is the best method to obtain needed financing?
  This covers the area of loan application, and combines the concepts of type and source of loans with that of project feasibility, because a lending officer is most concerned that the potential borrower understands, and has a sense of control over, all of the economic variables of his particular real estate project. The main financial marketing job is to convey to the lender in words or calculations that it is a good project and makes economic sense in its particular environment.

Lending institutions differ tremendously in policies and practices nationally, and sometimes within the same community. The type of proposal which will be acceptable to one lender, another may not even consider. The ownership entity's ultimate success with the project will depend upon the knowledge it possesses of the policies, practices, and procedures of various types of lending organizations in that particular market. It is important to recognize that shopping the market definitely is not the technique to be used to obtain industrial project financing. In the first place, lenders spend a great deal of time processing loan submission proposals; and, they should not be asked to undertake this work frivolously. Secondly, communication channels in the finance industry are open. If word gets around that the ownership group is shopping the market, financing the development under these circumstances will be difficult. Hence, knowing how to choose the correct lender, and where the project will be received most favorably, will aid in getting the project financed at the first submission.

Sources of Lendable Funds

There are four primary lending sources for industrial development: life insurance companies, commercial banks and savings banks, real estate investment trusts (REITs), and a final category which includes individuals, state or local agencies with bond authority, pension funds, savings and loan associations, and recently, foreign money. Since these are the primary sources for lending money on industrial development projects, it is important to know their capacities, resources, and organizational structures. The more that is known about the potential lending sources, the more intelligently the developer is able to deal with them.

Insurance Companies

Within the United States, there are more than 2,000 insurance companies, approximately 100 to 150 of which will consider lending on industrial development. Insurance companies have broad real estate investment powers, and huge resources which make them a good source for a multitude of financing transactions. They are a key source of long-term permanent mortgage financing. Most insurance companies are organized either as mutual companies which are completely owned by policy holders, or as stock companies which are privately owned by stockholders who receive dividends on operating profits. Except for REITs, insurance companies generally have broader lending powers than other lending sources. They operate with greater flexibility and their financing activities cover a wider range. These companies frequently have financed projects which are not attractive to—or which may not be permissible for—state or federally chartered banks, and the savings and loan associations. Being less conservative than commercial banks, insurance companies are interested sometimes in taking an equity or ownership position as well as a mortgage. Insurance companies are also interested in joint venture financing whereby they provide the debt financing, receive a return on their investment, and participate in the profits of the development.

Commercial Banks

Commercial banks represent the primary source of temporary short-term land loan or construction loan financing. They also exert a strong influence on other phases of mortgage financing. Commercial banks are either federally or state chartered institutions; their organization is complex and their lending policies vary greatly. Commercial banks are always business corporations operated by and for the stockholders.

For specific building financing, banks lend primarily for construction financing contingent upon the builder or developer's take-out commitment, or his permanent mortgage loan being obtained from some other lender. Except for small numbers of banks located in the nation's largest financial centers, the commercial banks providing funds for industrial development will have a local focus. Various state laws limit banks to state, county, municipal, and even single-office locations. They usually maintain close contacts with the local communities in which loans are made. Banks can and do make short-term loans for development purposes when a favorable economic climate is present. However, lending practices for commercial banks can be expected to vary widely since these banks adapt their practices to the competitive position they occupy in the marketplace for lendable funds.

Real Estate Investment Trusts (REITs)

Another major source of lendable funds is the real estate investment trust. There are four specific types of REITs. Although they can and do perform hybrid activities, they sometimes, as a single trust, engage in all four activities.

The most common type of REIT, known as the short-term trust, focuses on construction and development loans (C&Ds) much like a commercial bank. These loans generally do not run longer than 18 months. They are tied closely to the short-term outlook of real estate and the real estate market as a whole.

Long-term mortgage REITs finance permanent mortgages. They function just like insurance companies, savings banks, and commercial lending institutions. They lend long and borrow for the short term. The difference between them and other fund sources lies in the fact that they obtain funds by issuing securities and stocks rather than by inviting deposits.

Equity REITs go somewhat further than the long-term mortgage trusts since they invest directly in depreciable property as an ownership group. Thus, the equity trust can shelter much of its income from taxation through depreciation of its ownership properties at accelerated rates. Some REITs specialize in land purchase leasebacks. In this case, they purchase the ground under a completed property and enter into a long-term leaseback. Usually the REIT, because of the higher interest rate it commands, is found in a second or junior financing position, playing a secondary role to the prime position occupied either by an insurance company or by a commercial bank. Lending territory has practically no limitations because some REITs have loaned abroad as well as in the United States. REIT lending policies usually are not well defined, although they do have specific goals and objectives in their stock offerings. Typical lending policy is that if the project fits the broad objectives of the trust and stockholders, the trust usually will consider lending on the project.

REITs have been known as the swingers in the real estate market. In 1973 and 1974, however, a number of the C&D REITs experienced financial difficulties and entered into bankruptcy proceedings. Due to the substantial downturn in most REIT portfolios, it is doubtful that they will be able to provide new higher risk development funds in the foreseeable future.

Other Lending Sources

Additional lending sources include individuals (partnerships or joint ventures); corporate, institutional, union, and government pension funds; public development companies (as a joint venture partner); mutual banks; savings and loan associations (both federal and state chartered); and syndications. Beginning in the late 1960s and into the 1970s, many of the corporate and institutional pension trust funds undertook substantial shifts from stock market investments into real estate ownership and lending. Probably, there will be a continuing propensity on the part of pension trust funds to mortgage rather than own real estate.

Another source of financing is larger publicly owned companies, many of which have entered the real estate field over the last decade, and are actively seeking out joint venture deals with undercapitalized developers who have optioned well-located parcels.

Public development companies have a substantial advantage in financing industrial projects through the use of low-interest-bearing corporate debentures, stocks, and other securities, combined with easier amortization schedules. Since the early 1970s there has been evidence, although it is hard to quantify, of a substantial number of European, Arab, and Asian sources coming into the continental United States real estate market. These funds are seeking both ownership positions and mortgages. Although overseas sources of financial capital for industrial development do exist, gaining access to them may prove difficult; much time may be wasted dealing with unknowledgeable intermediaries. Lastly, individuals can provide a source of lendable funds or equity participation through the creation of limited partnerships or syndications. In any case, Dennis Kealey of Cabot, Cabot and Forbes advises careful examination of financing rates or equity participation from unusual sources which are significantly below market.

Selection Process

Certain specific data should be considered to help identify potential lending sources. Most lenders do not make loans on all kinds of properties. For the most part, they restrict themselves to property types if they have staff expertise and facilities to adequately service and handle the loan submissions. Some lenders prefer to make loans on single-family homes only; others will lend on apartments, shopping centers, and office buildings; still others will provide funds for industrial land development; or, a lender may participate in a combined portfolio. Usually, however, individual lenders and real estate investment trusts (REITs) will take greater risks than such institutional lenders as insurance companies and commercial banks.

Other lending sources are limited by state and, in some cases, federal laws as to the maximum amount of loan that can be made on any one property. Within this legal ceiling, these institutions generally will establish their own maximum and minimum loan limits. These limits are set at a range within which it is determined that the institution can operate profitably with reasonable security. Regardless of project economics, the credit and past record of the development principals will weigh heavily in determining this amount. It is often difficult for the developer to ascertain the size of a loan that can be made on his property since this depends upon the portfolio requirements and policy of the lender. Occasionally, lenders will limit their operations to a particular geographic region which they consider to be their lending territory. Moreover, they also may restrict the number of loans made in a region. Interest rates by lender type may vary as well. Information about how the lender treats interest and the variations over time from property to property will be important.

Financing Techniques

A financial manager, or some other individual representing the ownership entity, is normally responsible for preparing the project budget. This budget should indicate the amount of capital required, at various points in the development process, to implement the project. A front-end working capital budget also should be designed so that it can be financed later as the development progresses. ULI Industrial Council member R. John Griefen notes, "the capital budget, however large or small, should be as accurate as possible—deviations will affect the lender's valuation of the capability and credibility of the borrower."

Essentially, on large-scale, long-term projects, there can be three different types of financing through the development period—initial standby commitment, land purchase and development loan, and permanent financing. In practically all cases, if the development is well planned and structured, the interest rate will decline with each successive stage of development. The initial standby commitment is based almost entirely upon the credit rating and net worth of the ownership group. This is usually at a high interest rate—three to four points above the prime interest rate. These funds are used to defray initial startup costs, and to provide working capital for good faith deposits on land acquisition and for hiring outside consultants. Once the development is initiated and land has been controlled, two or three different financial sources may participate in the funding of a project.

Generally, an insurance company or commercial bank will take the prime position—giving as much as 75 percent of the land value and financing up to 100 percent of the land development costs, depending on the economic feasibility of the project. This is a standby commitment and is drawn upon when needed. Total funds for the land, however, are usually provided at closing of land purchases, with invoiced withdrawals given for the land development costs through the initial development period. Possibly a REIT is brought in, holding a secondary or junior position to the commercial bank or insurance company, and lending the balance of the needed funds at an interest rate substantially above prime. The life insurance company or commercial bank will lend usually at one and one-half to two and one-fourth points above prime (perhaps with an adjustment for compensating balances); whereas, the junior position would require as much as three to five points above prime. In this manner, there can be as much as 100 percent financing on the land acquisition, land development expenditures, and development expenses (selling costs, commissions, advertising, etc.), through the initial development period. This situation may last from 18 months to as much as two to two and one-half years. Usually, this kind of joint participation by two or more lending organizations is tied to a single prearranged financing agreement.

Once a sufficient length of time has passed, the lending consortium is capable of evaluating the project's track record and is able to envision an absorption period. Then, the opportunity arises to obtain permanent financing for the balance of the development. Generally, an appraisal of the unsold inventory is made at this point. Hopefully, it will have a current value substantially in excess of costs, and some specified loan will be made based on the appraisal with the unsold inventory as security (a nonrecourse loan). This constitutes a longer term loan with amortization payments negotiated at simple interest rates, which may be only one or two points above prime. This long-term or take-out loan carries the project through to completion. The multitier financing approach, characterized by three different risk rates—each associated with a specific step in the development process—is usually applied to those projects exceeding a five-year development time period. Smaller-scale projects with a rapid absorption pace may require only a two-tier financing arrangement. Since initial and midterm financing are so expensive, it is important to convert as quickly as possible to a permanent financing situation in which the cost of lendable funds is reduced materially.

ULI Industrial Council Chairman Marshall Bennett raises the caveat that there is extreme financial danger in financing development for any length of time at interest rates based on points of interest over a floating prime rate. He cites the eight-month period in 1974 when the prime rate rose a full seven points which meant that a developer paying three-to-five points over prime, as suggested above, would have been paying from 15 to 17 percent for money. He states emphatically that no industrial development project can exist for more than a few short months at these rates.

Financial Planning and Business Structure

Prior to implementation of the development, the project's organizational business structure should, to some degree, be formalized. The scale of the development, the number of partners, and the complexities of the project, may dictate either a loosely drawn or tightly defined framework for doing business. Identification of responsibilities and communication channels, however, is necessary for even the most loosely structured organization. The development team, and the framework within which that team guides the project to implementation, must operate with maximum efficiency. Moreover, the business structure and forms of ownership used will have substantial federal and state tax implications.

The Development Team

Members of the group who should be involved in the planning and development of the industrial project are: 1) the ownership group, the key member of which is the developer; 2) the economist or real estate adviser; 3) the site planner/engineer; and 4) the marketing specialist. The developer member of the ownership group is the pivot around which the entire project management evolves and the person(s) in whom key coordination and decision-making responsibilities rest. As previously explained, the development team is set up at the inception of a project and should operate continuously throughout the planning and construction period.

Experience has shown that in large-scale industrial developments, the development team functions most efficiently when its members continuously coordinate their respective functions. In addition to the key members of the team, specific aspects of overall development will require the assistance of specialists. Some of these needs might be met by the in-house capabilities of one of the development team members; others might be referred to independent outside consultants. Such temporary associates might include a leasing manager, a public relations specialist, independent brokers, attorneys, appraisers, accountants, and tax counselors, as well as the specialized engineering capabilities of surveyors, structural engineers, soil and foundation experts, traffic engineers, and landscape architects. The costs incurred for retention of these outside counselling services must be taken into consideration during the financial planning phase.

Figure 5-5 exemplifies the internal organizational structure which many private investment groups have found to be effective in large-scale industrial development operations. The actual organizational structure of any given individual development group will vary undoubtedly somewhat from the chart. Nonetheless, the relationship among the various functions is clearly illustrated and defined even though there might be differences in structure and/or title in a specific organization. The top position in the organizational pyramid—labeled project ownership/developer—refers to the ownership interest in the property to be developed. Its various forms can range from corporate or joint venture to partnership or any other legal entity. The developer/member of this ownership body obviously would exercise overall control of development and make final decisions with regard to the development team, in all matters affecting the property. The developer, while not necessarily involved in the day-to-day operation of the development process, must retain strong managerial control and is responsible for basic management decisions. The ownership entity, assuming that equity positions other than the developer exist, probably functions like a board of directors in setting policy and business objectives under which the developer's managing partner operates.

Ownership Entities

The ownership group's methods of acquiring and developing real estate property can have a material tax impact on the financial success of the development. No attempt has been made here to provide definitive legal advice pertaining to the best or most efficient business structure to hold and/or develop real estate. This is the task of legal counsel and other advisers. Forms of ownership change from project to project as well as within the various time periods of a single project. The most common forms are described briefly, with emphasis on the tax implication for each form and with some mention of liability. Before any investment analysis can be meaningful, basic decisions on tax advantages and investment objectives must be made. No one organizational structure can solve all problems for all investors, or serve all purposes. Rather, the specific goals, objectives, and tax positions of individual developers, must be balanced with the advantages, disadvantages, and implications of

5-5

GENERAL ORGANIZATION FOR INDUSTRIAL REAL ESTATE PROJECTS

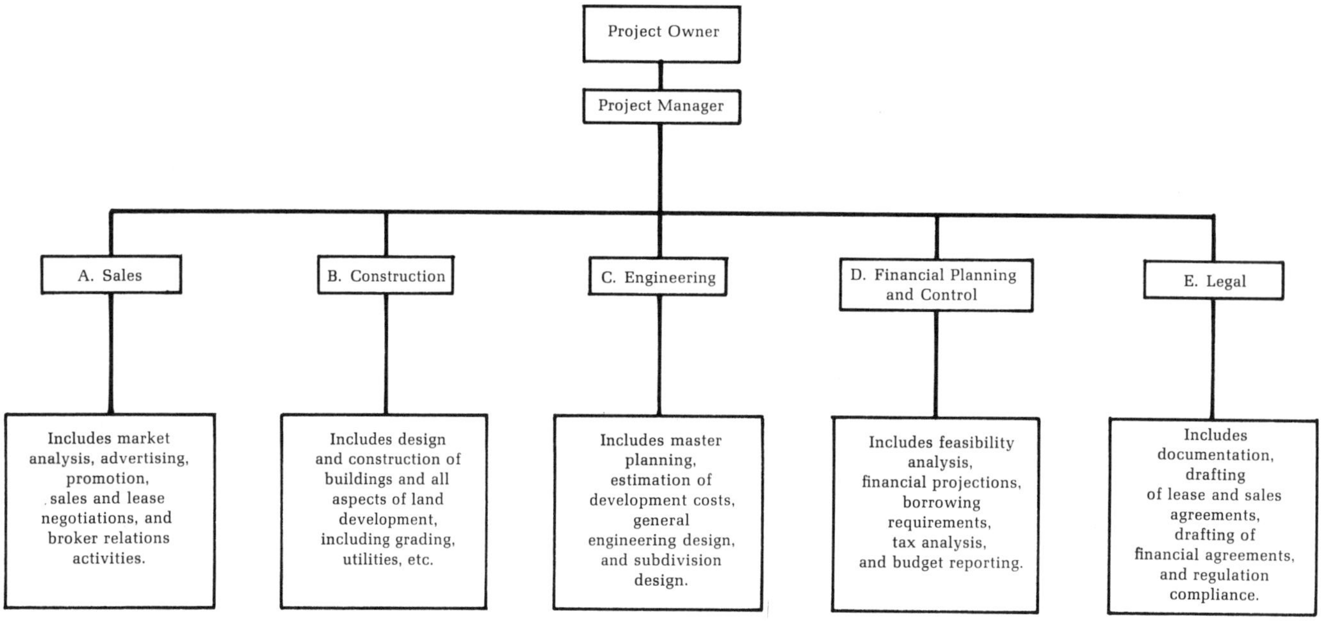

Provided by Cabot, Cabot & Forbes

each of the organization forms.

The key to any of the forms that might be chosen is the need to assume that the decision-making process is highly centralized in the developer member of the ownership group.

Partnerships

Partnerships are unincorporated businesses involving more than one individual. They can take the form of either general or limited partnerships. In the general partnership, each individual has a voice in management and can bind the business by his acts. In the limited partnership, usually one partner has full management control. Based on the need for developer control, the limited partnership is the recommended form.

In either the general or limited partnership, the partnership functions only as a conduit for income, losses, or tax benefits. In other words, the partnership entity is not a taxpayer; it merely computes income or loss during its accounting year, distributing to each partner his own share of the larger entity's profits and/or losses and tax shelter.

Joint Ventures

A joint venture is an association of two or more persons, partnerships, or corporations, or some combination of these, to carry on a development venture. A joint venture is most often described as a form of special limited partnership. The joint venture agreement establishes the partners' duties in the development process and specifies how ownership and profits are divided. The joint venture form of doing business can provide access to financing for under-capitalized firms and also aids in spreading the risk of loss. More important, however, through the joint venture approach real estate investors with no expertise in development can obtain ownership and income interests in professionally conceived and well-managed real estate projects. Utilization of tax benefits by some members of a joint venture provides, in some cases, a more attractive investment package. The joint venture form has allowed institutional investors such as insurance companies and REITs to obtain equity interest and even the kicker on mortgage loans, which aids in keeping the amortization dollar equivalent to the current dollar and in avoiding usury law restrictions at the same time. In other cases, the joint venture form has provided a source of management, consultation, and performance fees.

Corporations

A corporation is a legal entity which can be created under law for the benefit of its stockholders or members, and has three main features: limited liability for its stockholders, perpetual existence (in most cases), and transferability of interests of stockholders by means of stock shares representing some proportional unit share of the assets of the corporation. Regardless of the well-known advantages of the corporate form of ownership, this form of ownership is not often used for real estate development. Corporations generally are expensive to create, subject to double taxation on both corporate earnings and dividends, and finally, cannot offer the shareholders the advantage of the flow-through of losses including soft money and tax shelter.

Conclusions

The ultimate ownership form which is eventually adopted in order to implement an industrial project depends to a great extent upon the developer and his overall position with respect to capital, knowledge, and experience. Also of primary importance are the physical attributes of the site and its location in the community. Undercapitalized developers without experience, but controlling a prime site with good development potential, have some leverage in their joint venture negotiations with non-institutional entities. Developers with a successful track record in excess of five years can usually meet their financial requirements via joint ventures and institutional borrowing without undue difficulty, assuming they are operating in a normally expanding economy.

Basic Development Process

While the exact order and timing of the various steps in the development of an industrial project may be modified as experience with development unfolds, the steps which the development team will follow generally are shown in Figure 5-6.

This chart briefly indicates the activities of the other members of the development team and focuses on those activities which relate to the financing process. The junction points where activities converge are the decision-making points.

This general program is organized to permit staged development decisions within the limitations of financial commitment at each stage. The experience of most industrial developers has been that this approach actually permits more rapid completion of the physical and economic planning for large-scale projects than is likely to be the case when no coordination or review points in the process are defined.

Total time from selection of the metropolitan area to initiation of phase 1 can range anywhere from a minimum of six months to as long as 12 to 15 months, depending upon the scale of the project. Working capital requirements during this period can range from a minimum of $25,000 for a small project of perhaps 75 acres upward to $300,000 for a 500-acre project. These monies would provide funds for escrow and earnest money, deposits, staff overhead, traveling expenses, and consultant fees.

Investment Decision

Establishment of Investment Criteria

Before proceeding with any real estate project, and especially before making any substantial commitment, it is necessary to make some form of analysis to determine whether the proposed investment makes economic sense. This type of study is generally known as a feasibility analysis. It may be simple or sophisticated, short or long, but it must be done. This is particularly true in the real estate investment area of land subdivision and development because of the extended time frames which are usually involved.

The feasibility analysis is the crux of the investment decision because it provides the answers to four all-important questions:

- How much should be paid for the tract of raw land and what are the on-going costs?
- How must the end-product (subdivided lots, etc.) be priced, and is this within the market structure?
- What kind of capital structure and financing will be needed?
- How much profit will be made, and is it enough in view of the risks being taken?

Thus, the process of feasibility analysis requires the compilation and analysis of all of the key components and knowledge of real estate development

5-6

NETWORK DIAGRAM SUMMARY
TYPICAL INDUSTRIAL DEVELOPMENT

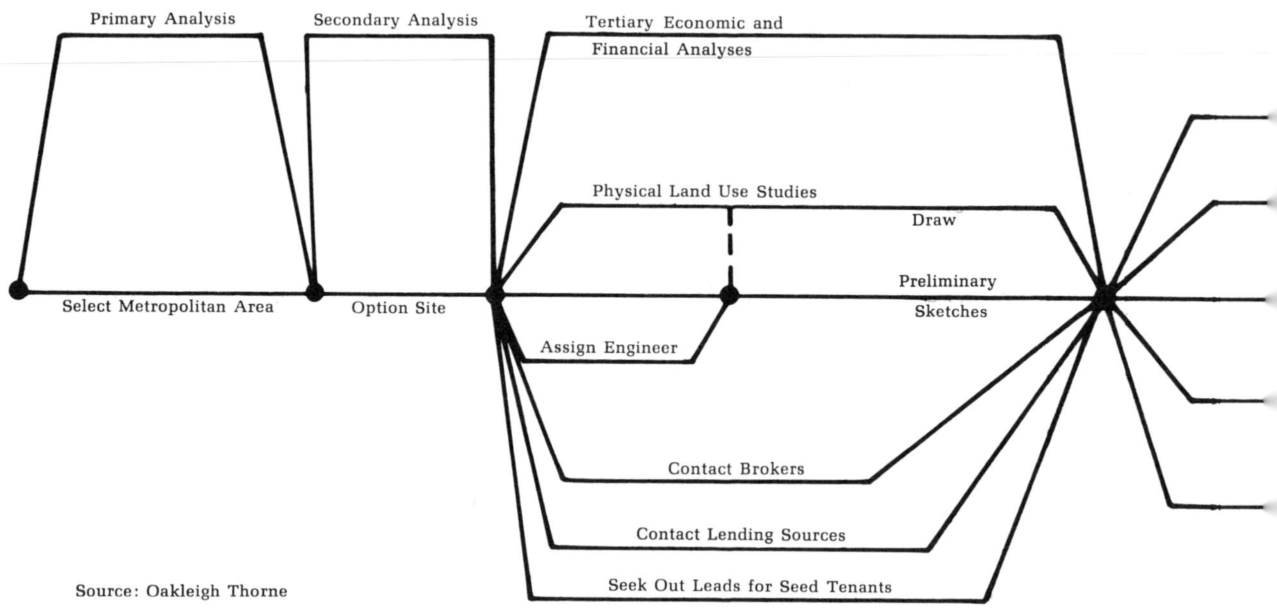

Source: Oakleigh Thorne

which have been discussed in detail in the other sections. It involves the coordinated efforts of the experts in marketing, land planning, engineering and contracting, financing, accounting, and sometimes law. The feasibility analysis, in a refined form, usually consists of detailed projections of sales, revenues, and costs, and is, therefore, the financial plan for the enterprise. It is vitally important that the operating organization be familiar with the plan and the goals which are defined. This group must participate in the preparation of the inputs which go into the original feasibility analysis.

After the financial model (feasibility analysis) has been prepared, it is possible to establish and apply investment criteria for that particular project. Every real estate project involves making a cash investment and putting it at risk. The investment is expected to be repaid with a profit, and projections are made of income, costs, and financing which demonstrates how the investment repayment and profits are to be achieved. The criteria for each investment becomes a function of the investor's evaluation of the risks inherent in each set of projections, so that a profit allowance can be determined which must be built in to compensate for taking these risks. The appropriate return on investment vis-a-vis risk is on a relative scale.

Secure bonds yield a return with no effort but offer no protection against inflation, and the income is taxable. Manufacturing companies' average after-tax profits are a percentage of total assets employed. Good investment real estate returns depend on tax characteristics and improvement opportunities. Where does this place an investment in a tract of raw industrial land or in a speculative warehouse building? The answer is different for each individual investor because the risks and alternative possibilities are viewed differently. However, it would seem that a normal-risk land development project with no tax advantages should earn at least equal to the average manufacturing company. A project offering some tax advantages (through leased buildings) could show a lower rate of return. A lower risk project, one which is a repeat of previous success or located in familiar territory with distinct tax advantages, might be justified by a rate of return somewhat lower than a manufacturing company. For a high-risk project (one in unfamiliar territory or where there is a hostile socio-political environment), the expectation of a higher rate of return before taxes is not unreasonable.

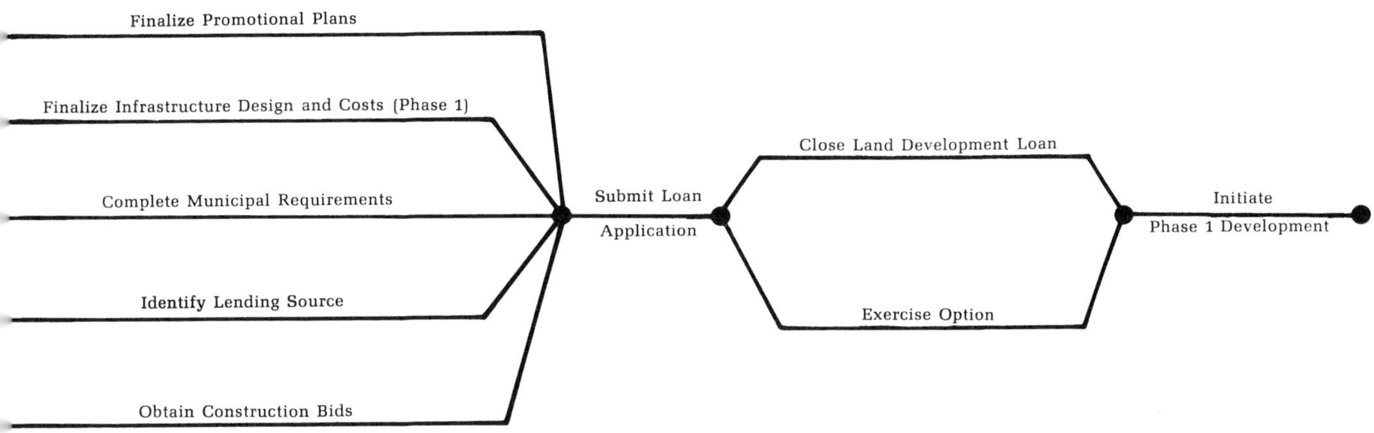

● Point of decision to proceed or stop

Rate of Return

The rate of return used to measure the profitability of industrial land development must take into account the timing of the cash flows. This concept is usually referred to as the *present value* or *discounted cash flow* approach, which rests on the simple premise that money invested must receive a profit calculated on the entire period of investment. This method also automatically provides identification of the rate of return of principal as well as profit. The discounted cash flow approach is an essential concept for evaluating land development projects because the cash flows may occur over a period of up to 10 years or more, and the cash flows tend to vary from year to year.

A simple illustration demonstrates the problem and the solution. Following are four examples of types of cash flows which can come from an investment. In each case, a $1,000 initial cash investment, and a project life of five years is assumed.

The annual cash flows and total cash flow are different in each case, but the calculated rate of return is the same: 10 percent per year on the investment of $1,000. All of these examples are solved with the use of tables showing "The Present Worth of $1.00" with income received at the midpoint of each year contained in many financial and appraisal handbooks.[2]

	TYPE 1	TYPE 2	TYPE 3	TYPE 4
INVESTMENT	$1,000	$1,000	$1,000	$1,000
CASH FLOWS				
Year 1	$ 264	$ 100	$ 105	$ (210)
Year 2	264	100	173	346
Year 3	264	100	159	380
Year 4	264	100	559	419
Year 5	264	1,100	346	461
Total	$1,320	$1,500	$1,342	$1,396

Type 1: *Level Annual Payments:* These cash flows are typical of a declining balance mortgage and can be solved most simply by finding the effective mortgage constant—$264 per year, divided by $1,000 is a constant of 26.4 percent. Mortgage tables (one payment per year) will show this to be a 10 percent, five-year constant.

[2] *Financial Compound Interest and Annuity Tables*, publication number 276 (Boston: Financial Publishing Co., 1968).

Type 2: *Level Payments with a Balloon:* These flows are typical of a rental property which is purchased for $1,000, pays rent of $100 per year, and is sold after five years for $1,000 (also identical to a five-year bond with a 10 percent coupon). This is clearly a 10 percent annual rate of return with full principal repaid at the end. If the end-payment (balloon or residual value) is not identical to the original investment, then the method of calculation shown under Type 3, below, must be used.

Type 3: *Irregular Annual Payments, All Positive:* These flows are typical of an industrial development having a five-year economic life. Present value factors must be used.[3]

	CASH FLOW		10 PERCENT PRESENT VALUE FACTORS		PRESENT VALUE
Year 1	$105	×	.953	=	$ 100
2	173	×	.867	=	150
3	159	×	.788	=	125
4	559	×	.716	=	400
5	346	×	.651	=	225
Total present value:					$1,000

Note that using 10 percent present value factors, applied to the actual cash flows, results in a total present value of $1,000, which is the amount of the investment. Thus, the cash flows represent a rate of return of 10 percent per year. The rate of return (interest factor in the present value tables) usually is not known at the start and has to be found by a trial and error process where the same calculations are made with different interest rate factors to find the situation where present value of cash flows equals initial investment.

Type 4: *Irregular Payments, Some Negative:* These cash flows are even more typical of an industrial development investment in which heavy expenditures in early years required for off and on-site improvements may not be met from lot sales. The method of calculation of the rate of return is the same as in Type 3, again using present value factors.

Here again, using 10 percent present value factors, the total present value equals the $1,000 initial investment.

The last example has been structured to demonstrate what the present value tables are doing. Receipt of the $380 cash flow in the third year has a present value of $300 today, assuming money earns 10 percent compounded. If the money is not to be

	CASH FLOW		10 PERCENT PRESENT VALUE FACTORS		PRESENT VALUE
Year 1	$(210)	×	.953	=	$ (200)
2	346	×	.867	=	300
3	380	×	.788	=	300
4	419	×	.716	=	300
5	461	×	.651	=	300
Total present value:					$1,000

received until the fourth year, it takes $419 to achieve the same return.

Finally, the above examples can be used to demonstrate the conceptual difference between rate of return and book (taxable) profit. The total cash flows vary from $1,320 in Type 1 to $1,396 in Type 4. The book profit is the amount by which the total cash flow returned to the investor exceeds his original investment of $1,000; hence the book profit is different in each example. However, the rate of return is 10 percent per year, and is the same in every case.

Appraisals

Shelves of books are devoted entirely to the subject of appraising, and anyone interested in the intricacies of the field is referred to the library. However, brief mention must be made here of the role of the appraisal.

The documented appraisal is required most often by a lending institution in order to satisfy statutory lending requirements. In this regard, some generalized observations are:

- Lending is a subjective process including a great number of intangible factors such as the lender's portfolio position, how much confidence he has in the borrower, and future business considerations.
- To a large extent, appraising is also a subjective process. Allow the lender to do his own appraising, he is usually the best one to match his sense of value to his lending objectives.
- Do not volunteer a formal appraisal unless or until the lender specifically asks for one. In some instances, the requirement may be waived.
- If the lender requires an appraisal, ask him for a list of appraisers who are satisfactory to the lender and who are familiar with his normal requirements.

[3] See Exhibit 5-7.

5-7

PRESENT WORTH TABLES—VALUES COMPUTED AT MIDPOINT OF YEAR

SELECTED RATES

Year	10 percent present worth	15 percent present worth	20 percent present worth	25 percent present worth	30 percent present worth	35 percent present worth
1	953 4626	932 5046	912 8709	894 4272	877 0580	860 6630
2	866 7842	810 8737	760 7258	715 5418	674 6600	637 5281
3	787 9856	705 1076	633 9382	572 4334	518 9692	472 2430
4	716 3505	613 1370	528 2818	457 9467	399 2071	349 8096
5	651 2277	533 1626	440 2348	366 3574	307 0824	259 1182
6	592 0252	463 6197	366 8623	293 0859	236 2172	191 9394
7	538 2047	403 1476	305 7186	234 4687	181 7055	142 1773
8	489 2770	350 5631	254 7655	187 5750	139 7735	105 3165
9	444 7973	304 8375	212 3046	150 0600	107 5181	078 0122
10	404 3612	265 0761	176 9205	120 0480	082 7062	057 7868
11	367 6011	230 5010	147 4338	096 0384	063 6202	042 8050
12	334 1828	200 4357	122 8615	076 8307	048 9386	031 7074
13	303 8025	174 2919	102 3846	061 4646	037 6451	023 4870
14	276 1841	151 5582	085 3205	049 1717	028 9578	017 3978
15	251 0765	131 7897	071 1004	039 3374	022 2752	012 8873
16	228 2514	114 5997	059 2503	031 4699	017 1348	009 5461
17	207 5013	099 6519	049 3753	025 1759	013 1806	007 0712
18	188 6375	086 6538	041 1461	020 1407	010 1389	005 2379
19	171 4886	075 3511	034 2884	016 1126	007 7992	003 8799
20	155 8987	065 5227	028 5737	012 8901	005 9994	002 8740

Source: *Financial Compound Interest and Annuity Tables,* publication number 276 (Boston: Financial Publishing Co., 1968).

The other main use of appraisals is by sellers and buyers to test their prices being asked or offered. Taking into account the subjective nature of the inputs, appraisals serve the function of identifying the upper and lower limits of the negotiations. Appraisals are not a substitute for a careful analysis conducted by the potential investor and his team of experts, who will be exposing themselves to the actual risks.

Feasibility Analysis

The feasibility analysis is a systematic approach to determining the profitability of a potential investment (or an investment in progress)—in this case, a tract of land for an industrial park. The analysis is usually conducted before any binding commitments have been made. Generally, the analysis requires some level of advance expenditures for options, preliminary market studies, and master planning and land improvement costing.

There are no universal rules of thumb when it comes to land acquisition feasibility; but clearly, the time to discover the peat and rock and the cost of removing it, is before large capital is committed.

Several layers of feasibility studies may be done. The first is a fairly rough analysis to determine whether it is even worthwhile to commit funds to extensive engineering. Additional studies follow as more precise information and data are developed. The rougher the input data used, the less accurate will be the answer on rate of return. The objective, therefore, is always to work with the most precise information possible. This section discusses only the types and steps of calculations involved. The discussion of financial projections which follows will describe in some detail the projection of income and expense.

Free and Clear Analysis

The free and clear analysis is a financial projection model which assumes that the investment is owned and developed free and clear of any debt financing. This type of analysis is recommended because it uses the basic economic data for the project, undistorted by any debt financing assumptions. These financing assumptions can sometimes cover up defects in the underlying business venture which will be identified easily in a free and clear analysis.

5-8

LAND FEASIBILITY ANALYSIS MODEL ($000s)

Given Data

Gross Acres: 450 acres
Net Salable Acres: 400 acres
Purchase Price: $6,000,000

Marketing Projections: Conservative: 100 percent sales

Financing Assumptions:

Mortgage: $4,000,000
Interest: 10 percent per annum
Releases: $13,000 per acre after 25 acres

Equity: $2,000,000

Years	1	2	3	4	Total	Line
Marketing						
Acres Sold	25	100	100	175	400	1
Price per Square Foot	$ .92	$ .92	$ .92	$ .92		2
Gross Revenue	$1,002	$4,008	$4,008	$7,012	$16,030	3
Less Sales Expenses (7 percent)	70	281	281	490	1,122	4
Net Proceeds from Sales	932	3,727	3,727	6,522	14,908	5
Expenses						
Land Development	1,600	800	800	0	3,200	6
Engineering	100	100	100	100	400	7
Contingency	100	100	100	100	400	8
Real Estate Taxes	100	125	80	40	345	9
Direct Overhead	75	75	75	75	300	10
Promotion and Signs	50	20	20	20	110	11
Other	20	20	20	20	80	12
Management Fees	25	25	25	25	100	13
Total Expenses	2,070	1,265	1,220	380	4,935	14
Net Cash from Sales, Free and Clear	(1,138)	2,462	2,507	6,142	9,973	15
Financing $4,000,000						
Release Payments	—	1,300	1,300	1,400	4,000	16
Interest	400	335	205	70	1,010	17
Total to Lender	400	1,635	1,505	1,470	5,010	18
Net Cash after Financing	(1,538)	827	1,002	4,672	4,963	19

Rates of Return Using present value factors where it is assumed that income is received at the midpoint of each year. See Exhibit 5-7.

Free and Clear: 18 percent on $6,000,000 purchase price.
20 percent justifies a purchase price of $5,660,000.

Equity Investment: 24 percent on $2,000,000 equity.

Generally, financing for a properly structured project always improves the rate of return on invested equity, but an analysis with financing should come only after the free and clear approach has been used to determine the proper structure.

Through the tool of feasibility analysis the developer answers the question, "How much can I afford to pay for this tract of land?" Or, if the acquisition price is fixed and nonnegotiable, "At what price do I have to sell the lots in order to make my required rate of return?" A feasibility analysis also allows the developer to experiment on paper with such business decisions as:

- Price structure: What is the effect on rate of return if retail prices are lowered to move the land faster?
- Front-end cost: What is the effect if greater amounts of money are spent at the start of the project on entrances, landscaping, amenities, etc., to permit a higher retail price structure?
- Sensitivity analysis: How sensitive is the rate of return to changes in projections? What would be the effect of a recession in the third year, etc.?

Using the best information available at the time, a feasibility analysis is made along a format similar to that shown in Figure 5-8. The example is a 450-acre tract, with a projected development program of four years. The key result is line 15—Net Cash from Sales, Free and Clear. In this example, the conclusion is that a rate of return of 18 percent per year will be earned on the $6,000,000 cash purchase price. This may be satisfactory to the developer. However, if his minimum rate of return requirement for this project is 20 percent per year, it is a simple matter to determine the highest purchase price he can afford to pay. The annual cash flows of line 15 are discounted at the 20 percent present value factors to arrive at a maximum price of $5,660,000.

Effect of Land Financing

Lines 16, 17, and 18, on Figure 5-8, depict the normal effect of a land financing loan. Here, it is assumed that $4,000,000 of the $6,000,000 purchase price is borrowed at a 10 percent fixed interest rate, and that the loan is to be repaid as land is sold by making release payments at the rate of $13,000 per acre. It has also been negotiated that the first 25 acres will be released free (see financing assumptions in Figure 5-8) as an incentive to the developer to complete substantial land improvements in the first year. Therefore, the initial minimum cash equity to buy the land is $2,000,000.

The key result is line 19—Net Cash After Financing, from which can be calculated the rate of return on the $2,000,000 equity. In this case, it is 24 percent per year, which is higher than the 18 percent rate of return on a free and clear basis. This is a proper relationship because, by leveraging the project, the developer is exposing his equity investment to greater risk. Naturally, when money can be borrowed at a cost which is lower than the free and clear rate of return, the rate of return on the leveraged equity will be higher than the free and clear. But consider the developer who undertakes a project on a 15 percent free and clear basis, then borrows at a floating rate of prime plus 4 percent in a 7 percent prime rate money market, only to find himself paying 16 percent for his financing when the economy goes into a tight money period and the prime rate rises to 12 percent.

Obviously, a financing analysis can be used to test the effects of varying amounts of financing and of various interest rates, including joint venture situations.

Both the free and clear, and the leveraged results on lines 15 and 19, highlight an extremely serious characteristic of land development programs: the need for adequate working capital. In this particular model, there is a large cash outflow in Year 1, because of the front-load improvement costs. This is a normal situation and the deficit must be planned for and met out of capital reserves or bank lines of credit.

Lease-Building Program

The preceding discussion assumes the entire land development program is realized by outright sale of all of the subdivided lots—i.e. 100 percent sales. In practice, it can be expected that anywhere from 10 to 80 percent of the land may be used for build-to-suit and for leased arrangements or on speculative buildings. The amount of this type of activity will depend on the market and the aggressiveness of the developer.

The effect on feasibility is significant. Figure 5-9 contains the same assumptions as Figure 5-8 except that 100 percent of the land is committed to a leased building program. It should be noted that the completed lease packages are sold off each year, the annual cash deficits are more severe, and larger working capitals are required. However, land values created in lease deals are generally higher than those in outright lot sales. This trade-off, if properly executed, will result in an improved rate of return.

Computer Analysis

In the days before computers came into common use, all of the copious and complicated calculations necessary for a financial model feasibility analysis had to be done by hand. In view of time, labor cost, and tediousness, it was normal to make only one or two analyses with various assumptions before making an investment decision. Now, many developers and consulting firms have developed computer programs, which permit virtually unlimited variations and print-outs of a given project, at a relatively low cost. Use of these computers hopefully results in more enlightened business decisions. Individual use of these programs usually can be obtained by inquiring at national real estate consulting firms, or sometimes at local computer time-sharing companies.

Conclusion

The objective, in this discussion of the investment decision-making process, has been to provide the

5-9

LAND FEASIBILITY ANALYSIS MODEL 100 PERCENT LEASING PROGRAM ($000s)

Years	1	2	3	4	Total	
Leased Packages						
Land Area—Acres	25	100	100	175	400	1
Land Value per Square Foot	$ .92	$ .92	$ .92	$ .92		2
Building Size	325,000 sq. ft.	1,300,000 sq. ft.	1,300,000 sq. ft.	2,300,000 sq. ft.		3
Building Cost—All Items	$ 12.00	$ 12.00	$ 12.00	$ 12.00		4
Rental Calculation						
Land Value	$ 1,002	$ 4,008	$ 4,008	$ 7,012	$16,030	5
Building Cost	3,900	15,600	15,600	27,600	62,700	6
Total Package	4,902	19,608	19,608	34,612	78,730	7
Cash Flows (see below)						
Net Flow from Buildings (a)	108	538	968	1,728	3,342	8
Net from Construction/ Financing (b)	500	2,000	2,000	3,500	8,000	9
Proceeds of Sale of Buildings (c)	—	—	—	17,280	17,280	10
Total Inflow	608	2,538	2,968	22,508	28,622	11
Expenses						
Land Development	1,600	800	800	—	3,200	12
Engineering	100	100	100	100	400	13
Contingency	100	100	100	100	400	14
Real Estate Taxes	100	125	80	40	345	15
Overheads	170	140	140	140	590	16
Subtotal	2,070	1,265	1,220	380	4,935	17
Net Cash Free and Clear	(1,462)	1,273	1,748	22,128	23,687	18
Detailed Calculations						
(a) Net Flow from Buildings						
Net Rents at 12 Percent of Lease Package	$ 588	$ 2,350	$ 2,350	$ 4,150		19
Mortgage Debt Service	480	1,920	1,920	3,390		20
Net Cash Flow	108	430	430	760		21
Cumulative	108	538	968	1,728		22
(b) Net from Construction/Financing						
Building Cost	3,900	15,600	15,600	27,600	62,700	23
Permanent Financing	4,400	17,600	17,600	31,100	70,700	24
Net Proceeds	500	2,000	2,000	2,500	8,000	25
(c) Proceeds of Sale						
Net Cash Flow at End of Project	$ 1,728					
Capitalized at 10 Percent	$17,280					

generalized framework and concept of investment criteria and feasibility analysis, along with the mathematical techniques which are used. The results of this type of sophisticated analysis are only as good as the data which goes in. Therefore, the objective is to use the most refined data available, at a cost which can be afforded. The more refined the data, the more refined will be the answers. As an example, if the results of an analysis show a 15 percent rate of return (and that is the investor's minimum), but he is only 50 percent certain of the accuracy of his rough input data, that means the actual rate of return could be as low as 7.5 percent or as high as 22.5 percent. The spread is too wide to make an intelligent business decision. The next section discusses in greater detail how accurate input data is developed and refined.

Financial Projections

Projections and Calculations

In industrial development, conditions are continuously changing due to internal and external events. The financial projection is an essential management tool which serves a number of purposes:

- determines the initial feasibility of the project;
- determines marketing and development strategy;
- obtains financing;
- checks on performance to date, as it affects future requirements;
- serves as a budget and a standard of performance;
- determines cash and borrowing requirements; and,
- provides a basis for tax planning.

While the compilation of data and inputs may be performed by one individual or department, it is essential that operating personnel be involved in the preparation of the inputs and the review of the final product. This requires a team approach, since the allowable pace of expenditures for land development (prepared by the project engineer) will depend on the rate of sales of subdivided lots (projected by the marketing manager).

The preceding subsection presented the basic format for a financial projection of a typical industrial park, and then discussed the procedures for making a feasibility analysis. This section provides greater detail on the derivation of the individual line items making up that basic format. The same sample data is used, and reappears as Figure 5-10. The line number references appear in the right hand margin of the exhibit.

Given data: Gross land area is 450 acres. After a review of the master plans, it is determined that the net salable acres are 400 acres after deducting streets, slope, unzoned land, etc. Local practice will determine whether rail and drainage easements are included in salable or nonsalable areas. This reflects approximately a 90 percent efficiency factor, which is not unusual in industrial development.

The purchase price is $6,000,000. This figure should include all costs incurred up to the time the land is actually deeded from the seller: optioning costs, preliminary engineering and economic studies, legal and brokerage fees, as well as the cost of demolishing any structures on the property.

Marketing projections: In feasibility studies, it is normal to make several runs using different marketing assumptions. Figure 5-9 was an example of a 100 percent leases building program.

Financing assumptions: These assumptions are discussed under lines 16, 17, and 18.

Marketing—lines 1 and 2: This indicates the projections as to annual rate of absorption of acres, and the retail selling price per square foot. Absorption of industrial land at a specific site will vary from year to year, based upon a number of economic variables interacting both regionally and nationally. The best basis for making these judgmental projections is the detailed market study which provides information on competitive projects, probable future needs in the community, and historical absorption rates for other industrial developments in the community. Pricing is directly related to the competitive marketplace, but can also reflect the developer's specific product and his investment objectives. Lower pricing will aid in a faster sellout period, whereas higher prices will defer an ultimate sellout. Depending on the market, a leasing program will affect the absorption rate. A lease-only policy will certainly extend it. For simplicity, line 2 shows constant land prices, however, financial projections usually indicate pricing increases over time, partly due to inflation and partly due to improving surroundings as development progresses. The ability to achieve projected price increases during the selling period is also affected by the supply and demand relationship between the project and its competition. The pricing indicated in the table represents averages for the acre-

5-10

LAND FEASIBILITY ANALYSIS MODEL ($000s)

Given Data

Gross Acres: 450 acres
Net Salable Acres: 400 acres
Purchase Price: $6,000,000

Marketing Projections: Conservative: 100 percent sales

Financing Assumptions:

Mortgage: $4,000,000
Interest: 10 percent per annum
Releases: $13,000 per acre after 25 acres

Equity: $2,000,000

Years	1	2	3	4	Total	Line
Marketing						
Acres Sold	25	100	100	175	400	1
Price per Square Foot	$ .92	$ .92	$ .92	$ .92		2
Gross Revenue	$1,002	$4,008	$4,008	$7,012	$16,030	3
Less Sales Expenses (7 percent)	70	281	281	490	1,122	4
Net Proceeds from Sales	932	3,727	3,727	6,522	14,908	5
Expenses						
Land Development	1,600	800	800	0	3,200	6
Engineering	100	100	100	100	400	7
Contingency	100	100	100	100	400	8
Real Estate Taxes	100	125	80	40	345	9
Direct Overhead	75	75	75	75	300	10
Promotion and Signs	50	20	20	20	110	11
Other	20	20	20	20	80	12
Management Fees	25	25	25	25	100	13
Total Expenses	2,070	1,265	1,220	380	4,935	14
Net Cash from Sales, Free and Clear	(1,138)	2,462	2,507	6,142	9,973	15
Financing $4,000,000						
Release Payments	—	1,300	1,300	1,400	4,000	16
Interest	400	335	205	70	1,010	17
Total to Lender	400	1,635	1,505	1,470	5,010	18
Net Cash after Financing	(1,538)	827	1,002	4,672	4,963	19

Rates of Return Using present value factors where it is assumed that income is received at the midpoint of each year. See Exhibit 5-7.

Free and Clear: 18 percent on $6,000,000 purchase price.
20 percent justifies a purchase price of $5,660,000.
Equity Investment: 24 percent on $2,000,000 equity.

age sold each year. Substantial variation can occur among individual sites depending upon size, aesthetics, and exposure.

Revenues—lines 3, 4, 5: This section converts the acreage and price marketing assumptions of lines 1 and 2 into dollar amounts. The provision for expenses of sale included as line 4 will depend upon the developer's marketing policy. If all sales are expected to be made through outside brokers, provisions should be made for full commissions. If an inside sales force is expected to obtain a certain level of business, the direct selling expenses should be adjusted.

Expenses: The projections of cash out-flow to develop, market, and manage the project, are shown in lines 6 through 14.

Land development—lines 6, 7, 8: Based on the detailed engineering plans, and in coordination with the rate of land sales, the costs of all physical improvements to the land are projected over the life of the project. Note that in common with most projects, the expenditures will be heavy in the first year and will tend to taper off thereafter. Line 6 is a summary of detailed projections for the various components of the site improvement requirements. Figure 5-11 provides a minimal suggested breakdown for land development expenditures. Line 7 is a reminder that expenditures for professional engineering services will continue throughout the life of the project in connection with specific parcel sales or lease deals and the almost certain need to revise the master plan a number of times. Line 8, contingency, is to provide an allowance for changes in costs due to poor estimating or redesigning. This category also can be used to make allowances for inflation over the life of the project.

5-11

LAND DEVELOPMENT COSTS
SUGGESTED DETAILED CHART
OF ACCOUNTS

- Clearing, Grading, and Fill Compaction
- Storm Sewers and Drainage
- Rail
- Utilities
  - Water Lines
  - Sanitary Sewers
  - Electrical Lines
  - Gas Lines
  - Lighting, Striping, and Signing
- Paving and Surfacing
- Curb, Gutters, and Sidewalk
- Landscaping, Trees, and Shrubbery
- Off-site Improvements and Miscellaneous
- Planning, Engineering, and Professional Fees
  - Site Concept Engineers
  - Soil Engineer
  - Topographic Mapping and Survey
  - Landscape Architect
- Professional Fees
  - Traffic Consultant
  - Environmental Expert

Real estate taxes—line 9: It is difficult to predict real estate taxes throughout a development program unless there is a clearly established historical pattern in the community for assessing tracts of industrial land. At an early stage, the developer should meet with the assessors to determine the probable taxes in the first year of ownership—which may be significantly different from the pre-acquisition level because the transfer price may form a new basis for assessment, and a re-zoning may occur which changes the character of the land. Taxes usually can be expected to increase in the early years as heavy site improvement expenditures are made. The increase in property value will be off-set to the extent that, as land is sold to third party users, the value of the inventory remaining in the hands of the developer goes down.

Direct overhead—line 10: This projection includes all of the management and staff who will be running the project, and their attendant expenses such as office supplies, rent, telephone, and automobiles. Therefore, the size of staff planned for the project should be determined in advance.

Promotion and signs—line 11: This category includes the cost of promotional materials and outside consultants involved in advertising and promotion of the project. Included will be promotional brochures, photographs, direct mailings, and publicity releases. The first year of the model includes a $30,000 item representing the cost of large signs to be erected on the site—assuming the location has exposure to major highway traffic.

Other—line 12: This is a miscellaneous category to provide for items such as auditors' fees, nonrecurring legal fees, general liability insurance, and any other items which have not been scheduled as direct overhead in line 10.

Management fees—line 13: Where this project is one of several owned by a single company with a headquarters management staff, there must be some method to allocate head office expenses and this is provided on line 13.

Net cash from sales free and clear—line 15: This is the difference between line 5, net proceeds from sales, and line 14, total expenses.

Financing—lines 16, 17, 18: In this model, it has been assumed that $4,000,000 of the total purchase price has been financed through conventional bank land financing, paying a fixed interest rate of 10 percent and with principal to be repaid through release payments as the land is sold. In a more complex financing arrangement, provision might be included whereby the bank would advance additional funds for land development expenses, but this has been omitted to simplify the presentation.

Release payments—line 16: Release payment formulas usually call for an acceleration factor since the lender will want to be paid back fully before the last acres are sold. The acceleration factor will usually be in the range of 10 to 25 percent, depending upon the negotiations. In this model, a factor of 18 percent has been used, which is calculated as follows:

$$\frac{\text{Total land loan}}{\text{Total repayment acreage}} \quad \frac{\$4{,}000{,}000}{375 \text{ acres}} = \$11{,}000 \text{ per acre}$$

$$\$11{,}000 \text{ per acre} \times 118 \text{ percent} = \$13{,}000 \text{ per acre release price}$$

Since the first 25 acres are to be released without principal payment, the loan has to be repaid from the proceeds of sales of 375 rather than 400 acres. The actual release prices are, of course, calculated by multiplying the acres in line 1 by the figure of $13,000 per acre.

Interest—line 17: Interest expense on the land loan is calculated on the average loan balance outstanding each year. It may be assumed that all of the sales release prices are made at the end of the year, or, that they are spread evenly over the year. In the latter case, the interest expenses would be lower.

Net cash after financing—line 19: This entry is the difference between line 15, net cash free and clear, and line 18, total to lender.

Rates of return: In Figure 5-10, the rate of return is calculated on a free and clear basis and also on the initial equity investment of $2,000,000 ($6,000,-000 purchase price less $4,000,000 land loan). The free and clear rate of return is calculated using the present value techniques discussed in the section dealing with the investment decision. On the $6,-000,000 purchase price, the rate of return is 18 percent per year. If it is desired to determine the maximum cash price which can be paid for the land if the developer is to meet a 20 percent rate of return objective, then the annual cash flows shown in line 15 are discounted using a 20 percent present value interest factor.

In order to calculate rate of return on the equity investment, present value factors are applied to line 19 in order to solve against the investment of $2,000,000 (see Figure 5-7).[4]

Cash Flow Projections

The format utilized in Figure 5-10 is the basis for preparing the details of the cash flow projections for the project. It would be normal to expect the cash flow projections to have a more detailed accounting breakdown and also to be made at least on a quarterly basis for the immediate 12-month period.

Profit and Loss Statements

The profit and loss statement used for determining taxable income each year employs the same data contained in the cash flow projections but the treatment of the items is somewhat different.

The most significant element is the cost of goods sold which is charged against the revenue from each sale. In its simplest form, only land acquisition cost and land development cost items are treated as capital accounts which are pro-rated equally over each acre. Thus, in the model of Figure 5-10, the pro-rata land cost is determined as follows:

Land Acquisition Cost	$ 6,000,000
Projected Land Development Costs	$ 4,000,000
Total Land Costs	$10,000,000
Land Cost per acre (400 acres) =	$25,000 per acre

Thus, for every acre sold, in this case at $40,000 per acre, there is a cost of sale of $25,000, leaving a gross profit of $15,000 per acre.

Other expense items can be treated in different ways depending upon objectives of the developer, who may choose between expensing all of the general overheads each year or capitalizing them into the land costs and thereby spreading them evenly against all of the sales. The former method will result in lower taxable earnings in the early years. But, a public-oriented company may wish to show book earnings and, therefore, adopt the latter method.

As a final refinement, allocation of the total land and land development cost of $10,000,000 on an uneven basis is permitted. Thus, the frontage land, which presumably will command a higher retail price, might be charged with a cost of goods sold of $40,000 per acre, and less desirable land may be charged at $10,000 per acre, but averaging to $25,-000 per acre. This procedure has to be adopted at the beginning of the project as the unit acreage cost zones are difficult to change once established. However, if properly used, this method can have material effect upon the timing of the impact of federal income taxes on the project.

Conclusion

While certain methods of projection and analysis have been suggested in this chapter, it is entirely up to the developer to determine the methodology which is most convenient and understandable to him and his organization. The most important thing is that a pattern be established of assembling financial projections on a routine basis and of using these to measure the performance of the project against the original standards under which the venture was justified.

---

[4] Ibid., *Financial Compound Interest and Annuity Tables.*

Elements of the Loan Submission Package

Obtaining financing can be the most important single requirement for a project and considerable care should be devoted to it. The application will contain complete project details, financial statements, and the recorded experience of participants in the proposed development. The developer and other participants also may be required to submit personal biographical material and experience as a part of the mortgage application.

The mortgage application document should be a persuasive presentation using all appropriate and favorable facts and figures as well as visual aids. In practically all cases, the mortgage application is prepared either by the developer's staff or by outside advisors. If a quality mortgage broker has been retained, he will prepare his own style of presentation but will depend heavily on the cooperation of the developer.

The market study, an essential ingredient, predicts pricing and absorption rates for the project. It also will have evaluated the overall market demand in sufficient detail, both with respect to future population and to employment trends to support the predicted absorption rate at the site. The detailed market study can either be summarized or condensed into the loan submission package. The results of the field investigation describing the competitive parks and their absorption rates also should be enclosed, perhaps only as a single table. It is also appropriate to supply aerial photographs not only of the competitive parks, but also of the site itself. These can serve as important visual aids in the presentation.

The importance of photos, graphics, maps, and engineering data cannot be overemphasized. A desirable effect can be obtained by packaging facts and figures with artists' renderings, and by preparing slides which portray proposed land use plans and building code restrictions. In situations where severe topographic problems prevail, scale models can aid in providing a three-dimensional image of land planning and architectural design.

Financial projections covering revenue expenditures and expenses will be reviewed very carefully by the potential lending institution. Undoubtedly, questions will arise as to the assumptions behind the numbers, and it is necessary to be able to support these assumptions in a convincing manner. For example, specific questions might be asked about staffing and about administrative and general costs.

The industrial developer should remember that the loan presentation is his sales tool upon which the potential implementation of his concept rests. The developer will use it as an aid in a personal meeting with the lending officer. However, the material submitted by the developer also must equip the lending officer to make an equally persuasive presentation to a loan committee composed of individuals who, in many cases, will not be familiar either with the geographical area or the technical aspects of industrial development. The material, supplemented by the lending officer's own analysis, must be sufficiently complete to enable him to answer any questions arising in committee. Failure to obtain financing on the first application can damage the project's image with any other lender which might be approached. Following is a checklist of the contents of the loan application. Order of insertion will vary with the emphasis required.

Introduction:
- Letter of transmittal
- Loan application (supplied by lending institution and filled out by the ownership group)
- Recordation of good faith deposits
- Mortgage loan submission summary sheet

Discussion:
- Narrative of the concept of the project, its general description, and rationale. Synopsis of advantages of key economic and site data

Economics:
- Survey of competitive developments
- Aerial photos of the competition (optional)
- Financial forecasts of the project
- Model lease forms for the rent of speculative built space
- Economic base analysis of the community
- Aerial photo of the city

Site:
- Land comparables (if a separate appraisal is not submitted)
- Overall development plan with detailed characteristics on phase 1
- Aerial photo of the project site
- Map illustrating location of the project site and competitive development
- Detailed map of the metropolitan area
- Complete engineering data covering soil tests, bearing strength, grading, road specifications, etc.
- Copy of purchase option

Buildings:
- Plans and specifications either of build-to-suit or speculative building, or architectural

drawings and preliminary plans for the initial tenants

Credit Information:

Dun and Bradstreet reports on individuals in the ownership group or partners in the joint venture and seed tenants

Copy of the purchase money mortgage (if any) and a financial statement of the land owner

Financial statement of the applicant(s)

Marketing and Management

Marketing Strategy

Marketing of the development starts almost at the time the option to purchase the tract is considered. Acquisition of seed tenants and advertising of the project begins quietly with informal contacts by the ownership group among major national companies and prestigious regional and local firms. Attraction of high quality companies at the outset is helpful to the success of the project. The initial marketing is not undertaken by a formal marketing team, but consists mostly of capitalizing on communication channels already open to the ownership group. Marketing materials used at this stage, including advertising literature, need not be highly polished but should illustrate the location of the property, its relationship to the total city, and the overall project identity being promulgated. In some cases, initial marketing may be indirect. A few of the brokers that the ownership group has been associated with extensively in the past should be contacted for leads as to possible seed tenants. This is particularly true if the local industrial real estate brokers are skilled and effective. In other cases, contacts are made directly to the corporations.

As development progresses, additional steps in the marketing program, including selection of advertising agencies and public relations firms, budget development, broker contacts, and brochure design, are undertaken. The marketing program or strategy must be coordinated with the conclusions drawn in the detailed market analysis and with the goals and objectives of the ownership group. The result will be a well thought out project identity and marketing strategy. ULI Industrial Council member Robert R. Walker, Jr. cautions: "Care must be taken to avoid excessive costs for advertising and promotion. For example, advertising in the general media is rarely effective, yet can be costly."

Marketing Documents

Two marketing documents should be a part of the program:

Technical Services Package

The first is the technical services package which is provided to brokers. It consists of statistical data and descriptive materials of interest to prospective tenants and includes information relating to population growth and change, statutory taxes, beneficial real estate taxes and sales taxes, and interstate commerce trucking zones and rates. A complete detailed description of the utilities available to the development must be outlined. This item should cover the typical water and sanitary sewer dimensions, capacity, static pressure, and design flows. Fire protection services and requirements should be summarized. The availability of electrical capacity should be indicated also, and any secondary services. Availability and frequency of rail service, together with the name of the line serving the development, should be included.

A summary of the protective covenants applicable to the project is essential. All development constraints, building setbacks, landscaping requirements, building exterior qualities—e.g., brick on three sides, or a brick front with a painted concrete block on the other three sides—must be outlined together with the procedures for architectural inspection of tenant-built structures, and the actual approval process. Parking and service areas should be outlined. Uniform sign standards must be explained, both with respect to location and design. Objectionable uses not permissible within the project, such as open storage, must be itemized together with procedures for dealing with and controlling objectionable situations such as noise, odor, vibrations, and smoke. Any adjustment for variation in the requirements should be indicated. These regulations should be set forth in a positive way rather than as negative controls. The potential tenants need to be convinced these covenants were created for their benefit—rather than as a way to prevent abuse. This data should be assembled in a small leaflet or pamphlet which the broker can give to possible clients.

Sales Brochure

The other document is the sales brochure. The professional, well-designed sales brochure usually is created with the help of a public relations or advertising consultant. It should carry a distinctive logo identifying the project and the ownership group. Most such documents consist of a nine-by-twelve-inch, fold-out, jacket-type brochure with single-page or staple-bound literature located in the jacket. This approach allows for updating as development proceeds without complete revision and reprinting each time new or revised information needs to be added to the sales package.

The lead piece describes the ownership group and its development experience, sets forth the theme or concept for the development, and hopefully, identifies the seed tenants. It is also appropriate to enclose an overall development plan of the entire project in a fold-out leaflet within the jacket itself. The map identifies the parcel configuration on a preliminary basis and shows interior road networks. Tenants already committed to specific sites can be located on the map. It should be scaled so user prospects can approximate distances and sizes. A location map showing the relationship of the site to the total community and the general road and rail networks which provide access to the property, should be included also.

The sales brochure jacket can contain a stapled brochure of several pages summarizing some of the data provided in the technical services package. Information about the Interstate Commerce Commission's trucking zone rates and the number of trucking companies which serve the area are probably of interest to all perspective buyers. Information on rail service should be included. Identification of airport facilities is important when air cargo may be an integral part of tenant operations. Water port information, listing the annual total of waterborne tons and some commerce statistics, should be indicated if it is relevant to the development.

Also included in the sales brochure jacket is a synopsis of community characteristics drawn from the market analysis studies. The track record of the ownership group, complete with photographs of previous developments, can be presented as the final piece of the handbook.

In all cases, the brochure design should have a professional and distinctive quality to it, identifying not only the project but the ownership entity. Complete factual accuracy is essential.

Marketing Budget

The total cost of the entire marketing program, including brochures, should not exceed 5 percent of total gross projected revenues. However, some projects may require a minimum of 3 percent while others need the 5 percent allocation. The stature and development expertise of the ownership entity and the nature of the competition will determine whether the program could be as low as 3 percent, or whether the project may require a 5 percent advertising budget. Thomas Wolff, Jr., of the Irvine Company and ULI Industrial Council, suggests that in addition to the general marketing budget each building built for speculative market should have its own supplemental marketing budget.

Marketing Procedures

The most inexpensive method of marketing the property is to deal honestly and openly with the brokers in the community. Some industrial development entities prefer to take 10 to 15 minutes at a local realtor meeting to discuss the project and then invite all the brokers to participate. In dealing with the brokers, it is important that the standard commission be maintained and that the development entity not get involved in commission cutting. After all, marketing expenses and brokerage commissions are a relatively inexpensive portion of the entire project. ULI Industrial Council member R. John Griefen of Boston points out, "The broker will follow the path of least resistance and therefore will stress those projects in which the ownership is open and cooperative and can provide the best services for the broker's clients." Robert R. Walker, Jr., a ULI Industrial Council member from Oakland, California comments that the question of exclusive representation by a single brokerage firm is sometimes considered if the developer does not have the sales skill on his own staff. He felt this method has merit but the effect of this approach on other brokers must be considered. "If undertaken, the exclusive agent must be held accountable for his promotion and sales activity, and his program and his performance must be reviewed periodically."

Direct solicitation of prospects by the project manager and his staff, in some cases, will motivate the brokers. If the brokers understand that the ownership entity is aggressive and has the internal capability of soliciting local or regional industries, the broker will be aggressive in making such contacts as are not made initially by the developer's sales staff. It is important that the broker knows he will make his commission as well. Paul Shepherd of Cabot, Cabot & Forbes further points out that regular sales meetings should be held with brokerage firms to inform them of new developments in the project, such as changes in the plan, pricing modifications, and current sales or leases which have been made. An occasional tour of the project to familiarize the brokers with the status of development is helpful. The ownership group also should engage in what is referred to as "missionary work", that is, mailing the brochures to their clients, industry contacts, and brokers known to them in other communities in the region.

It is also expected that the ownership entity will provide guidelines for continuing communication procedures, such as courtesy calls to industrial development agencies and to public departments within the community and region as well as to utilities and railroads which are operating in the territory. Planning commissions, redevelopment agencies, and other public or quasi-public entities should be contacted regularly. These groups frequently can be a source of leads on prospects for the project.

Spot advertising can be placed in trade journals or professional publications associated with industry; however, this type of advertising generally is not very productive except to establish image and visibility. Brian Hogg, vice president of sales and marketing of the Irvine Industrial Complex, suggests that with a new project, it is important to have initial, heavy exposure in the media and then fall back to a steady, consistent marketing program.

Brochures sent directly to prospects and cooperation with brokers provide the best avenues for marketing the property. Naturally, for the marketing strategy to be most effective, the entire thrust of the program must be directed toward the potential clients identified in the market analysis.

Not to be ignored as a potential marketing tool are the signs on the project site announcing its development. If highway visibility is good, these signs can be an asset. The developer should design his signs to convey the quality of the development. They should be consistent with the sign standards planned for the development, and should be designed in accordance with local regulations. If a suitable site is available, the signs should be set in landscaped areas which will further convey the planned character of the development. Signs can be costly—the financial projection model discussed earlier in this section had a budget of $110,000 for signs and promotion, of which $25,000 was allotted for signs.

Marketing flexibility during the course of development is critical also. In the event a given strategy is proving ineffective, the developer must be prepared to modify his approach. For example, parcelization and lot sales must be organized to permit adjustments in parcel size should demand prove to be different than anticipated at the outset. If this dictum is ignored, the result may be an extremely slow absorption period. Therefore, planning, engineering, and marketing must be designed around flexibility. ULI Industrial Council member R. John Griefen reinforces this concern for flexibility, "If a developer does not have a flexible program of lot sales, speculative buildings for sale or lease, and a build-to-suit capability, he should consider selling sites to other developers who can provide these programs. However, as vital as this flexibility is, the developer must maintain control in order to assure the correct amount of each type of product and the quality of the product is maintained at all times." Those industrial developments which are concentrating on providing a product mix with a substantial portion of outright sales together with some leasing, and which can accommodate almost any site requirement, will have a more rapid absorption period—everything else being equal.

On-Going Management and Operations

Management of industrial development is a combination of public relations, standards administration, and delicate coordination of development activities. If an industrial project is subject to failure, there are generally two internal areas responsible for the nonprofitability. The first area is financing of the development (previously discussed in this section). The second area is management. Many good projects have turned sour because of mismanagement.

Management of industrial development is a specialized field; it is divorced almost completely from traditional management concepts applied to the operation of commercial and apartment buildings. Traditional management occupies itself with the operational and maintenance aspects of building improvements. Management in the industrial development field, on the other hand, is more sales or leasing oriented and public relations assumes a greater importance. Most buildings, if they are leased, are managed and maintained by the tenant. The project manager must be attuned to changes in processing and distribution technology, as well as be knowledgeable about building design criteria for housing both processing and storage activities.

The project manager also becomes the industrial promoter and public relations man for the project both as the development takes form and after it has become part of the community. Maintenance of a harmonious relationship between the project and the larger community is an essential ingredient in the expertise of the project manager.

Management of large-scale, long-term industrial developments generally follows a three-phased program which features different priorities and goals for each phase. During the initial project implementation phase, coordination and timing goals are created to establish the park's character and reputation and to initiate the sale of land to seed tenants. The second phase covers the maturing stages of development and is primarily concerned with sales pace and price levels. The last phase of development concerns itself with the residuals (leased buildings) which are owned by the ownership group. At this stage, management is involved only peripherally except for periodic inspections to see that the properties are being operated and maintained properly.

Embryonic Period

The project manager's tasks are concerned primarily with coordinating the process of infrastructure development and attracting seed tenants to the development. However, at this point community response concerning environmental factors is critical and, consequently, a public relations role may take priority. The scale of the overall development and its impact frequently will influence the nature of the community's response to the development. To the degree that the developer has incorporated measures designed to insure the implementation and maintenance of a high quality industrial environment, this problem can be mitigated. Nevertheless, the project manager would be ill advised to ignore the probability of negative community reaction. The implementation of recommended procedures found in environmental impact statement codes adopted by such states as Florida, California, Minnesota, and Illinois, can be expected to expend a considerable amount of the project manager's time. The construction process with its noise, dirt, and increased traffic flow caused by construction equipment and material delivery vehicles, can be a significant source of community discontent.

The success of the "good neighbor" image of the development may hinge on the diplomacy with which these problems are resolved. Failure to respond quickly to community concerns can result in time consuming, costly delays. ULI Industrial Council member Thomas Wolff, Jr., feels that *on-site* management, during the embryonic period, is vital at the start of development and should be continued as long as feasible. Timely completion of infrastructure is vitally important to the seed tenant who will establish overall project identity. If the seed tenant experiences delays in occupancy due to what he considers poor management of problems by the developer, it is highly likely he will convey these dissatisfactions to other potential tenants to the extreme disadvantage of the project. The project's tenants and owners are the best salesmen a project has. Once initial development is underway and the pace quickens, management efforts should be focused toward coordinating the financing and development planning during the major period of market absorption.

Management During the Absorption Period

The coordination and timing necessary in the initial stages of an industrial development project are also necessary during the maturing process, but with a different emphasis. Primary emphasis at this later stage is placed upon tenant selection, occupant relations, enforcement of project standards, financial management, and maintenance of the development's reputation.

The detailed market analysis will have suggested an optimal tenant mix for the industrial project. If this study has been carefully prepared, and the advertising campaign is directed toward the desired segments of the economic market, then the actual occupancy pattern should fall in line with the projected occupancy pattern. Parcelization, locational relationships, and compatibility of potential tenants with those already committed, requires continuous attention and reevaluation. The ownership group and the manager must keep the long-term goals and objectives firmly in mind and must not sacrifice them for any short-term considerations.

It is particularly important to preselect a seed tenant who will fit the optimal market mix for the development. A conflict between the goals of the seed tenant and of the development may result in the creation of a problem which will have a deleterious impact throughout the entire absorption period. If the ownership group and the developer have done their homework properly, the selection of tenants and their compatibility, and the achievement of the desired mix, should fall in line with relatively few conflicts. At the same time, Dennis Kealey of Cabot, Cabot and Forbes restates that "industrial development is a dynamic process. Thus, there is a continuing need for flexibility and accommodation of potential tenants and for finding the solution to their particular problems."

It is particularly important to remember that principles of agglomeration appear to have an almost universal application to industrial real estate. That is, similar or linked industries tend to locate together. Often, the seed tenant will set the entire reputation and character of the project and of the tenant mix. In addition to the type of tenants attracted by the seed tenant, the size of seed tenant buildings seems to have some relationship to the pattern which follows. If the first tenant occupies a 10,000-square-foot building, it frequently develops that most of the remaining buildings will be in the category of 10,000 to 25,000 square feet. If the first seed tenant occupies a building of 100,000 square feet, most of the users attracted later will range from 50,000 square feet upward.

The continuation of an industrial development's marketability is assured through the establishment and enforcement of covenant restrictions concerning architecture, outdoor storage, loading, and parking requirements. The concept of positive encouragement, backed by controls and continuing supervision, must be handled on an equal and impartial basis. The greater the supervision and control of the development as manifested in its architectural features and especially exterior yard controls, the easier financing becomes during the mature stages. Such a program illustrates to the lending institution that the ownership group is interested in the maintenance of its reputation for future projects. A well-nurtured understanding among tenants of the value of complying with the covenants will be far more effective than waiting to enforce noncompliance.

Financial management, including cost accounting, pricing, and new leasing and sales information, almost becomes a day-to-day task of the developer's staff during the absorption phase. Whereas the initial objective for the management's staff was timing, coordination, and acquisition of a seed tenant for the implementation of the project, the basic objective of industrial development management at this stage is to maximize the long-term financial potential inherent in the development. This stage includes planning and budgeting both expenses (staff salaries, advertising, legal, accounting, and office space rent) and development expenditures (infrastructure costs). Either quarterly or semiannual updating of the original projections with actual project experience is required.

Throughout the absorption phase, management must work to maintain the reputation of the ownership group and to continually enhance the project's identity in the community. The coordinative aspect of management's role with respect to the financial aspects of the park, tenant selection, and administration of standards, comprises the project's reputation as development proceeds. A project's reputation can create a substantial markup for remaining lots in the development which are most difficult to sell because of location, topography, or parcelization factors. During this period, it is important to maintain continuous contact with the local governmental and regulatory agencies. Their requirements and concerns are continually changing. New or changed requirements may be detrimental to the pace of absorption if not given management attention.

Final Management Phase

There are two methods of handling the management of a multi-tenant project once it is sold. The first is through an occupant association, which operates much like a homeowners association. In this instance, the occupants play an integral role in the maintenance of the park and in enforcement of the standards ascribed to by the ownership group. If it is to be a voluntary association, it can be created as the developer begins to phase out toward the end of the development period. If it is to be a mandatory association, then its legal responsibilities and purposes must be established at the beginning of development so it can grow as the development grows. In this way, it will be a mature organization with enthusiastic support from its tenant members and, therefore, capable of taking over when the developer relinquishes control. ULI Industrial Council member Robert R. Walker, Jr. points out, however, "in either case there are inherent problems with these associations. They can be a source of expense and problems for the developer if they become a focal point through which the occupants can press for services or benefits to which they are not entitled."

At this stage, the ownership group is concerned only with the residuals, or those buildings leased either for long or short-term periods. Generally, semiannual inspections are sufficient. The project's staff and its office space will have been vacated from the premises, and the semiannual inspections are carried out by administration from headquarters locations or via administration of the ownership group. It should also be recognized that the bank or the lending institution which has provided the financing for the buildings also will be making its own inspections as well.

It has been customary, once the development is sold out, for the developer to depart, leaving the enforcement of standards to the municipality's zoning department. This would continue to be the case if a voluntary association is created which relies primarily on social pressure and lacks a legal mandate to maintain the development. With a mandatory association, however, the responsibility for enforcement should rest very significantly with the tenants, allowing them to act independently of the public agency to enforce covenant requirements which are not legally enforceable by the public. While the creation and nurturing of an effective association of tenants places another burden on the industrial developer's management team, the long-term protection of the initial development objectives yields benefits to the tenants, to the community as a whole, and to the developer's reputation when initiating the next project. From this standpoint, the benefits should more than equal the effort required.

6. Future Trends

The Nature of Change in Industrial Development

Introduction

The nature of the industrial environment is evolving in company with the broader scale of change occurring throughout society. In this discussion of future trends, the industrial environment is taken to mean all the aspects—physical, social, regulatory, technological—which define industry's and therefore industrial development's function and role in society. Change is continuous. Current innovations give way to newer concepts, techniques, and development strategies. Old ideas once discarded are revived, modified, and reapplied. The rate at which new and modified ideas will be incorporated in future developments is difficult to predict. Certainly, in the industrial sector, a key factor in acceptance of new ideas and change will be economic viability. The discussion which follows is projective, perceptive, and predictive. Projective in the sense that emerging trends are identified and projected into the future. Perceptive in that phenomena which might result in change and their importance are offered for evaluation. Predictive in that some attempt has been made to go beyond the mere projection of trends and the perception of forces for change to predict the probable nature and extent of change.

Evolution of industrial development during the twentieth century was treated in Section 2—Background and Characteristics. During the years the modern industrial park evolved from the original central manufacturing district, innovations were introduced slowly and were adopted nationally over a period of decades. More recently, however, the pace of change has quickened. The industrial developer of today faces a much greater variety of problems and challenges and requires a substantially more sophisticated approach than that of his predecessor. Even with these increased difficulties, innovations have made it possible for the industrial developer to create working environments far superior to those commonplace only a few years ago.

The industrial developer has been somewhat slower than his counterparts in commercial and residential development in identifying and directing new trends. The idea of a controlled and planned industrial environment is of much more recent origin than the idea of planned and protected residential and commercial settings. Many of the physical planning ideas now being applied to the industrial environment are adaptations of concepts first tried and then refined in residential and commercial development. This divergence is now closing. Within the next few years, industrial land use will be treated with as much professionalism as any other land use type.

The purpose of this section is to make some judgments about the future directions of industrial development and to apprise the developer and his professional associates of the opportunities available so that they may take full advantage of the potential in their projects. In addition, public officials at various levels of government will be better able to anticipate the activities of the private sector.

The industrial park, with its comprehensively planned environment, has been the focus of much discussion in this Handbook. The majority of innovative industrial development has taken place within the framework of the planned industrial park where the nature and extent of change can be observed most readily. Nevertheless, the majority of land developed for industrial use is not found within such planned environments.

Industrial development takes place within a profit framework. For the developer, the purpose of innovation is to increase project profitability either by reducing costs, capitalizing upon an unmet need, or enhancing competitive position. This is true now and will remain true in the future. ULI Industrial Council member O. G. Linde points out that innovative development can and has been a generator of profitability. "Upgrading by one developer breeds upgrading by others—it is contagious." Innovative development is good for the community and can be beneficial for the developer unless economic feasibility is adversely affected. Innovation, therefore, is rooted in the profit motive. For a change to receive acceptance, it must be directed either toward reducing costs or increasing income. The average developer needs encouragement to initiate a worthwhile innovation. He will do so only if he is confident that the change will enhance his project's prospects for economic success, or if the change has been required through legislative mandate. Within this framework of economic feasibility and changes in public policy, the following trends have been identified as reflecting the greatest degree of innovation to be found on the industrial development scene.

Planning, engineering, financial, and marketing innovations—large and small—have been touched upon throughout the Handbook. This section, however, focuses on key trends which will be influential over the long term. Projections of the future must be approached with caution. All too frequently, an apparent trend will disappear or change dramatically as new forces appear in society.

Current Trends in Industrial Development

Environmental and Energy Considerations

Concern for environmental preservation and energy conservation is the most influential current trend affecting industrial development. It also will be an influential force for change for the next several decades.

The industrial developer is facing increasingly stringent environmental controls over the development and use of his property. The objective of these governmental restrictions is to insure a high standard of environmental quality within the project and throughout the larger community. Concurrently, rising energy costs and shortages have become a major national concern. In some instances, energy conservation in industrial development will be complementary with environmental protection. However, these two concerns are not totally complementary and therefore not totally mutually achievable. These circumstances will be a major cause of innovative responses on the part of industrial developers and industrial land users.

Integration of Land Uses

Initially, industrial parks were planned as specialized employment areas, with all other activities rigorously excluded because of a presumed deleterious impact. A strong trend is now emerging toward meeting some of the commercial, recrea-

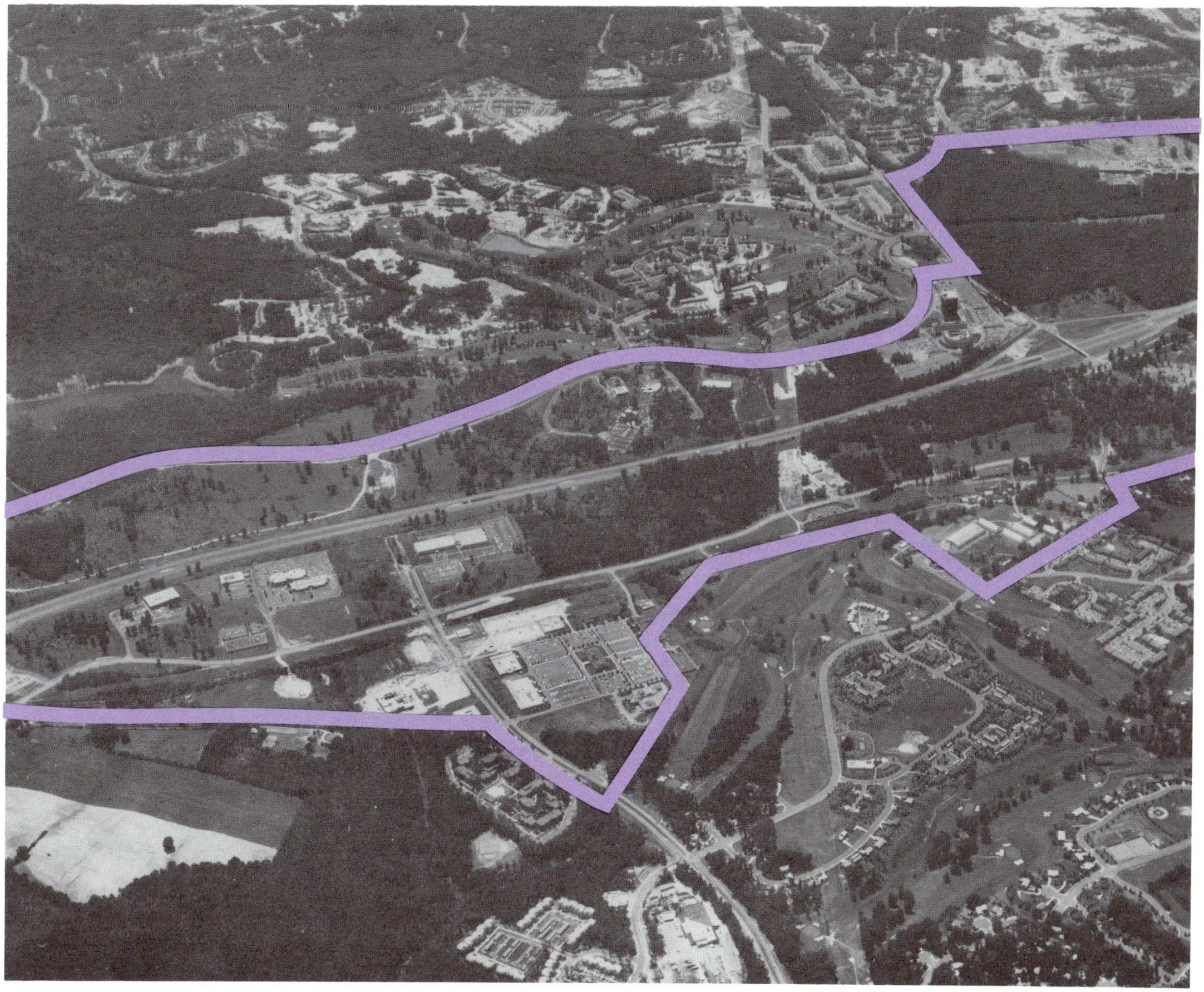

6-1 Increasingly, industrial development is taking place in a comprehensively-planned environment. Shown is an aerial view of Reston, Virginia's primary employment area. It is situated on either side of the Dulles Expressway which connects Washington, D.C. with the Dulles International Airport.

tional, and housing needs of employees within the industrial developments themselves.

Specialized Industrial Aggregations

An awareness of the advantages of clustering compatible industrial uses is growing. This will take the form of subclusters within an employment center as well as in single-purpose groupings. When the specialized requirements of an industrial subgroup such as meat processing, computer technology, or automotive services, are incorporated in the project planning and design from the outset, the result is the provision of special features and solutions to problems peculiar to that subgroup which are not easily or economically obtained by individual users.

Employee Social Needs

In the past, the primary emphasis in industrial park planning has been placed upon the efficient and orderly function of the industrial process to maximize productivity. Now, increasing consideration is being given to identifying and meeting the social and psychological needs of the employee during the working day. This changing trend in the organization of work tasks will be reflected in industrial development requirements at all levels of scale from individual buildings to the regional scale. However, as ULI Industrial Council member Cyril Herrmann points out, "the competitive posture of the United States in world markets requires increased productivity as a concomitant benefit of improved social and psychological needs of employees."

The nonworking hours of the employee increasingly have become the employer's concern. If productivity is related to personal satisfaction as well as job satisfaction, as has been suggested by social psychologists, the future industrial user will be seeking an environment for his employees which will give access—as an integral part of the work environment—to convenient commercial, recreation, and health services. The forces of social change will continue to impact industrial development.

Innovations in Industrial Marketing Techniques

Innovations in other areas of development practice lead inevitably to new industrial sales and marketing techniques. These techniques include adoption of new ownership types such as the industrial condominium, utilization of residential marketing techniques such as furnished model areas, creation of foreign trade zones, provision of inventories of speculatively built buildings, and increased staff specialization to understand and anticipate the needs of potential industrial users.

Technological Change

Along with considerations for environmental protection and energy conservation, the force which will most influence future trends in industrial development will be technological innovation. New user groups will appear as a new technology is applied. The rapid growth of the aerospace industry in recent years presents a dramatic example. A strongly held belief among ULI's Industrial Council members is that communications technology more than any other technological advance will cause dramatic changes in the current patterns of industrial land use. Innovations and modifications in transportation modes including highway, rail, water, and air will occur both for the movement of goods and people. Because the very life blood of industrial activity is tied both to efficient and economic movement of goods and to convenient access for employees, transportation will be an important area of responsive innovation among industrial developers.

Shifting Patterns of Industrial Location

Two broad national trends in industrial location have strong implications for the industrial developer. First is the continuing major shift of employment and production away from New England, Middle Atlantic, and Great Lakes states toward the rest of the country. Second, is the strong shift of new industrial facilities toward nonmetropolitan areas. These two trends, discussed in greater depth later on, have important implications in identifying future markets for the industrial developer.

Countering these present trends is the continually suggested public policy objective of inducing industrial development and redevelopment in the inner cities of major metropolitan areas. A few instances of success and certain ongoing attempts have been identified. Nevertheless, a variety of impediments and constraints have tended to outweigh such positive factors as ready access to a skilled and semi-skilled labor force. Although no trend is yet visible, the rise of social concerns, the increasing scarcity and rising cost of suburban land, and a more definite understanding of the cost of servicing land both in terms of energy conservation and environmental protection, should provide the impetus for a serious reconsideration of inner city industrial development. ULI Industrial Council member Donald Johnson feels that major industries are already reevaluating the reuse of existing sites as an alternative to pioneering new sites.

Financing Development

Underlying all the trends described above will be the impact of changes in the method and availability of financing industrial development. Indications are that during the next decade and beyond, capital for development will be in short supply and more expensive to obtain. Methods of financing also will change as mortgage terms differ from past practice. Higher rates or variable rates, and various indexing techniques which account for inflation or changing money market rates and other mechanisms, are beginning to appear as alternatives to long-term fixed rates. These new trends are as yet ill-defined and can be only the subject of speculation.

Summary

The trends suggested above will be validated in the following discussion. In many instances, projects currently in the process of development which are in some way responsive to these trends will be identified. The actual strength of these trends will become apparent only with the passage of time. Trends which are most responsive to identifiable public needs will be the ones which gain eventual support. The industrial developer can no longer satisfy only the industrial user but must be prepared to make a positive contribution to broader societal needs.

Evaluation of Trends

Environmental and Energy Considerations

Increased interest in preservation and enhancement of the natural environment, coupled with the recognized need to efficiently utilize scarce energy resources, constitute the two most dramatic developmental constraints to have arisen over the past five years. Their impact has been as substantial in the industrial field as in the other segments of the development community. The environmental protection issue was raised first by environmental groups concerned with the preservation of areas of great natural beauty or areas of unique or fragile ecological structure. On a broader basis, there is also a growing national awareness of the hazards to our very existence from pollution of our air and water.

Public action has taken place at all levels of government, but most dramatically at the federal level. The National Environmental Policy Act of 1969, the Clean Air Act of 1970, the Noise Control Act of 1971, the Federal Water Pollution Control Act Amendments of 1972, the Coastal Zone Management Act of 1972, renewed enforcement and interpretation of the Rivers and Harbors Act of 1899, and the creation of the Environmental Protection Agency have placed the federal government in a powerful position as the watchdog responsible for protecting the Nation's environment. It is likely that national land use legislation will be enacted in the near future—probably in support of planning at the state level but clearly involving the federal government in land use.

At the state and local level, a variety of parallel actions has taken place. Perhaps the most significant actions are those intended to control, restrain, delay, or manage growth. In many instances, these actions are taken for fiscal or political reasons but almost always with the declared purpose of environmental protection. This latter phenomenon is complex and evolving so rapidly, it serves no useful purpose to cite specific state and local actions. They have been compiled as a separate three-volume publication by the Urban Land Institute.[1]

ULI Industrial Council member Donald E. Johnson suggests that "these impedimenta will, or may have already, reached a peak and that controls will not be eliminated but will be applied in a more balanced way."

For several decades, concern for energy conservation was an issue discussed only in academic circles. Suddenly in 1974, this issue achieved major national importance as a result of international political actions. One of the results was creation of the Federal Energy Administration in 1974. A significant amount of funds was committed for energy research.

The long-term impact of these two areas of concern—environmental protection and energy conservation—will not be assessed for at least a decade. Administration of federal, state, and locally enacted legislation is changing on almost a daily basis. A variety of litigation is in process. Therefore, the particular impact on industrial development must be determined at the time development is initiated.

The Environment

The day when most communities indiscriminately sought industrial growth has passed. City officials now weigh potential costs such as pollution, increased traffic volume, and higher tax revenues, in their attempt to insure that benefits will outweigh drawbacks. As a result, the industrial setting increasingly reflects a recognition of and concern for the environment.

The industrial land user increasingly is seeking to create an environment which will be attractive to his employees and to the larger community. The industrial park has been one response to these circumstances. Landscaping, careful placement of buildings, sign controls, and use of compatible building materials, are now commonplace features of the more prestigious industrial parks. Initially, such environments were thought to be justifiable only when research and development activities, branch offices of national companies, and prestige conscious local firms were the primary occupants. The environment conscious perspective began emerging at the time of the initial Dartmouth-Arthur D. Little conference on industrial parks in

[1] *Management and Control of Growth—Issues • Techniques • Problems • Trends* (Washington, D.C.: Urban Land Institute, 1975).

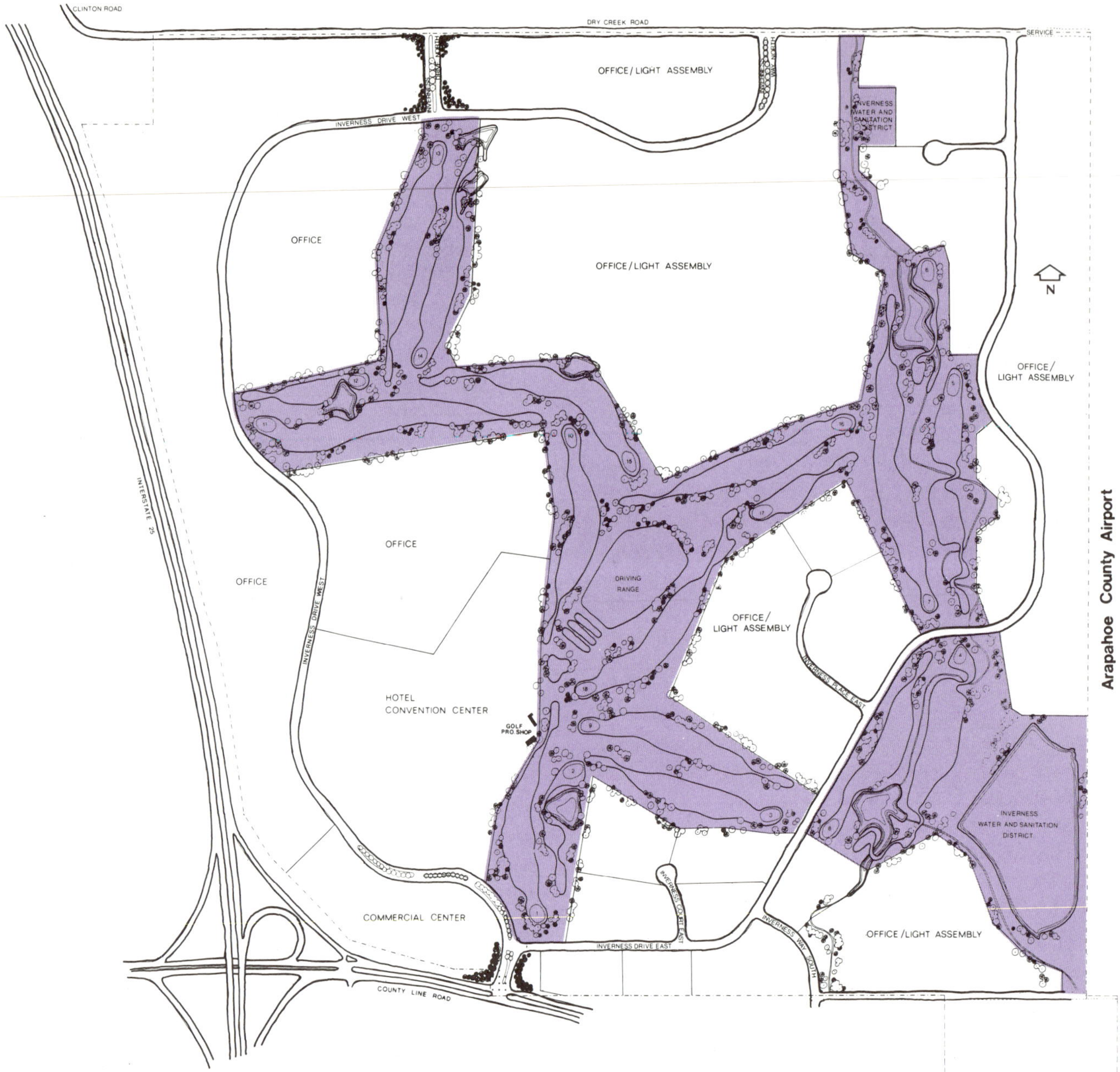

6-2 Inverness site plan. The growing variety of land uses found within the industrial environment is indicated by Inverness, Arapahoe County, Colorado. An 18-hole golf course provides the setting for industry. A hotel-convention center and commercial center are also planned.

1958. Now, most office/warehousing parks incorporate codes, covenants, and restrictions which specify limitations on site coverage, delineate landscaping requirements, list permissible uses, and outline performance standards and similar devices intended to insure maintenance of high environmental standards.

Several major companies are redesigning existing heavy industrial plants to make them more compatible with the environment. One of the largest such projects now underway is Dow Chemical Company's 2,000-acre chemical production complex in Midland, Michigan. Within the site, overhead power lines were buried, a compatible color scheme for tanks and towers was selected, peripheral parking areas were bermed and landscaped to provide a green buffer from outside properties, sediment ponds were converted to naturally shaped lakes, and new parking was located in the landscaped center strip of a new parkway created by closing two streets. When this project is completed in 1980, the plant will have been transformed into a visually attractive element within the community. Presumably, the pollution control problem of this industry will receive similar attention in response to legislative requirements and industry self-policing actions. This project has widespread application throughout the American industrial landscape. Similar undertakings by enlightened companies can be expected in the future.

At Inverness in Denver, Colorado a multi-use project by Bennett & Kahnweiler, an employee-oriented open space recreation area has been incorporated within the 602-acre project area. An 18-hole, double-fairway golf course with returning nines, driving range, and water reservoir, occupies almost one-third of the project area. Such a design was adopted because the developer believed the value of the remaining property would be sufficiently enhanced to make setting aside over 200 acres as permanent open space economically justifiable. Aesthetics has been presumed to be an economic value, even in an industrial setting.

The economic benefits derived from this approach can be demonstrated in various ways. At Inverness, it was possible to locate the open space and the golf course within the natural drainage framework on steeply sloping and flood-prone terrain difficult to develop for industrial use. In this manner, costly grading and channelization improvements have been avoided while contributing materially to the overall appearance of the park. Likewise, the reservoir created by damming the stream can be used to water the golf course. The linear course configuration opens up expansive views of the Rocky Mountains, creates a maximum number of parcels having golf course frontage, and is the location of an extensive pedestrian walkway system tying together the various activity areas found within Inverness. Developer Marshall Bennett indicates the golf course has generated substantial use from the development and its presence has been an attraction to new customers considering location in Inverness.

These two examples of a creative approach to environmental design represent a positive response to environmental concerns on the part of the industrial developer and user.

One issue of growing environmental concern is the location of industrial uses within flood plain areas. Historically, industrial development followed the course of railroads along the river valleys. Because most of these rail lines are being used and interest in the availability of multi-modal transportation is increasing, flood plains remain attractive to the industrial developer. Rapid growth

6-3 Earth City, St. Louis County, Missouri raises the issue of development in a flood plain.

in many cities bypassed the flood plains in earlier years; pressure now is growing to fill in the gaps. In some instances, it may be economically justifiable to undertake the extensive engineering required to floodproof these flood plain areas preparatory to converting them to industrial and other urban uses. However, the environmental protection implication of this strategy is obvious. From an ecological point of view, river valleys are particularly sensitive. They reflect an intricate balance between regional precipitation patterns and runoff characteristics. Unplanned interference in such systems often can have unexpected effects.

ULI Industrial Council member Ben Dyer points out that in many locations, environmental pressure is growing for total preservation of flood plain areas. In Prince George's County, Maryland, local legislation has been introduced to prohibit any structures within the 100-year flood plain regardless of the present uses permitted under the existing zoning category.

Resolution of these conflicting interests has taken various forms. At Earth City in suburban St. Louis County, Missouri, more than 1,000 acres of flood plain have been enclosed by levees. The hydrological impact of these levees has been mitigated, at least in part, by developing an artificial lake and designating open space areas within Earth City itself. ULI Industrial Council member R. John Griefen points out that Cabot, Cabot and Forbes' industrial park in Phoenix, Arizona, which is developed on land diked off from the Salt River Flood Plain, has successfully resisted flooding since its inception in 1965 and has successfully attracted industry despite its location in a floodplain. The need to quantify the trade-offs involved in such an undertaking suggests the desirability of developing a measurement or evaluation system which can be used as a basis for making well-considered decisions.

Energy Constraints

The sudden and dramatic shift from having relatively cheap energy in abundant supply to having a real or artificially induced shortage at a much higher cost has meant that future industrial development activity will reflect increasing efforts to minimize energy costs during initial development

6-4 Computerization of materials handling and heating and cooling processes is an important way of enhancing the energy efficiency of industrial processes and buildings. This picture shows the control room at Burns Harbor.

and continuing operation. In the past, capital costs of buildings, mechanical equipment, and production machinery, were the most important cost control areas. While energy costs in production processes were evaluated carefully, relatively low energy costs did not justify the same attention in plant building design. Highly efficient patterns of energy consumption in building designs will become a key concern in the future. ULI Industrial Council member Thomas C. Wolff, Jr. suggests the possibility that government will establish energy consumption maximums for new development. These specific types of new trends are easy to identify but broader trends may not yet be clear.

Recognizing patterns of industrial development will undergo modification as a result of the energy crisis and its aftermath. In March 1974 a ULI Energy Ad Hoc Committee evaluated the long-term development implications of the energy crisis. The following probable changes in industrial development were identified:

- Current patterns of large-scale industrial parks and districts will continue. However, the mix of uses may expand. More office/commercial use and the likelihood of related residential use—either within or adjacent to industrial use—will occur, reducing transportation requirements for the home/work trip.
- The need for rail-access sites and a concern for the reliability of energy supply for manufacturing operations will continue to be emphasized.
- While a shift to multistory industrial building configurations might suggest an opportunity for industry to return to smaller city-center sites, the general opinion is that operational constraints, coupled with the continuing availability of suitable and well-located suburban locations, will preclude this from happening on more than a modest scale.
- Increasingly, industries will seek locations in warmer climates, particularly in areas able to offer a reliable fuel supply and a trained labor force.
- Smaller cities located within 100 miles of major metropolitan areas, which have a surplus labor force, may prove attractive to new industry as higher energy costs cause industry to seek savings in less expensive labor markets.

Many of these trends were emerging before the energy crisis. The forces which create trends are multi-faceted; and, the concern for energy conservation will reinforce some of these trends and weaken others.

For example, access to public transportation increasingly will become an important consideration for industrial park developers. This trend is in marked contrast to the typical post-war industrial park which relied primarily on employee access by private auto. Public transit tended either to be limited or entirely absent. At Hayward Park in Alameda County, California, primary employee access is via automobile, however, a station of the Bay Area Rapid Transit system is a short distance away. At Park Forest South, the joint venture developer I.C. Industries is extending the South Suburban line of the Illinois Central-Gulf Railroad to Bennett and Kahnweiler's 1,200-acre Governor's Gateway Industrial Park.

In recent years, many public and corporate groups have been upgrading bus system access to existing and recently developed employment centers. At Peachtree Corners in Atlanta, for example, the developer initiated the Parkway Coach system—a series of busses which operate on regular schedules throughout the project with connections to the Atlanta public transit system (MARTA), shopping centers, and points in between. Approximately 40 to 50 trips per day are provided. The planning for this system was initiated by the developer and is financed by tenant subscription in the development. Tickets are available through the subscriber, who either gives them away or sells them. Also, part of the planning process is an intracommunity transportation network which provides pathway connections for golf carts and bicycles to all portions of the project and the county and state park system. In Flint, Michigan, an effort was made to connect General Motors' assembly and manufacturing plants with residential areas by means of minibuses for call on an individual basis.

While these efforts have met with varied success, their utility in an era of high and increasing energy costs becomes readily apparent. ULI Industrial Council member Otto Pongrace notes, however, that the major problem with these efforts has been the negative reaction of the employees whose life style has been patterned on the use of a personal vehicle. "A particularly vexing problem for the transportation planner is the employee's preference for combining personal business and pleasure trips with the work trip."

The trend toward integration of uses within the industrial development environment also will pro-

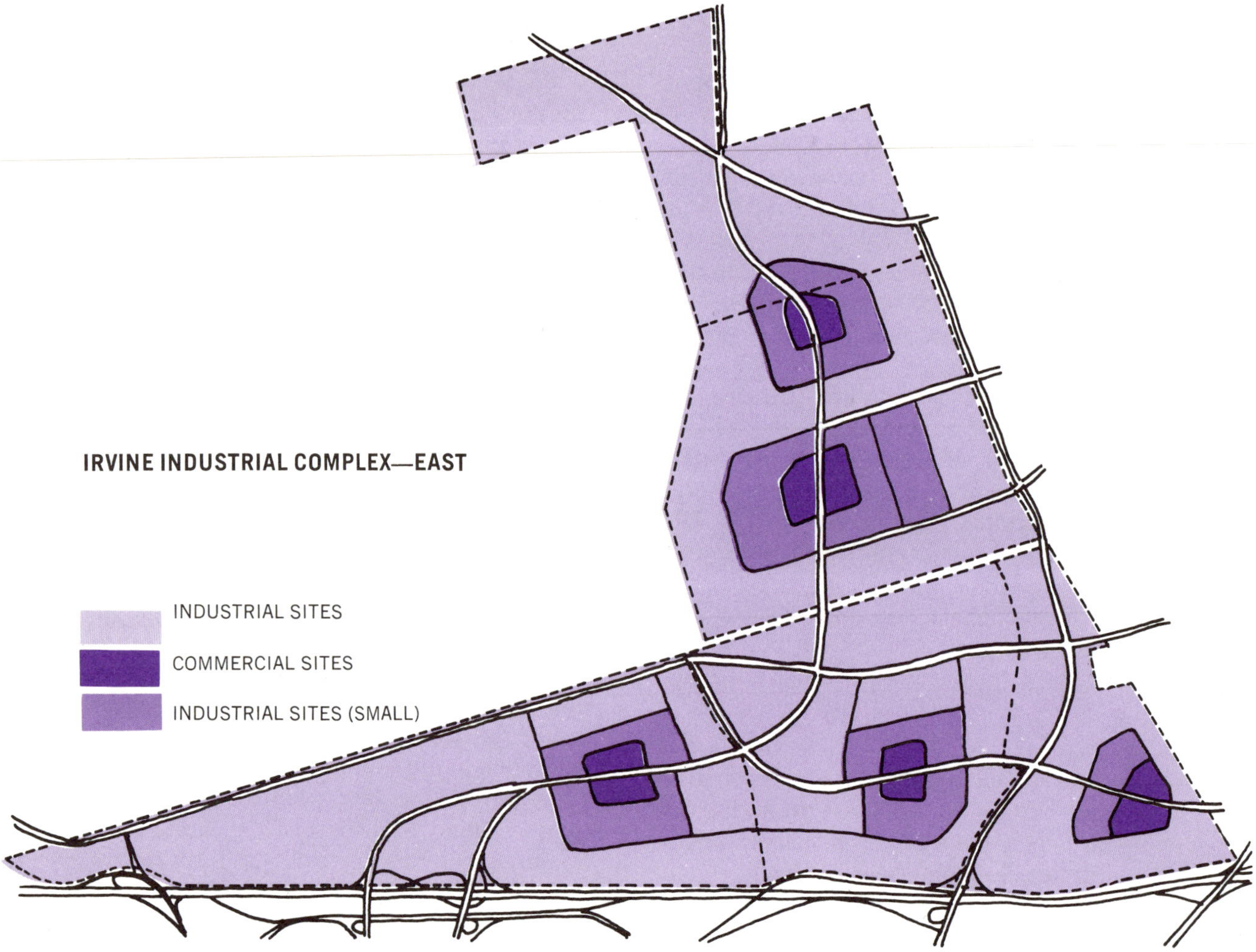

6-5 Irvine Industrial Complex East—this project features an innovative integration of supporting uses into the industrial environment.

vide more efficient use of energy by reducing employee travel needs during the work day. When basic shopping facilities such as restaurants, banks, and drug stores are available within walking distance, it is not necessary for employees to patronize more distant commercial areas using private automobiles.

Most of the basic criteria considered to be essential elements of a restructured industrial environment have been incorporated into Irvine Industrial Complex East. Here, a series of industrial neighborhoods are planned with dimensions of roughly one-quarter mile per side, centralized parking areas, and a radial road system. Each neighborhood, comprising 200 to 300 acres of light industrial and/or office warehouse facilities, is to be oriented around a commercial center. The periphery of each neighborhood will be within easy walking distance of the commercial center where the park employee may have lunch, do convenience shopping, and leave his car for the day. Ultimately, a mass transit stop will be located at the center permitting even further reduction of auto use. A system of landscaped pedestrian paths and bikeways will constitute the framework of an open space network within the overall employment center and will link it with the rest of the Irvine Community.

Overall Implications

In terms of project design, environmental and energy considerations are requiring a more comprehensive and sophisticated approach to industrial development. Higher densities, integration of various land uses on a single site, provision of several transit modes, and redevelopment in existing urbanized areas, are all ways which may be used to accommodate the new constraints. Industrial developments will be likely projects for application of experimental energy conservation techniques in-

6-6 Neighborhood Commercial Center—Irvine Industrial Complex East. Each center is designed to provide for the commercial needs of a 200-to-300-acre industrial "village."

cluding solar heating, central heating, air conditioning, and technologies designed to utilize a variety of fuels and to recycle wastes.

In addition, the developer will need to reevaluate his short-term interest in the project. Since it is becoming increasingly difficult to assemble suburban sites for industrial use, obtain necessary approvals, and provide essential utilities, the developer will intensify his search for practical ways to increase values and extend the economic life of his investment. As ULI Industrial Council member R. John Griefen suggests, "A significant percentage of venture capital may be required to meet environmental or other regulatory requirements, and these must be built into the cost of development." The implementation of various environmental protection acts along with local actions to manage growth, regardless of the extent to which unreasonable specifics are eliminated by administrative reconsideration or adjudication, will make approved projects more valuable. The industrial developer must be responsible. He must seek ways to assure that approvals, when obtained, provide maximum flexibility for intensification of use over time even if not appropriate with initial development.

At Miami Valley Industrial Park near Cincinnati, ULI Industrial Council member William Robinson has used the technique of attaching covenants to individual deeds rather than filing covenants with the plat. This method allows the developer to be flexible and to meet technological changes, while keeping within the spirit of the covenants.

At the Denver Technological Center, protective covenants drawn up by the developer insure a low-density building profile and an open space environment. Nevertheless, the option exists to increase density by decking over at-grade parking lots. This option may not be available for many existing parks, but provision for it should be made for those being designed now. ULI Industrial Council member John O'Donnell, however, suggests extreme caution with regard to higher densities. He believes that even with the new constraints, public officials will be reluctant to allow higher densities. Furthermore, "high cost and problems with appearance may deter use of parking structures except in special circumstances."

This last example tends to focus on a design detail which at first glance would seem insignificant to the broader issues. However, this very kind of attention to detail, and the exploration of every opportunity for wise and efficient first use and planned redevelopment (or second-phase use of land) should be the industrial developer's concern. This longer range view represents a new trend which is responsive to environmental concerns and energy conservation.

Integration of Land Uses

In physical terms, the most noteworthy trend in industrial development over the past half-dozen years is the increasing integration of other land uses within the industrial environment. This trend is exemplified by the increasing number of activities found both within the parks themselves and with respect to the placement of industrial areas within the larger community. Such integration has been prompted by recognition of the fact that the traditional industrial park, while carefully providing for efficient operation for industrial purposes, typically has ignored many fundamental employee and tenant needs and certainly has not included any residential uses. In most older parks, commercial, professional, business service, and recrea-

6-7 Decked over at-grade parking lots represent one way of increasing density in industrial projects. This view is of Denver Technological Center.

tional facilities have been excluded completely or provided on an individual basis as when a company operates a cafeteria or lunchroom/canteen or operates a check-cashing service. The typical location of industrial activities was on land for which there was only marginal demand for residential uses.

The older industrial park design philosophy which advocated tightly controlled uses represented an important advance from the early industrial districts with their casual intermix of commercial, residential and industrial uses in unattractive and polluted environments. With the industrial park developer now exercising stringent control over every aspect of the industrial park design and operational processes, it has become possible to create a better environment for employees as well as industries without sacrificing overall environmental quality. Careful design control now makes it possible to locate many industrial activities close to the houses of employees.

New Uses Within the Industrial Employment Center

A variety of nonindustrial land uses is being found with increasing frequency within industrial parks. These uses include offices; convenience commercial; professional and business services; public and quasi-public uses; food service; and transient lodging facilities. To a lesser degree, residential uses are being incorporated in, or situated in proximity to, industrial facilities. This latter situation has not occurred in a sufficient number of projects to qualify as a trend. Nevertheless, the success of large-scale, mixed-use, urban complexes with a mix of residential, office, and commercial uses suggests that the concept is equally logical for developments in which industry is the basic use. The degree of integration versus coexistence will vary with each use category and with the size of the project.

Public and Semipublic Uses

Public and semipublic uses including fire stations, fire training facilities, and police stations represent the logical extension into the industrial environment of important public services required by industry. Little or no incompatibility exists between these uses because of the mutuality of benefits derived by both developer and community. Police and fire services were the first nonindustrial uses to be provided within the park setting. In the case of the Los Angeles Industrial Center (LAIC), the city of Compton built a fire station and a police station on land in the park. To the city, LAIC represented an important addition to the tax base. To the developer, the availability of fire and police services within the park represents an important asset to the project and is a valuable selling point to prospective land purchasers. Hospitals, health clinics, state employment agencies, motor vehicle licensing and driver testing centers, are a few examples of public and quasi-public uses which might be located appropriately in planned industrial employment centers.

6-8 City of Compton Fire Station—Los Angeles Industrial Center. Cabot, Cabot and Forbes donated the land for this public facility so that it would be located within the LAIC.

Recreational Uses

Provision of recreational facilities, developed both publicly and privately, represents another growing element in multi-use industrial developments. To date, most efforts have consisted of private developers providing an array of recreational facilities for the use of park tenant employees. Facilities include golf courses, tennis courts, health clubs, basketball courts, and other amenities intended to introduce a new dimension to the industrial setting. It is rare to find public open space and recreational facilities in industrial areas. Even though industry is a major source of tax base, industrial employees rarely are provided with a nearby public park.

One major corporation, Anheuser-Busch, Inc., has sought to combine industrial production with a leisure time recreational complex open to the general public. Growing out of its experience with a rathskeller hospitality center at its St. Louis brewery, and its Grants Farm estate, the company has expanded its entertainment concept to include the development of major tourist attractions called Busch Gardens at several of its new plants. Size of these attractions has grown over time since the first was built at the 85-acre Los Angeles plant site. Most ambitious of all has been the theme park called The Old Country, located in Williamsburg, Virginia on 600 acres adjacent to the brewery. Opened in mid-1975, the park features re-creations of English, French, and German villages together with a broad array of theme rides and concessions in a landscaped, garden setting. These facilities provide substantial publicity for the manufacturer's product as well as constitute successful recreational attractions in their own right.

6-9 An archway frames a restaurant entrance in the French village portion of Anheuser-Busch's theme park, The Old Country.

Commercial and Office Uses

Commercial facilities now being placed within industrial employment centers consist primarily of restaurants and overnight accommodations. The availability of these facilities within an industrial park represents an important convenience to firms, employees, and visitors who might otherwise have to travel some distance to find them. In the early years of exclusive industrial zoning and industrial park planning, not only were the needed service commercial uses conspicuous by their absence but frequently they were excluded by legislative mandate. Currently, a trend toward integration of service commercial uses for the employees is emerging. Fast-food services, banks, service stations, auto repair facilities, barber and beauty shops, dry cleaners, commercial recreation facilities, medical and dental offices, and health clubs are obvious uses for inclusion in an industrial setting. At the Nonconnah Park in Memphis, for example, the presence of a hotel was recognized as vital from the outset. The hotel provided rooms, meeting facilities, restaurants, and limited shopping facilities both for employees and for visitors to the park. The hotel was situated at the entrance of the project to draw patronage from the nearby Memphis International Airport as well as from the development itself.

At the Irvine Industrial Complex East, six commercial sites have been planned within the 2,068-acre industrial area. Each site is 30 acres or less in size and has been designed to meet the limited shopping and restaurant needs of adjacent industrial uses. A further example of this trend is found at Irvine's original 2,310-acre industrial area. Restaurants and other limited commercial activities in the original Irvine park are being located now on parcels bypassed during the initial phases of development.

Centex Industrial Park, in the Chicago area, has three hotels, a commercial indoor tennis facility, fifteen restaurants, a bank, six service stations, a convenience shopping area, an industrial health clinic, and a series of office buildings.

The expansion of office buildings or office uses in industrial parks or employment centers is proceeding rapidly. Traditionally, the planned office

6-10 Brookhollow Hilton Inn, Houston, Texas—this 200-room hotel with restaurant and club facilities is located on a 3.5-acre site at the Brookhollow Business Park. The facility was designed to accommodate both visitors to Brookhollow and transients. Brookhollow Business Park has 170 acres and 1.3 miles of freeway frontage.

park has been considered a commercial land use and for that reason has not been covered in this text. Nevertheless, office functions typically occupy 10 to 25 percent of most industrial buildings. Many industrial parks are becoming hybrid employment centers with up to half, or more, of the total site area occupied by offices. The nature of this trend is suggested by the experience of Crocker Industrial Park, San Mateo County, California. The park was begun in 1959. Initial tenants had only small office requirements. By the late 1960s, the developer had experienced increasing demand for additional office space from existing tenants and had potential *office-only* tenants seeking space. At Cross Westchester Park in Elmsford, New York, approximately one-third of total building area is occupied by office users. Denver Technological Center in Englewood, Colorado, probably represents the best example of the bridge between the industrial park and the office park. The tenants are drawn from high technology industries. The center has more office users than warehousing or manufacturing users. In the future, drawing the line between an industrial park and a combined office/industrial development clearly will become increasingly difficult.

A particularly sophisticated example of an integrated land use pattern combining office/retail with industrial functions is to be found at Bennett and Kahnweiler's Inverness in Arapahoe County, Colorado. Separate parcels have been set aside for a commercial center with freeway access and for a hotel convention center at an interior location. The industrial uses themselves are divided into office and office/light assembly categories. A golf course ties together the entire 602-acre project and provides a focus for employee activities. This project has all the elements of the multi-use developments referred to earlier; however, industrial is the primary land use.

Residential Uses

Probably, the pre-industrial revolution historical relationship between industrial and residential uses, perhaps best exemplified by artisans and craftsmen living and working in a single structure, will not become viable except in certain unusual circumstances. For this phenomenon to occur, major shifts in employer and employee attitudes about the production process would be required. Nevertheless, relating residential uses to employment centers in various ways appears to be a future trend. Most ULI Industrial Council members feel that residential uses will coexist rather than integrate with industrial developments.

For example, one situation requiring integration

has been the incorporation in many industrial developments of apartments for project managers and guard personnel. Each mini-warehouse developed by Public Storage, Inc. of El Monte, California, includes a 900-square-foot apartment built into one of the basic buildings. Because security is a major consideration for mini-warehouses, the rationale for housing a full-time manager on the property is very great. At the other extreme is the example of Anheuser-Busch, Inc. in Williamsburg, Virginia, which has set aside 2,900 acres of its brewery property for development of Kingsmill on the James—a residential community which will ultimately house 12,000 to 15,000 persons. Because of the price structure at Kingsmill, few residents are expected to be employed at the plant. Under construction since 1973, the development

6-11 Winster Fax townhouses, the first residential development undertaken by Busch Properties on its 2,900-acre Kingsmill on the James property in Williamsburg, Virginia.

has been planned as a PUD combining a variety of residential unit types, commercial areas, and an array of recreational facilities including a golf course and a marina. Forty percent of the site is to remain permanent open space.

ULI Industrial Council Chairman Marshall Bennett notes that at the Centex-Schaumburg Park west of Chicago, 40 acres were deleted from the original 675 acres planned for industrial use. This land was rezoned to multifamily use. One thousand units of middle-income housing are being constructed with funding by the Illinois Housing Authority. In this instance, many of the residents probably will be employed in the adjacent industrial area. Similar undertakings of this nature appear likely. However, actual provision of housing for employees will be dependent upon demonstrable economic benefits to be derived from housing employees in immediate proximity to their place of work. ULI Industrial Council member Otto Pongrace is quick to point out that these ideas are not new, and that the company town and the company store were commonplace during the initial phases of industrialization in the United States but subsequently lost favor. The idea of the company town is being recast in new form.

Large-Scale, Multi-Use Projects

Increasingly, PUDs, new communities, and other large-scale, multi-use developments have incorporated industrial areas within their boundaries. The purpose of this incorporation has been to generate additional revenue from the sale of land and to create more balanced communities.

The policy of ULI's New Communities Council has been that, "Provision of a wide range of employment opportunities for new community residents should be a primary objective." Two well-known new community developments in the United States—Reston, Virginia and Columbia, Maryland—each have designated 17.6 and approximately 18 percent, respectively, of their total land area for industrial use. With these and similar projects, residential and industrial areas have been placed in close proximity to each other and frequently are interconnected by means of pedestrian and bicycle

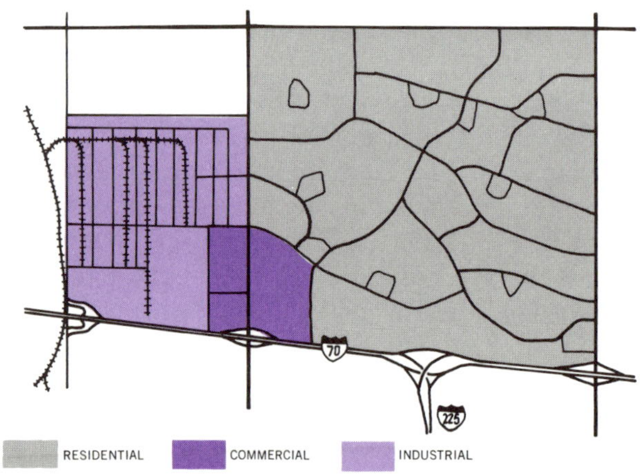

6-12 Montbello, Denver, Colorado—many residents live in close proximity to their places of work.

trails. Enforcement of stringent performance standards and use of careful building placement and landscape design have mitigated potentially deleterious impacts.

6-13 Photo depicts use of parkland to buffer residential and industrial areas.

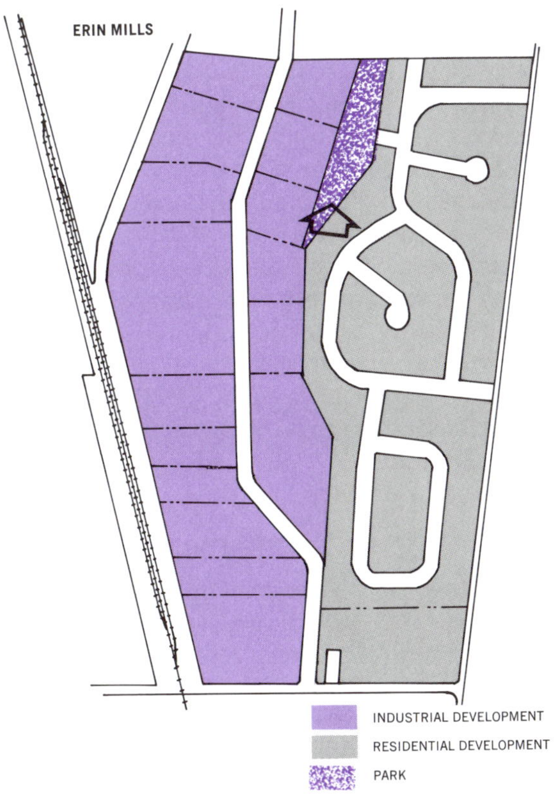

6-14 Site plan—Parkland Buffer. Arrow on the site plan indicates the perspective in the photograph above.

Montbello in Denver affords one of the best examples of a project designed to house much of its work force within the development area. The industrial portion of Montbello—600 acres—is adjacent to an area planned for 7,000 moderate-income, single-family units and for 6,000 multifamily units, plus attendant commercial areas.

At the 6,000-acre Elk Grove Village, west of Chicago, 2,250 acres are devoted to Centex Industrial Park, 2,700 acres are in residential and commercial uses, and the remaining 1,050 acres have been set aside as recreation and open space areas. The percentage of employees working in the industrial park while residing in Elk Grove Village cannot be precisely determined, but the proximity of the two land use areas suggests the likelihood of a reasonably high percentage.

Frequently, buffer zones are used to separate and define the perimeters of industrial areas in large-scale projects, At Cheyenne Mountain Ranch, a 3,500-acre new community in Colorado Springs, these buffers have taken the form of wide open space strips which follow areas of steep topography and natural drainage courses. At Erin Mills in Toronto, Canada, the buffers are narrow landscaped pedestrian pathways. In both instances, an attempt has been made to unite divergent activities on the basis of a single common denominator—green space.

A countervailing point of view takes the position that industrial areas should remain isolated from residential and commercial sections of the new community. One argument in favor of this approach has been the desire to maintain maximum project flexibility. An independently developed industrial district accommodates change more easily than an industrial district which is joined to residential and commercial areas. The more fine-grained the development plan and the more detailed the level of interaction, the less flexible the plan. Vandalism is a problem when industrial activities are placed too close to residential neighborhoods. Isolation can be reinforced by using road systems, natural barriers, and open space buffers to separate various activities. Coexistence, as opposed to integration, is exemplified by the initial section of the Irvine Industrial Complex. The Complex's 2,300 acres of industrial land have been developed with only a limited effort to establish a close relationship with the residential villages under development nearby. Access to and from the complex is exclusively by auto.

Despite differences of opinion as to how employment centers should be incorporated into large-scale community development projects, the economic and social advantages of a variety of land uses within an industrial development will remain a valid and growing future trend.

Specialized Industrial Complexes

Concurrent with the trend toward greater internal and external integration of other uses into industrial environment has been the development of industrial parks designed for highly specialized and inter-related uses. Among the more significant examples of industrial developments designed to serve specialized segments of the market are warehousing and distribution parks, research and technology parks, and business parks. In addition to industrial developments catering to a single specialized market, a variety of industrial subgroupings have unique requirements that are being accommodated as well. In both instances, the validity of specialized industrial developments as a future trend hinges on the following important points:

- The potential market or the design must be flexible enough to meet several needs and must be sufficiently large to support a specialized aggregation;
- To justify the provision of specialized engineering and design features, the development's characteristics must be unique to the aggregation;
- Some apparent advantages must be derived by locating the uses in close proximity to one another; and,
- The specialized user group must be willing to pay for special improvements.

The specialization feature can be a very valuable marketing tool in and of itself. Disregarding other factors, a specialized market subgroup can be attracted simply on the basis of a strongly targeted marketing program.

Several problems associated with this kind of development need to be kept in mind. Concentration on one aspect of the economy exposes the developer to the fluctuations of that particular market. As a result of this circumstance, such projects probably will be developed as sections of a much larger development rather than as wholly self-sustaining undertakings which would be adversely affected in the event of change or decline of the industry or subgroup. To this point, ULI Industrial Council member John O'Donnell adds that special improvements required in projects of this kind are, by definition, costly. "Any users moving into a development, but not needing these facilities, probably will be unwilling to help amortize them in the form of higher land prices or higher rent."

ULI Industrial Council member Ray Sydansk suggests that "in contrast to specialization and specialized design, many industries are turning away from custom-designed buildings for specific operations. This approach resulted in odd sized and shaped single-purpose structures. While this custom-design practice is still common, the direction is toward construction of general purpose buildings that are constructed as a protective shell or envelope for any type of operation—not just the one currently contemplated."

The simplest way to describe this phenomenon and its future prospects is to identify several of the most common types of specialized industrial complexes which have evolved already or might be expected to develop.

Warehousing and Distribution Centers

Many of the early industrial parks fell into the warehousing and distribution center category because these activities were either the sole or the predominant use. Such parks place a heavy emphasis upon a well-developed internal circulation system including both highway and rail transportation modes. A strategic location with ready access to the region is essential. Crocker Industrial Park in San Mateo County, California; Los Angeles Industrial Center, California; and, Great Southwest Industrial Park in Dallas are typical examples.

Technology and Research Centers

Parks with a campus-like atmosphere have been developed in areas in which the local economy is

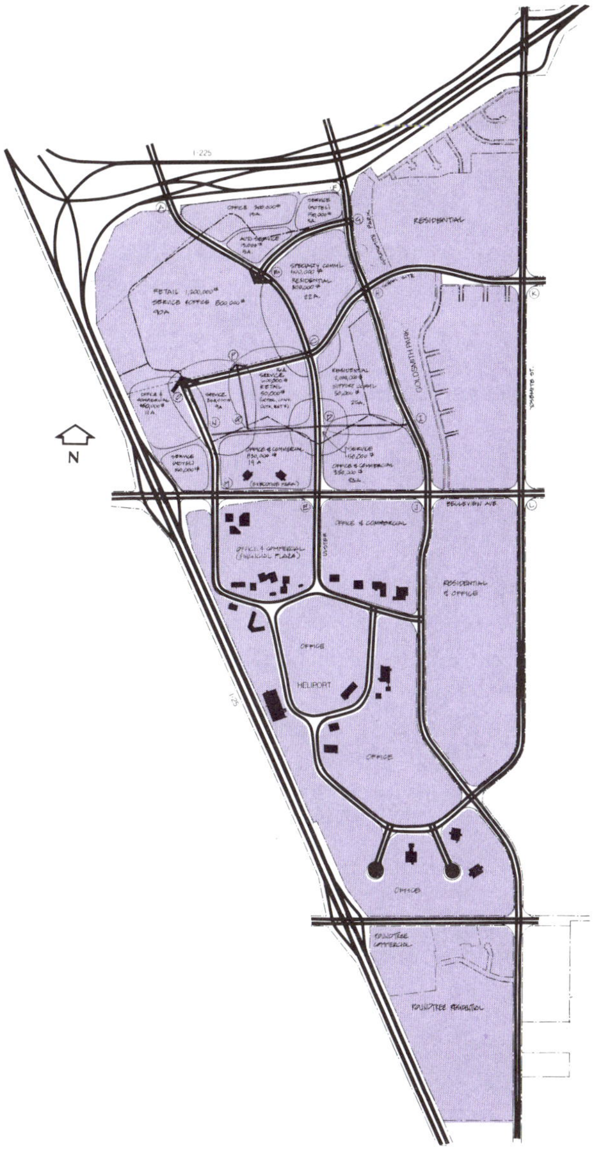

6-15 Denver Technological Center—an outstanding technology and research park.

oriented heavily toward scientific research and high technology industries. These parks are characterized by spacious and extensively landscaped settings, excellent architectural quality, rigorous exclusion of incompatible industrial uses, and heavy emphasis on aesthetic and environmental considerations. Examples include Stanford Research Center, Palo Alto, California; Denver Technological Center, Denver, Colorado; and Westgate Park, McLean, Virginia. ULI Industrial Council member John O'Donnell notes these examples are the exception rather than the rule. From his perspective, many research parks first announced in the late 1950s never developed as originally planned. Those which succeeded, did so because the scope of the research park was broadened to include a wider range of users.

Special Needs Subgroupings

Certain industrial activities have specialized requirements—utility, lot configuration, and locational—which if designed into a park from the outset, can be utilized to attract a major share of an industry subgroup. Denver Technological Center, for example, has set aside a portion of the park for exclusive use of firms placing a heavy reliance upon computers. This particular group needs a very reliable power supply with very strict limitations on wattage fluctuations. By providing major on-site energy generation facilities featuring precise power output controls and standby generation, the developer has offered a feature that is attractive to tenants in the computer industry. If possible, highly specialized users should be committed before project initiation. Any speculative development for a specialized market should be approached with great caution.

The key element of a first attempt at industrial condominium redevelopment conversion of a loft building in New York City was the perceived benefit of aggregating printing and printing-related industries in close proximity to one another. These benefits include the proximity of binders, graphic services, paper supplies, and other technical and intermediate production services.

The Brookhollow Home Improvement Center, developed by the J. C. Nichols Company, occupies a portion of the Brookhollow Industrial Park in suburban Kansas City. The development consists of 105 acres specifically designed to accommodate the special display, office, and warehouse needs for retail and wholesale distribution of furnishings, equipment, and materials for residential decorating and construction. The wholesale/retail sales operation, pioneered in the furniture business, requires the use of a warehouse/showroom where the buyer selects merchandise and loads it directly into his car or truck. At the Brookhollow Home Improvement Center, a curving landscaped boulevard surrounds a series of centrally located reflecting pools. Commercial display and office areas within individual buildings are oriented toward the boulevard and reflecting pools while truck loading and rail access is provided at the rear of each structure. This campus-like environment is enhanced by use of a uniform architectural style.

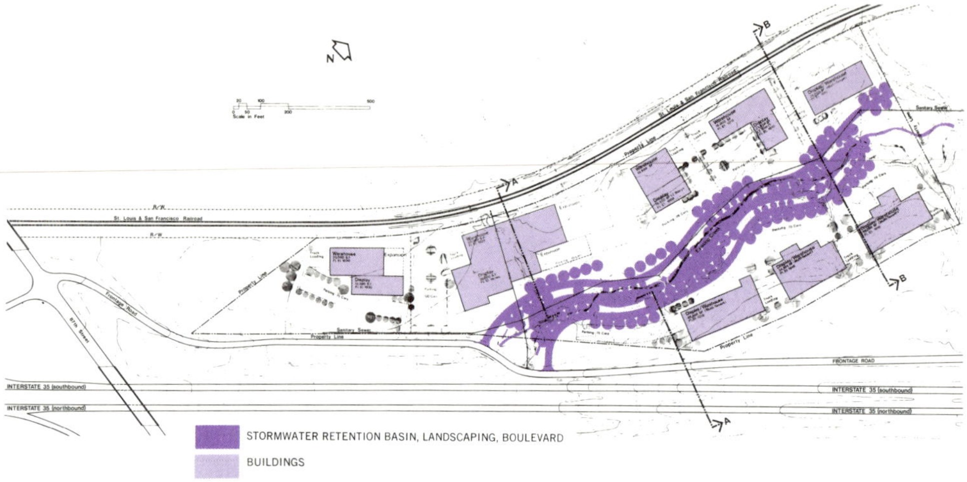

6-16 Brookhollow Home Improvement Center, Johnson County, Kansas—a park designed for the specialized needs of furniture and building materials wholesalers and retailers.

The rail service requirement for this type of operation precludes wholesale/retail warehouses from locating in conventional retail centers. This subgrouping can be considered either industrial or commercial in terms of land use categorization. However, its functioning process makes it more compatible with industrial land uses than with traditional retailing activities.

At Pacific Intrapak, food processing, warehousing, and distribution functions have been integrated on a comprehensively planned 110-acre site. Located in Brea, California, the project features a central energy plant, an office building for firms wishing to locate on site, 1.2 million square feet of processing floor space, 800,000 square feet of cold and dry storage, and 50-foot-high storage areas utilizing automated stacker cranes. The primary advantage of projects of this kind is the efficiency gained through the joint and flexible use of facilities required in common.

Industrial subgrouping as a development technique is at an early stage. The potential inherent in this approach might be applied to a substantially wider range of industrial activities in the future. The meat packing industry is an example. Over the past 30 years, this industry has shifted from the great packing centers of Chicago, St. Louis, and Kansas City, to the meat producing areas themselves. Meat packing is now concentrated in the smaller cities and towns of the Great Plains. The by-products subgrouping of the meat packing industry represents another example of an industry subgroup with problems or needs which might be solved jointly. Hide and tallow processors, and many others in this subgroup, could cluster in one community provided a developer would solve the industry's common sewer and solid waste problems. A central pre-disposal treatment facility would be the sales incentive.

ULI Industrial Council member Otto Pongrace feels very strongly that this specialized clustering concept will be important in the future. He points out that subgrouping has always occurred but notes the reasons were different—the need for proximity of interdependent functions was the factor. In contrast, he uses the example of a group of foundry operations that could solve its environmental and appearance problems by joint action thus allowing this type of development in areas or on sites that might otherwise be refused public approval. The environmental and energy concerns discussed previously may identify a significant number of specialized subgroups whose needs can be accommodated by aggregation. As previously discussed, the future strength of the specialized clustering trend will depend upon not only the willingness of industry subgroups to precommit to participate in such an effort but also upon the developer's ability to design an offering which is attractive on a cost-saving and problem-solving basis and allows for equitable distribution of costs of special improvements.

Meeting Employee Social Needs

Employee social needs can be viewed from two perspectives—the work functions which encompass the work day itself and the remainder of the employee's time. Both perspectives have implications for the industrial developer because an em-

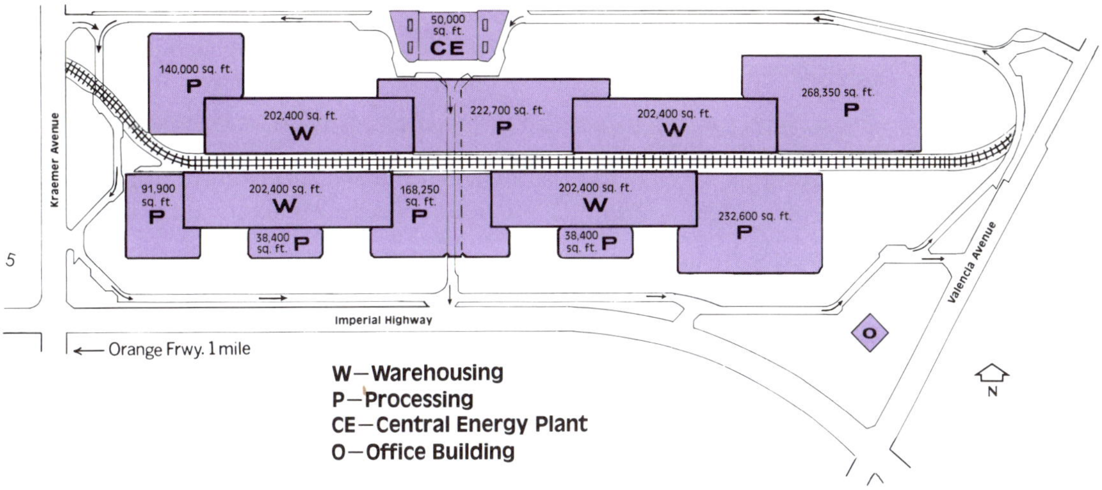

6-17 Pacific Intrapac, Brea, California—integrates food processing, warehousing, and distribution functions.

ployee's well-being is dependent, in large measure, upon his satisfaction both with his working and leisure environments. To the degree that working and leisure environments can be affected positively by industrial developers, steps are being taken to enhance these factors.

The Work Function

Since the beginning of the industrial age, the prevailing approach to worker productivity has been characterized by the application of two basic concepts—the assembly line and labor specialization. Hand craftsmanship—an individual creating an entire product—has been replaced by assembly line methods—an individual responsible for only a small portion of the production process. While this approach is considered to be responsible, in large measure, for world economic growth in the past century, it has been criticized for eliminating important human values from the production process. For many workers, it is no longer possible to look upon something which he himself created with pride. However, a shift away from specialization toward a return to the concept of craftsmanship currently is underway. The implications of this trend for industrial development are not yet clear, but surely both industrial building design and servicing will evolve to reflect this new direction. Entirely new systems of materials handling and industrial production may well be required if this concept takes hold.

Nonwork Activities

The nonwork activities of the industrial employee are similar to those of the rest of the population. Traditionally, however, these activities have been either excluded entirely or severely controlled within the employment environment. While many industrial parks are both attractive in appearance and highly efficient in operation, only recently have they been considered as more than just a place to work.

Historically, the first to recognize the nonworking needs of employees were employers who assumed a paternalistic responsibility for the welfare of their employees. Many such companies were owned privately by individuals who felt responsible for the welfare of their employees. Examples of company concern with employee well-being may be found as early as the 1840s with the development of the textile industry in Lowell, Massachusetts. This paternalistic view fell into disrepute during the labor movement days. A revived manifestation of this perspective is the recreation area—including golf course and clubhouse—owned and maintained by the S. C. Johnson Company of Racine, Wisconsin for the benefit of its employees. However, this facility and others like it are typically located at some distance from the industrial site itself. The modern concept of integrated recreation and other services is exemplified by the Connecticut General Life Insurance Company's suburban headquarters complex in Bloomfield, Connecticut. Built in the late 1950s, the company provided a wide range of employee amenities on site, including outdoor recreation facilities for field sports, tennis, basketball, and indoor facilities for bowling, billiards, cards, a smoking lounge, and a boutique shop for personal services. Many other examples might be cited. In

Minneapolis, for example, the 3M Company, Control Data Corporation, and Honeywell have all developed major recreation facilities for the use of their employees.

The Nonconnah Corporate Center in Memphis, Tennessee provides an example of this approach as applied to the physical layout of a multi-tenant industrial park. Developed by a group of local investors—Harris, Long, and Gates; and Hunt, Daily and Bagley—this park was designed to encourage maximum individual group interaction. Freestanding office and office warehouse buildings are interconnected by means of pedestrian paths which converge on grouped facilities used in common by all park tenants, including restaurants, hotels, shops, auditoriums, multimedia forums, exhibition spaces, health clubs, and meeting rooms. The plan is designed to require both project resident and visitor to pass by, if not to use, common facilities which provide a social focus for the entire development. Paying attention to the social needs of employees hopefully will result in greater acceptance of the project and will contribute to its economic success. Other projects incorporating similar features are now in the development stage. The industrial neighborhood center concept planned for Irvine Industrial Complex East in California is a particularly ambitious example.

These concepts are being applied to internal building design as well. Normally, industrial office buildings have no social focus; the isolated tenants have little opportunity to meet one another. At Nonconnah, individual multi-tenant buildings have atriums which perform as subcenters for social interaction. Filled with tropical plantings which provide a pleasant view from adjacent offices, these spaces provide circulation (elevators, stairs, corridors), utility functions (toilet rooms, service spaces), and socialization (dining, visiting, observing). As design elements they are more than just lobbies.

The provision of commercial and service facilities such as service stations, auto repair, shops, banks, cleaners, barber and beauty salons, and health care facilities in or near industrial developments has been discussed previously. This trend has evolved in response to employees' social needs, and represents an important step forward in enhancing the convenience of the working environment.

Innovations in Industrial Marketing Techniques

Perhaps no other aspect of the business of buying, improving, and marketing industrial land has changed as dramatically over the past decade as marketing techniques. Until recently, the industrial land sales business was conducted by a small number of experienced professionals who relied upon their intuition and experience to identify product demand and to market improved property. This informal approach has been supplanted by totally different tactics. Industrial developers now are adopting the sophisticated marketing and analytical techniques which have been used successfully in commercial and residential development.

6-18 The atrium, with areas of tropical planting, is an attractive element in many new industrial park buildings.

6-19 At Koll/Irvine Center, industrial condominiums have been merchandised like houses. Finished models show the various configurations available and will be sold later. Note the extensive landscaping and the standardized design.

The Model Complex

Currently, some developers are building the industrial equivalent of the residential model home complex—buildings which are erected and furnished, and grounds which are landscaped to demonstrate the actual appearance of space being made available in the project. The project manager's office is located within the model area. Once the rest of the project has been marketed, the model units are sold. This technique has applicability particularly when the developer is marketing a relatively small product in a limited number of unit configurations.

In the case of the Koll/Irvine Center industrial condominium, being developed by the Irvine Company in Costa Mesa, California, six different unit configurations—ranging in size from 4,000 to 8,000 square feet and located both in attached and freestanding buildings—are clustered at a highly visible location near the entrance to the project. Prospective purchasers may tour the model complex, and then select their own unit from the site plan which indicates both the location and type of each of the 101 units to be incorporated within the 51.4-acre site.

This technique has evolved in response to the increasing sophistication of industrial property purchasers and is most in evidence in highly competitive locations such as southern California. The technique may or may not receive widespread acceptance throughout the rest of the country.

The Industrial Condominium

Conventional lending institutions have been reluctant to make funds available for industrial condominium developments in the past because of the traditional risks inherent in small business activity and the relatively new and untried nature of the concept itself.

Many lenders do not feel that a small businessman, the typical purchaser of an industrial condominium, will have the same degree of commitment to his business mortgage as to his home mortgage. Lenders prefer to limit their industrial lending to larger and more established firms. The recent success of some office condominiums provides a track record against which the potential success of industrial condominiums may be compared. According to many persons active in the field of industrial marketing, industrial condominiums is an idea which gained prominence in 1975 and which will receive widespread application in the future. Many industrial property marketers feel this technique was about to receive widespread market acceptance when high interest rates in 1974 virtually dried up all available large source funds and delayed implementation of projects under more traditional financing.

The Irvine Company, which financed Koll/Irvine Center internally, was not tied to the money markets. Individual units in the project require a minimum 15 to 25-year mortgage period. With the success of this project, the company plans to initiate additional industrial condominium projects in the future.

Speculative Building Programs

Many industrial park developers now maintain an inventory of speculative industrial buildings available for immediate occupancy. This trend has grown because of an increasing awareness that many firms require space on short notice. Unless the market is overbuilt, the developer who is ready to meet that need will find that the prospective tenant is willing to pay premium prices and to modify his space requirements, should that be necessary, in order to obtain quick occupancy. This strategy requires the maintenance of a full product line so that a prospective tenant for any scale facility planned for accommodation within the park can be located immediately. Since the Irvine Company adopted this policy in 1972, it has kept approximately 400,000 square feet of unoccupied office and warehousing space in the Irvine Industrial Complex available for lease or purchase on short notice. These buildings are constructed as shells and the interiors are finished to a client's specifications. As soon as a category of industrial property is exhausted, the company initiates work on a new

6-20 North Anaheim Dunn—an example of a small-scale standardized industrial park.

section of the complex. At its 278-acre Pacific Gateway Center in Los Angeles, Cabot, Cabot and Forbes initiated the development process by constructing an inventory of 35,000, 42,000, 65,000, 72,000 and 94,000-square-foot buildings ready for immediate occupancy. Located on the former site of a synthetic rubber plant, Pacific Gateway Center is planned as a parklike setting for offices, light manufacturing, warehousing, and distribution facilities. At Centex Industrial Park in Chicago, Bennett and Kahnweiler maintains an inventory of approximately one million square feet of various sizes of "spec" warehouse space.

"Spec" warehousing programs are designed to accommodate the increasing diversity of tenants found in modern industrial parks. Such a program requires an increased capital commitment on the part of the developer but also helps speed up the project build-out time. In an increasingly competitive environment, the prospective tenant who cannot find what he wants in one park—whether it be combined highway/rail access, a specific parcel size, or a specific building size—will turn quickly to another park.

The Standardized Industrial Project

Standardization, the repetition of a preplanned industrial park design concept at a number of different locations, is an innovation in the field of industrial development. One of these innovators is the Dunn Properties Company of California which packages, develops, and markets small-scale industrial projects in the 10 to 20-acre range. Each Dunn project contains a basic combination of office/warehouse space in a variety of sizes including restaurant and office space. As soon as a facility is built out and leased, it is sold to an investor.

Miniwarehouses were designed to satisfy the demand for small amounts of secure space for storage purposes. Public Storage, Inc. offers miniwarehouse projects ranging in size from 144 to 723 units. Individual storage units range from 5 feet by 10 feet to 10 feet by 30 feet. Project sites average two acres in size.

Despite inflexibilities inherent in this approach, standardization has proven successful when combined with in-process adjustment of building design based upon special needs of the local market. For instance, a building in southern California does not need the snow-load design that would be required in Denver. Likewise, the same building built in Seattle and Anchorage, Alaska will require different amounts of insulation.

The Foreign Trade Zone

The creation of noncoastal foreign trades zones prevents opportunities for industrial development. A foreign trade zone permits the importation of foreign goods for processing or assembly into a designated location where they are not subject to customs duty. The goods are taxed only after they leave the trade zone. Moreover, goods or material components of finished products imported and subsequently exported without entering U.S. Customs territory also are exempt from customs duty. In addition, a company using the zone may process its goods in such a way as to obtain lower duties and/or lower freight charges; or save on customs duty by exercising various options permissible within the zone.

Because of clear economic benefits to importers, exporters, and the locale within which the foreign trade zone is located, 15 such areas have been designated in the United States. One area is the Greater Kansas City Foreign Trade Zone. The Kansas Zone occupying 3.5 million square feet of space at three locations, is the largest Foreign Trade Zone in the country. It is unique because it is the first

6-21 Underground storage at the Greater Kansas City Foreign Trade Zone.

such zone to operate at some distance from the nearest sea or river port. Over 3.2 million square feet are located in two underground limestone cavern sites—one in Missouri and one in Kansas.

Given the attractiveness of the Foreign Trade Zone to date, an increasing proportion of industrial development probably will be undertaken in conjunction with this concept in future years.

Inverness

Central Development Group
Southeast Venturers
Arapahoe County, Colorado

Consultants:

Planning and Landscape Architecture
bde associates

Civil Engineering
KKBNA

Water and Sewer Engineers
CH^2M-HILL

Transportation Consultants
Alan M. Voorhees & Associates, Inc.

Golf Course Architect
J. Press Maxwell

Graphic Design/Marketing Communications
Genesis, Inc.

Soil Engineers
Zeff, Cogorno and Sealy, Inc.

Marketing:

Bennett & Kahnweiler Associates

6-22 Listing of participants in the Inverness development program gives some idea of the range of professional skills which must be utilized in undertaking a large industrial project. In this instance, heavy reliance was placed upon consultants.

Industrial Developer's Staff

Increasingly, the industrial developer's staff includes an expanding range of professional expertise. Highly trained personnel, each expert in a specific portion of the business, are needed to develop, construct, and manage the complex projects now being built. For example, Watson Land Company in Los Angeles has a resident community relations representative whose sole function is the maintenance of developer/community relations.

The trend, however, has not been entirely toward larger permanent staffs, but toward a more flexible arrangement for putting together a team of consultant experts. While some prominent exceptions exist, such as the Irvine Co., the practice over the last five years has been to concentrate ownership expertise on the business of development. ULI Industrial Council member Robert R. Walker, Jr. points out "many developers joint venture or sell land to subdevelopers with special expertise not available in the developer's organization. This is clearly a recognition that industrial development is just as cyclical as other real estate development—a circumstance which makes it very difficult to justify a large payroll and high overhead expense during periods of low demand." Therefore, the likely trend in the future with some exceptions will be for industrial developers to draw on outside expertise as needed rather than to build large in-house staffs.

Technological Change

Technological innovation and change has provided the base upon which U.S. economic growth has been grounded. Recently, however, many persons have begun to consider this reliance on technology as excessive. Regardless of the merits of this viewpoint, it seems clear that the United States is too deeply committed to technology to discard or even substantially alter the foundation upon which its progress and prosperity is based. Continued reliance on technology, therefore, seems inevitable but within a more humanistic context. Efforts are now being made to minimize the undesirable byproducts of industrial development such as pollution and waste of irreplaceable natural resources. Likewise, the industrial environment is being restructured to enhance its attractiveness to both employees and communities.

In the industrial development field, three areas of technological change can be expected to induce new trends—transportation, materials handling, and communications.

Transportation

Use of alternative transportation modes is based on the relationship of cost to utilitarian value. These relationships, however, are never fixed and can be expected to shift with changing technology. Transportation variables include speed, route flexibility, and cargo carrying capability and cost. In the recent past, for example, the flexibility inherent in truck transport has meant this transit mode supplanted railroads to a substantial degree. Thus, the Interstate Highway system caused a revolution in industrial location. Currently, however, new factors including higher fuel and labor costs have been introduced into the transportation equation which are having the effect of creating a more favorable economic climate for rail transportation. The future will see more careful consideration of all potential transportation modes.

Highway Access

If there has been any clear locational trend over the past two decades, it has been the location of industrial development away from water and rail and toward highway transportation as the major transportation access to be optimized. The development of the Interstate Highway System has been a highly influential factor in this trend. The need for good highway access will not diminish; however, the need for flexibility and alternate model options will be a major adjustment in this trend. As ULI Industrial Council member Otto Pongrace suggests, "The new locations will be where the Interstate Highway systems intersect with each other *and* with rail service."

Rail Service

Cost and availability of various fuels will have the most dramatic impact on transportation systems in the foreseeable future. ULI Industrial Council member William Dillard feels, "Until 1974, the trend was away from providing rail service in industrial parks. The truck permitted greater flexibility and the Interstate Highway system had dramatically improved the time and distance of truck movements as contrasted with rail. Then, with fuel costs rapidly rising, and the prospect of limited availability in the future, potential customers for industrial sites began insisting on rail service as a mandatory option."

In a period of uncertainty, an increased demand for rail service—even on the part of customers who have never used rail service—is to be expected. Rail service, however, has its own limitations not the least of which is the questionable financial viability of many of the nation's railroad companies. Resolution of the financial problems of railroads may result in the reduction of elimination of rail service to many sites presently suitable for industrial development. ULI Industrial Council member James R. Scott points out, "A railroad cannot discontinue service at its own discretion. However, new service—particularly when the service railroad is in financial difficulties—may be more difficult to obtain. Likewise, coming decades may see rather dramatic changes in the ground rules under which railroads are regulated." Good rail-served land will then become even more precious.

ULI Industrial Council member Otto Pongrace points out, "To a great extent, optimal flexibility in transportation of goods is constrained by a variety of institutional arrangements and regulatory agency regulations. For example, common ownership of truck and rail lines, which would make possible the combined provision of multimodal transportation by a single transportation company are currently precluded or constrained by regulation." The probable direction of future trends suggests that the United States can no longer afford built-in inefficiencies in its transportation systems.

In addition to increased concern for transportation options for the transport of goods, the availability of various public transit options for employees also is becoming a matter of concern. Many people feel the private automobile cannot remain the primary mode of travel for the work trip. The implications of this viewpoint for the location of industrial development are highly significant. As ULI Industrial Council member R. John Griefen suggests, "Despite the costs of providing rail service, it will be a mandatory feature of industrial development in the future. Rail likely will be used more for employee transportation as well as goods movement."

Air Transportation

This transportation mode continues to have a great potential for industrial use. Numerous instances of industrial development are occurring around commercial and general aviation airports. The need for actual linkages between commercial airports and air cargo operations, however, remains unclear. Industrial development has tended to occur around airports, probably because it is a compatible land use for communities to encourage in such areas rather than for any other reason. With good highway access and relatively cheap motor transport, immediate access to airports has not been demanded so far. In the absence of a demonstrated need for air freight availability, it is difficult to ascertain the direction of future change in this area. At least one indication may be found near O'Hare Field in Chicago. Sears Roebuck and RCA have created extensive distribution centers relying on air transport to provide distribution from a centralized location. These facilities rely upon the exceptional distribution flexibility provided by O'Hare but are not actually located within the airport itself.

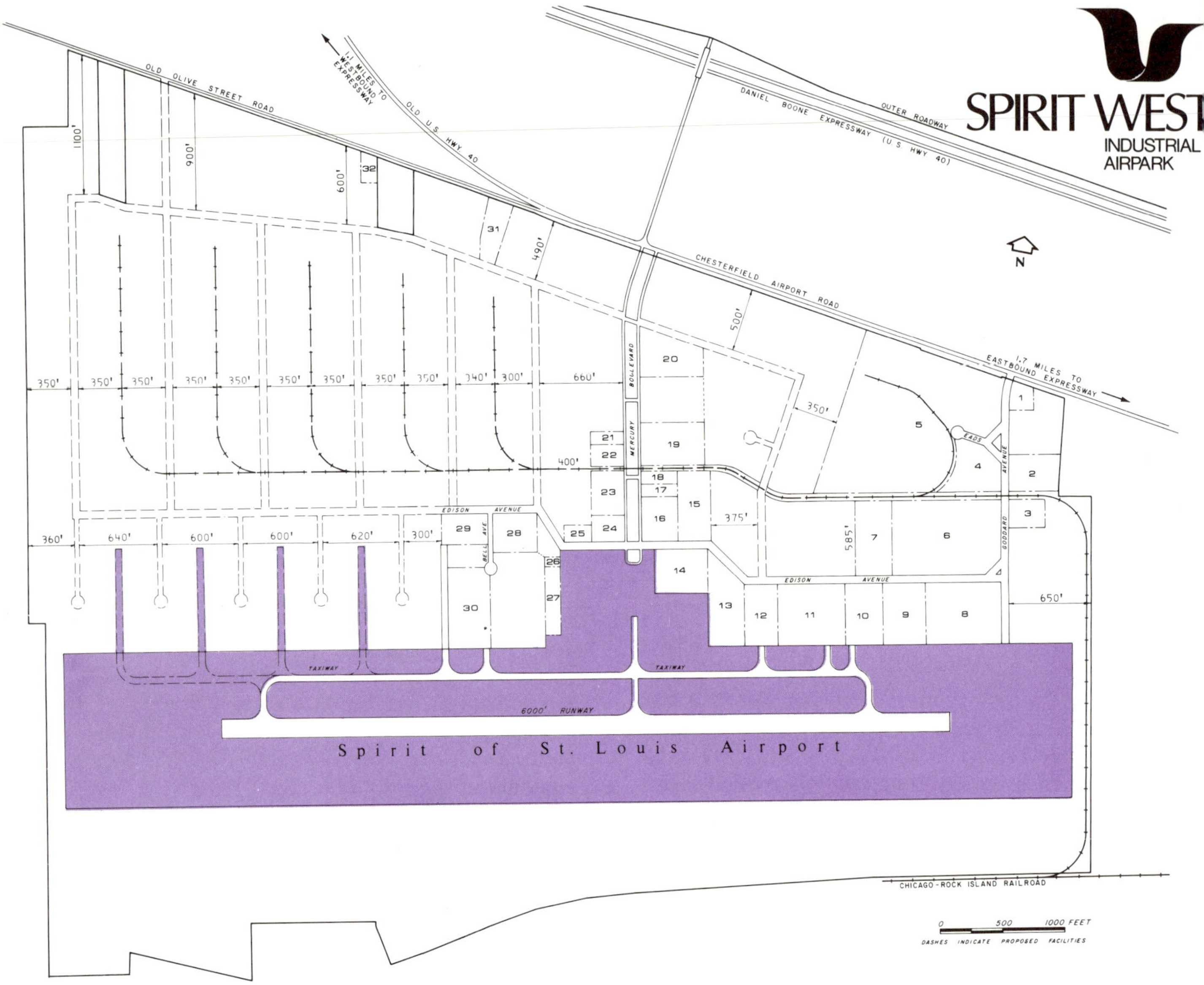

6-23 Spirit West Industrial Airpark site plan—note taxiway easement providing air access to individual properties within the park.

Airport-oriented industrial parks are being built at general aviation airports. Bennett and Kahnweiler's Spirit West Industrial Airpark in St. Louis County, Missouri has a light aircraft strip with taxiways extending into the park interior. Cabot, Cabot and Forbes' Hayward Airpark, in Hayward, California has easements between buildings for taxiways leading to the main runway, Airparks cater to firms producing lightweight high value products shipped by air freight. For industrial developments to become more closely linked to airport access, air transportation technology will need to be developed to the point that load capacity and costs are competitive with ground transport. Likewise, a materials handling breakthrough will be required to eliminate the need for motor transport as the link between the warehouse or manufacturing operation and the aircraft itself. This breakthrough would be an alternative approach to constructing a system of secondary taxiways which permit the aircraft to move to the goods.

Water Transportation

Historically, warehousing and industry have located adjacent to port or river facilities. Prior to the advent of the railroad, shipping by water was the only cheap, high-volume transportation system available. Industrial/port relationships continue to remain important and numerous innovations are being introduced. The major changes are concentrated in the areas of ship size, harbor configuration, and material handling techniques. The super tankers and super cargo ships now being built require different harbor facilities, thus the proposals for off-shore superports. Sea/land containerization techniques currently in operation permit direct inland movements by truck or rail from docking areas. This sea/land containerization technique has changed the warehousing requirements at the ports themselves.

6-24 Port of Portland—an important link between the Pacific Northwest and the Orient.

At least three port-related industrial developments have been undertaken successfully during recent years: Rivergate Industrial District in Portland, Oregon; Watson Industrial Park in Carson, California; and Bayport in Clear Lake City, Texas. In each instance, these developments have attracted not only warehousing facilities, but manufacturing operations with shipping requirements as well. ULI Industrial Council member R. John Griefen points out, "The success of both Watson Industrial Park and Los Angeles Industrial Center (see case study) has been due, in large measure, to their location within the Port of Los Angeles and the Port of Long Beach special trucking rate zones."

Inland and Great Lakes water traffic, which is dependent on variables such as climatic conditions and seasonal shipping patterns, continues to be allied closely with specific industries—those requiring bulk shipment of raw materials such as coal, grain, iron ore, and fertilizer. Unit/barge shipping and containerized packaging techniques, however, are improving substantially the efficiency of inland water transportation. The total tonnage of barge traffic is substantial, but it is not possible to predict whether this mode of transport will grow to the point that it rivals either rail or truck in terms of volume shipped. Such projects as the Tulsa Transport Terminal, being developed by the Seltzer Organization, are envisioned as being basically river-related but are tied to rail and highway systems as well.

Pipelines

Whether pipelines should be called a transportation mode or materials handling is difficult to determine. However, in recent years, many innovations have occurred in the movement of goods—including suspended solids—by pipeline. Also, it is possible to move a variety of different products—one behind the other—without contamination. The potential role of pipelines, for transporting goods other than liquids, is open to speculation.

Materials Handling

Some significant breakthroughs in materials handling have taken place in recent years and further changes can be expected. Sea/land, truck/rail, and truck/air containerizing concepts have revolutionized intermodal transfers of bulk quantities of goods over the last two decades. Most current innovations are moving in the direction of coupling containerization concepts with computer-controlled inventory, sorting, and routing programs. ULI Industrial Council member O. G. Linde forecasts, "LCL (less than carload) freight forwarding will increase and handling will become more sophisticated." He cites the example of "partial carload containerization with direct destination drop off" as an alternative to LCL freight forwarding.

New systems—combining high-stack storage using palletized goods and high-reach fork lifts, computerized inventory procedures, modular product configuration, and modular building design—have improved dramatically warehousing operational efficiency. ULI Industrial Council member Donald Johnson points out, "Ceiling heights of 12 to 14 feet, common 20 years ago, have been sup-

6-25 Aerial view of Hallmark Cards, Inc. distribution center in Liberty, Missouri.

planted by heights of 25 to 30 feet and higher, with the result that low ceiling buildings now are considered obsolete." However, ULI Council member Raymond Sydansk notes, "While this is true in warehousing, low ceilings of 12 to 16 feet are still very usable for many forms of manufacturing."

Improved material handling systems for multistory industrial buildings have been conspicuously absent. A breakthrough in this area will be needed to prompt greater interest in high-density urban locations for industrial development.

The general direction in which materials handling methods are moving is illustrated aptly by the Hallmark Cards, Inc. distribution center in Liberty, Missouri. Located on a 38-acre suburban site, this 1.6 million-square-foot warehousing complex includes a sophisticated materials handling system, storage space for 163,000 pallets of mer-

6-26 Distribution system for Hallmark Cards, Inc. Graphic is oriented the same as the photograph on the opposite page. Computer operated system reduces manpower needs to a minimum.

chandise, and a distribution capability for shipping 30,000 cartons daily to Hallmark dealers across the country.

Designed on a "shell" or "envelope" concept which encloses the automated operating system, the center consists of a 900,000-square-foot, 15-foot-clearance, operations building with 48 shipping docks, and four separate warehousing modules enclosing an additional 600,000 square feet. The warehouse modules have a 62-foot ceiling clearance and house the six-story, computer-controlled, automated stacker cranes. Similar facilities are being built in various parts of the country by Sears Roebuck and other large merchandisers which handle a multiplicity of items on a very high-volume basis.

Communications Technology

One of the most difficult factors to assess in relation to industrial development is the impact of rapid advances in the field of communications technology. ULI Industrial Council member Cyril C. Herrmann, has identified this area of technology as one which ultimately will have a substantial impact on industrial development patterns. Entirely new industrial subgroups have emerged already. More importantly, Herrmann suggests, "The organizational structure and physical location of both manufacturing and service distribution industries could undergo dramatic changes." For example, a distribution firm with a national or worldwide, computer-controlled, inventory system with capabilities of automatic search, determination of least cost for handling from various locations to destinations, and simultaneous billing, could revolutionize the distribution industry. One implication of such a system would be a reduced need for regional warehousing facilities to maintain substantial local inventories of less-called-for items if such items could be furnished with greater efficiency from a more remote location.

One reason for the past concentration of industrial development in and around metropolitan areas has been the historical superiority of current modes of vital communication—mail, telephone, and direct wire services—at these locations. The advent of non-wire media, plus direct automated transmission, receiving and printing from principals or from computers, plus the rapidly accelerating cost of writing and forwarding written communications, will continue to diminish the advantage of metropolitan areas in terms of communication efficiency and may increase the advantage of rural locations when improved service can be assured.

Among service industries, the theory that employees need to be in physical proximity to their employers during the work day is open increasingly to reevaluation. Given increasingly sophisticated communication systems, workers whose products are ideas, words, or numbers, can frequently perform these tasks just as easily at remote locations as in the industrial facilities themselves. "In the manufacturing industry," as ULI Industrial Council member Andrew Evans points out, "communications technology and data handling equipment has already seen application in the remote control of manufacturing operation by computer facilities." The impact of innovation in communications technology upon industrial locations is just beginning to be felt, but seems to be substantial.

Shifting Patterns of Industrial Location

Future trends in the pattern of industrial location may well be a combination of continuing trends of past decades along with some rather dramatic shifts. For example, the dramatic growth of suburban industrial development may slow as issues of energy conservation and access to transportation make some suburban sites less attractive.

ULI Industrial Council member Ben Dyer adds, "The total economic impact on local government is being assessed much more thoroughly than ever before. Not only are tax revenue benefits being examined, but the off-setting costs of providing public facilities are being evaluated." Also, the developer needs to be aware of the salary range of the jobs which new industry will create and, consequently, of the economic potential of new residents in the community in terms of housing needs and spending ability.

A greater effort to make industrial development and redevelopment in the core cities feasible is gaining momentum. In addition to modifications in location within metropolitan areas, there are two other broad trends. One, is a shift toward non-metropolitan areas by manufacturing industries and, the other is a nationwide shift of both employment and production away from the East Coast and Great Lakes areas to other parts of the country. For the industrial developer, awareness of these trends will be important in future site selection decisions and in identifying new or growing market areas.

Industrial Locations in the Inner City

One outgrowth of increased environmental concerns, energy conservation issues, and an unwillingness on the part of many suburban jurisdictions to accommodate new industries, may be a renewed interest in long-established, inner-city, industrial locations. Among the advantages of such sites are availability of necessary public utilities, a fully developed transportation network, proximity of a skilled blue collar workforce, and a favorable attitude on the part of the center city governments themselves, particularly if new or expanded industrial development will provide employment opportunities for minorities.

Few statistics are available to document either

the reality or potential of this trend. Nevertheless, the central city clearly has a major and continuing role to play in industrial development. James H. Boykin's *Industrial Potential of the Central City*, published by the Urban Land Institute, documents the advantages inherent in locating in center city. This report concludes that such a location has particular appeal to industrial establishments serving intraregional as opposed to out-of-region markets as well as to those seeking multi-tenant rather than free-standing buildings, and to warehousing establishments. Boykin also found that taxes per square foot on industrial property are lower in the city.

One example of an industrial firm successfully locating a manufacturing plant in an inner-city location is provided by Brown Shoe Company. In 1968, this firm built a shoe manufacturing facility in a low-income area on the near north side of St. Louis. The 60,000-square-foot, two-story, brick and concrete structure occupies one square block of land. Currently, employment stands at 270 persons and productivity is comparable to the company's other plants throughout the United States. Vandalism has not been a problem even though the plant is located across the street from the abandoned, and partially demolished, Pruitt-Igoe housing project.

Brown Shoe Company cites the existence of a viable citizen organization—Jeff-Vander Lou, Inc.—as crucial to the project's success. Comprised of local residents, Jeff-Vander Lou Inc. fulfilled a number of liaison functions in working with the company to bring a plant to the area and to staff it. Strong and effective local leadership, together with the availability of labor, were the fundamental considerations upon which the locational decision was based.

The primary problem associated with establishing the plant was training the labor force. Initially, skilled workers were brought in from other plants to train the new employees, most of whom were not used to factory work. Absenteeism and labor force turnover have been brought within acceptable limits. Company executives emphasize that difficulties associated with moving into an inner-city location should not be minimized; however, these difficulties can be overcome.

Nevertheless, the disadvantages inherent in inner city locations, which precipitated the major shift to suburban sites during the 1950s and 1960s, still remain. Among them are prevailing high land costs of inner-city locations and the difficulty in assembling parcels of sufficient size. While many government programs have been devised to make available such property for industrial reuse, red tape and continuing shifts in official policy have reduced the actual level of participation on the part of industrial developers. However, more effective federal policies and programs could improve the level of participation. Additional deterrents to selection of inner-city sites for location of industry include higher insurance costs, and the undocumented propensity of executives to select sites within convenient commuting distance from their suburban homes.

Railroad-owned properties represent an exception to the problem of finding suitable sites for inner-city industrial development. Most major cities have abandoned rail terminals, old passenger station terminals, and switchyards occupying strategic locations. These parcels present excellent development opportunities, although, in many instances, urban land values have risen above the acceptable level for feasible industrial construction. An example of this situation is the 1973 ULI Panel which recommended to the CBD Study Group in Toledo, Ohio, that the Middleground—an abandoned Penn Central switching yard adjacent to the Maumee River—be redeveloped as a middle to high-density residential area.

If a substantial effort to expand industrial activity in inner-city locations ever takes place, much of it will occur on railroad-owned land, simply because these city properties tend to be large and well-located. As a countervailing trend, railroads have begun to consolidate and relocate switchyards and sorting tracks to the perimeter of metropolitan areas. This tendency for rail to follow industry to the suburbs might further reinforce the process of suburban industrialization, even though new areas for industry within the city might be opened up concurrently.

Outside of certain heavy industrial applications which utilize gravity flow process systems, the demand for multi-story industrial buildings remains limited and is not likely to increase in the near future. This circumstance suggests that the frequently advanced schemes for multi-floor industrial rehabilitation on inner-city sites will have limited applicability. Land and labor cost savings do not appear to offset physical constraints for buildings of this type. This factor tends to limit the demand for high cost, inner-city, industrial land as well. Likewise, the advantages of the urban industrial labor pool may be more illusionary than real. Consisting of poorly educated and unskilled minority groups, these potential employees may be less attractive than the labor pools available in rural and small town environments which are substantially unskilled but evince a better record for productivity, low absenteeism, and a less labor strife.

In summary, inner-city industrial development will continue to attract only a limited portion of all new industrial locations. Nevertheless, the industrial developer no longer categorically rejects inner-city sites as may have been the case in recent years. ULI Industrial Council member Donald E. Johnson suggests, "Major industry may consider acquiring and redeveloping existing facilities to be more attractive than pioneering new sites—not because of dollar savings but because there will be less time required for approvals, fewer environmental issues, less labor conflicts when replacing lost industry, and other factors which may make this approach an attractive alternative." In certain circumstances, as when responding to governmental inducements, and in relation to specific industrial activities, inner-city locations will have a continuing and possibly growing appeal.

Industrial Development in Nonmetropolitan Areas

Statistics reveal a strong shift toward nonmetropolitan areas as a location for new industrial facili-

6-27

EMPLOYMENT CHANGES IN THE UNITED STATES BY METROPOLITAN STATUS, 1960 - 1973

Location of employment and industry	Employment			Average annual change: Number		Average annual change: Rate per 100 persons	
	1960	1970	1973	1960-70	1970-73	1960-70	1970-73
	Thousands					Pct.	Pct.
Metropolitan areas[1]							
Total employment	47,527	60,409	62,591	1,288	727	2.4	1.2
Nonfarm wage and salary	40,953	54,183	56,428	1,323	748	2.8	1.4
Goods producing	15,440	17,599	17,099	216	−167	1.3	−1.0
Manufacturing	**13,248**	**14,887**	**14,214**	**164**	**−224**	**1.2**	**−1.5**
Mining	241	227	226	−1	[3]	−.6	−.1
Construction	1,951	2,485	2,659	53	58	2.4	2.3
Service performing	25,513	36,584	39,329	1,107	915	3.6	2.4
Trade	8,579	11,579	12,375	300	265	3.0	2.2
Service groups	5,667	9,149	10,034	348	295	4.8	3.1
Finance, insurance, and real estate	2,237	3,113	3,382	88	90	3.3	2.8
Transportation, communications and utilities	3,153	3,590	3,600	44	3	1.3	.1
Government	5,877	9,153	9,938	328	262	4.4	2.7
Other nonfarm[2]	5,474	5,396	5,368	−8	−9	−.1	−.2
Farm	1,100	830	795	−27	−12	−2.8	−1.4
Nonmetropolitan areas[1]							
Total employment	18,483	21,232	22,895	275	554	1.4	2.5
Nonfarm wage and salary	12,157	16,213	17,896	406	561	2.9	3.3
Goods producing	4,578	5,971	6,501	139	177	2.7	2.8
Manufacturing	**3,651**	**4,909**	**5,265**	**126**	**119**	**3.0**	**2.3**
Mining	450	379	386	−7	2	−1.7	.6
Construction	477	683	850	21	56	3.6	7.3
Service performing	7,579	10,242	11,395	266	384	3.0	3.6
Trade	2,450	3,112	3,603	66	164	2.4	4.9
Service groups	1,402	2,127	2,360	72	78	4.2	3.5
Finance, insurance, and real estate	359	478	552	12	25	2.9	4.8
Transportation, communications and utilities	857	868	928	1	20	.1	2.2
Government	2,511	3,657	3,952	115	98	3.8	2.6
Other nonfarm[2]	2,773	2,609	2,641	−16	11	−.6	.4
Farm	3,553	2,410	2,358	−114	−17	−3.9	−.7

[1] Metropolitan areas essentially as defined in January 1972.
[2] Self-employed, private household workers, and unpaid family help.
[3] Less than 500 persons.

SOURCE: Adopted from state employment security agency estimates by Economic Research Service, USDA.

ties. This trend was visible during the 1960s when manufacturing employment in nonmetropolitan counties was increasing at a rate of 3 percent per year and only 1.2 percent per year in metropolitan areas. In actual numbers gained, however, the more heavily urbanized areas retained an edge. ULI Industrial Council member Edgar McCaffery suggests, that "The movement of industry to rural areas may have positive encouragement from government rather than past policies which encouraged movement of rural population to urban centers."

Since 1970, the metropolitan areas have experienced a net loss in goods-producing employment, with decreases in manufacturing and mining being only partially offset by increases in construction. Nonmetropolitan areas, by contrast, are gaining goods-producing employment at an accelerating pace with new jobs per year amounting to 119,000 in manufacturing, 2,000 in mining, and 56,000 in construction.

In part, this fundamental shift in locational preferences reflects the general movement of industry from the heavily urbanized Northeast to the more rural South. Specific attractions include lower taxes, abundant nonunion labor, high employee productivity, moderate living costs and wage scales, a relaxed life style for employees, and a cooperative attitude on the part of local government. To these specific attractions must be added the growing difficulty of doing business in metropolitan centers, many of which are beset by congestion, escalating costs, deteriorating physical environments, and a frequently militant labor posture. Improved transportation networks also have contributed to the geographic dispersion of industrial activity. Modern highways now make it possible for nonmetropolitan area residents to live and work in uncrowded rural locations while having access to the attractions of nearby metropolitan areas. Little evidence suggests any major reversal of this trend for the rest of the 1970s. In fact, the attraction of rural areas may become even stronger. Small and medium-sized SMSAs (standard metropolitan statistical areas) may experience an even more dramatic expansion of industrial development than has occurred already.

In the area of distribution of goods, nonmetropolitan areas may become more attractive locations. ULI Industrial Council member Otto Pongrace gives the example of the Kresge distribution facility in Sparks, Nevada which serves a 12-state area from a nonurban location.

Geographical Distribution of Industrial Employment

Geographically, both employment and production are moving away from the older and heavily urbanized areas of New England, Middle Atlantic, and Great Lakes states toward the rest of the country in general and the less urbanized states in particular. The primary beneficiary of this trend has been the South. Between 1969 and 1973, the 16 states and the District of Columbia which comprise the South Atlantic, East South Central, and West South Central geographical divisions of the U.S. Bureau of the Census, realized a net gain of 404,000 industrial jobs in the face of a nationwide decline in industrial employment of 301,000 over the same period. Within the South, largest gains were registered in the states of North Carolina, Tennessee, and Texas. Elsewhere in the country, smaller but proportionally significant increases in industrial employment took place in the West, North Central, and Mountain states which have relied traditionally upon agriculture and mining as the predominant elements of their economic base. ULI Industrial Council member O. G. Linde suggests the likelihood, "The future will see an even greater increase in manufacturing employment rather than just distribution, in the west."

6-28
MANUFACTURING EMPLOYMENT AND VALUE ADDED—PERCENT DISTRIBUTION BY GEOGRAPHIC DIVISIONS 1967, 1972

	Total U.S. Manufacturing Employment		Value Added	
	1967	1972	1967	1972
New England	8.1	7.2	7.2	6.4
Middle Atlantic	22.6	20.9	21.9	20.0
East North Central	26.7	26.1	28.6	28.2
West North Central	6.2	6.4	6.4	6.7
South Atlantic	12.9	14.4	11.2	12.4
East South Central	5.7	6.6	5.2	6.0
West South Central	5.6	6.6	6.3	7.0
Mountain	1.6	2.0	1.7	2.1
Pacific	10.6	10.4	11.3	11.3
	100.0%	100.0%	100.0%	100.0%

Source: Statistical Abstract of the United States, 1974.

Nationally, most categories of employment are expanding, although at differing rates. Largest gains are taking place in wholesale and retail trade, government, and service employment. This phenomenon is believed to be characteristic of a maturing economy when production is increasingly balanced by service and administrative functions.

EMPLOYEES IN NONAGRICULTURAL ESTABLISHMENTS (in 000s)

By States, Annual Averages — States 1969, 1973

State	Total 1969	Total 1973	Total Percent Change	Manufacturing 1969	Manufacturing 1973	Manufacturing Percent Change	Wholesale & Retail Trade 1969	Wholesale & Retail Trade 1973	Wholesale & Retail Trade Percent Change	Government 1969	Government 1973	Government Percent Change
United States	70,141	75,567		20,121	19,820		14,644	16,288		12,227	13,657	
New England:												
Maine	330	353	+ 7.0	116	104	−10.3	64	74	+15.6	65	70	+ 7.7
New Hampshire	258	295	+14.3	98	96	− 2.0	49	61	+24.4	35	43	+22.9
Vermont	145	162	+11.7	44	41	− 6.8	27	33	+22.2	24	39	+20.8
Massachusetts	2,239	2,340	+ 4.5	682	619	− 9.2	474	522	+10.1	296	343	+15.9
Rhode Island	343	362	+ 5.5	127	124	− 2.4	66	75	+13.6	53	54	+ 1.5
Connecticut	1,195	1,241	+ 3.8	475	422	−11.2	218	245	+12.4	148	168	+13.5
Mid Atlantic:												
New York	7,181	7,121	− .8	1,874	1,622	−13.4	1,438	1,463	+ 1.7	1,175	1,258	+ 7.1
New Jersey	2,582	2,759	+ 6.9	897	837	− 6.7	517	600	+16.0	356	419	+17.7
Pennsylvania	4,370	4,480	+ 2.5	1,581	1,476	− 6.6	816	886	+ 8.6	609	651	+ 6.9
East North Central:												
Ohio	3,891	4,112	+ 5.7	1,468	1,422	− 3.1	761	857	+12.6	545	596	+ 9.4
Indiana	1,871	2,019	+ 7.9	747	756	+ 1.2	363	417	+14.9	288	304	+ 5.6
Illinois	4,368	4,353	− .3	1,404	1,321	− 5.9	945	945	.0	612	655	+ 7.0
Michigan	3,064	3,253	+ 6.2	1,170	1,164	− .5	583	647	+11.0	520	534	+ 2.7
Wisconsin	1,521	1,656	+ 8.9	518	529	+ 2.1	323	364	+12.7	255	276	+ 8.2
West North Central:												
Minnesota	1,298	1,437	+10.7	330	331	+ .3	307	352	+14.7	222	255	+14.9
Iowa	880	976	+10.9	225	241	+ 7.1	205	228	+16.1	173	183	+ 5.8
Missouri	1,668	1,754	+ 5.2	459	455	− .9	373	402	+ 7.8	282	305	+ 8.2
North Dakota	158	182	+15.2	9	12	+33.3	43	51	+18.6	49	50	+ 2.0
South Dakota	171	205	+19.9	16	20	+25.0	45	52	+15.6	53	59	+11.3
Nebraska	471	536	+13.8	86	90	+ 4.7	118	137	+16.1	97	111	+14.4
Kansas	684	753	+10.1	146	159	+ 8.9	157	179	+14.0	151	169	+11.9
South Atlantic:												
Delaware	208	236	+13.5	73	73	.0	43	51	+12.6	30	35	+16.7
Maryland	1,277	1,415	+10.8	281	256	− 8.9	293	347	+18.1	245	273	+11.4
D.C. (3)	681	691	+ 1.5	20	17	−15.0	85	72	−15.3	360	375	+ 4.2
Virginia	1,434	1,730	+20.6	370	400	+ 8.1	295	354	+20.0	293	390	+33.1
West Virginia	513	559	+ 9.0	131	128	− 2.3	93	107	+15.1	95	103	+ 8.4
North Carolina	1,735	1,962	+13.1	713	770	+ 8.0	309	357	+15.5	254	279	+ 9.8
South Carolina	812	990	+21.9	339	376	+10.9	135	175	+29.6	140	173	+23.6
Georgia	1,522	1,783	+17.2	477	494	+ 3.6	320	393	+22.8	285	331	+16.1
Florida	2,079	2,708	+30.3	329	373	+13.4	539	707	+31.2	379	463	+22.2
East South Central:												
Kentucky	295	1,038	+16.0	247	285	+15.4	181	209	+15.5	167	197	+18.0
Tennessee	1,312	1,539	+17.3	470	522	+11.1	255	316	+23.9	214	244	+14.0
Alabama	999	1,133	+13.4	324	346	+ 6.8	185	222	+20.0	205	226	+10.2
Mississippi	569	676	+18.8	182	218	+19.8	104	132	+26.9	129	139	+ 7.8
West South Central:												
Arkansas	532	620	+16.5	168	201	+19.6	105	126	+20.0	101	110	+ 8.9
Louisiana	1,044	1,161	+11.2	181	183	+ 1.1	228	270	+18.4	207	235	+13.5
Oklahoma	754	849	+12.6	130	150	+15.4	165	195	+18.2	185	195	+ 5.4
Texas	3,611	4,151	+15.0	749	796	+ 6.3	853	1,023	+19.9	651	745	+14.4
Mountain:												
Montana	197	226	+14.7	24	24	.0	47	56	+19.1	52	55	+ 5.8
Idaho	199	245	+23.1	40	47	+17.5	47	61	+29.8	47	56	+19.1
Wyoming	107	126	+17.8	7	8	+14.3	24	27	+12.5	28	32	+14.3
Colorado	709	899	+26.8	114	135	+18.4	168	219	+30.4	165	190	+15.2
New Mexico	285	345	+21.1	20	27	+35.0	60	77	+28.3	86	99	+15.1
Arizona	514	702	+36.6	93	108	+16.1	117	164	+40.2	113	147	+30.1
Utah	348	417	+34.2	54	64	+18.5	78	99	+26.9	99	107	+ 8.1
Nevada	189	243	+28.6	8	11	+37.5	36	47	+30.6	36	41	+13.9
Pacific:												
Washington	1,125	1,151	+ 2.3	279	245	−12.2	247	260	+ 5.3	238	259	+ 8.8
Oregon	704	816	+15.9	179	197	+10.1	159	190	+19.5	141	160	+13.5
California	6,909	7,635	+10.5	1,655	1,648	− .4	1,494	1,719	+15.1	1,387	1,522	+ 9.7
Alaska	85	109	+28.2	7	9	+28.6	14	18	+23.6	33	41	+24.2
Hawaii	276	316	+14.9	25	24	− 4.0	63	76	+20.6	71	78	+ 9.9

Source: *Statistical Abstract of the United States: 1970, 1974.*
Technical Publications Division, ULI-the Urban Land Institute.

(1) Combined with services (2) Combined with contract constructi

Services			Transportation and Public Utilities			Finance, Insurance, Real Estate			Contract Construction			Mining		
969	1973	Percent Change	1969	1973	Percent Change	1969	1973	Percent Change	1969	1973	Percent Change	1969	1973	Percent Change
,103	12,266		4,448	4,611		3,359	4,053		3,411	3,648		628	625	
41	54	+31.7	17	18	+ 5.9	12	13	+ 8.3	15	20	+33.3	(1)	(1)	NA
41	53	+29.3	11	13	+18.2	10	13	+30.0	13	17	+30.8	(2)	(2)	NA
27	32	+18.5	8	8	.0	5	7	+40.0	10	11	+10.0	1	1	.0
453	491	+ 8.4	115	123	+ 7.0	125	134	+ 7.2	94	109	+16.0	(1)	(1)	NA
53	62	+17.0	15	15	.0	15	17	+13.3	14	15	+ 7.1	(1)	(1)	NA
175	209	+19.4	53	56	+ 5.7	70	84	+20.0	56	58	+ 3.6	(2)	(2)	NA
330	1,430	+ 7.5	500	470	− 6.0	595	592	− .5	262	279	+ 6.5	8	7	−12.5
400	458	+14.5	176	184	+ 4.5	115	131	+13.9	118	126	+ 6.8	3	3	.0
670	742	+10.7	267	267	.0	184	206	+12.0	203	213	+ 4.9	39	40	+ 2.6
544	648	+ 7.4	222	224	+ .9	152	174	+14.5	180	167	− 7.2	20	23	+15.0
203	258	+27.1	100	104	+ 4.0	74	85	+14.9	89	89	.0	7	7	.0
671	701	+ 4.5	289	280	− 3.1	229	244	+ 6.6	196	184	− 6.1	23	23	.0
400	491	+22.8	150	153	+ 2.0	114	127	+11.4	114	127	+11.4	13	13	.0
217	264	+21.7	80	84	+ 5.0	59	68	+15.3	67	68	+ 1.3	3	3	.0
206	255	+23.8	88	91	+ 3.4	62	71	+14.5	68	67	− 1.5	14	14	.0
142	166	+16.9	51	55	+ 7.8	41	46	+12.2	41	45	+ 9.8	3	3	.0
252	289	+12.5	126	125	− .8	89	96	+ 7.9	73	75	+ 2.7	9	8	−11.1
29	35	+20.7	12	13	+ 8.3	7	8	+14.3	8	12	+50.0	2	2	.0
30	42	+40.0	10	12	+20.0	7	8	+14.3	8	10	+25.0	2	2	.0
79	96	+21.5	37	39	+ 5.4	28	32	+14.3	26	30	+15.4	2	2	.0
101	116	+14.9	52	54	+ 3.8	30	34	+13.3	36	34	− 5.6	12	10	−16.7
29	39	+34.5	11	12	+ 9.1	9	11	+22.2	13	16	+23.1	(1)	(1)	(1)
227	273	+20.3	79	80	+ 1.3	66	78	+18.2	85	107	+25.9	2	2	.0
134	143	+ 6.7	31	29	− 6.5	32	33	+ 3.1	18	22	+22.2	(1)	(1)	(1)
200	259	+24.5	95	106	+11.6	65	83	+27.7	94	121	+22.7	14	16	+14.3
64	74	+15.6	41	41	.0	15	18	+20.0	26	35	+34.6	47	52	+10.3
201	246	+22.4	90	103	+14.4	67	85	+26.9	97	119	+22.7	4	4	.0
83	115	+38.6	35	42	+20.0	28	37	+32.1	49	70	+42.9	2	2	.0
177	240	+35.6	103	117	+13.6	72	92	+27.8	82	108	+31.7	7	7	.0
387	527	+36.2	148	184	+24.3	124	180	+45.2	165	266	+61.0	8	9	+12.5
123	156	+26.8	60	62	+ 3.3	35	40	+14.3	55	58	+ 5.5	27	32	+18.5
176	227	+29.0	66	72	+ 9.1	56	67	+19.6	68	84	+23.5	7	7	.0
127	156	+22.8	56	61	+ 8.9	41	48	+17.1	54	65	+20.4	8	9	+12.5
66	79	+19.7	30	36	+20.0	20	25	+25.0	33	42	+27.3	6	6	.0
72	82	+13.9	32	37	+15.6	21	27	+28.6	30	34	+13.3	5	4	−20.0
151	188	+24.5	93	97	+ 4.3	49	57	+16.3	83	80	− 3.0	52	53	+ 1.9
108	130	+20.4	53	55	+ 3.8	36	44	+22.2	37	45	+21.6	40	36	−10.0
585	688	+17.6	252	279	+10.7	184	236	+28.3	232	275	+18.5	105	108	+ 2.9
32	42	+31.3	18	19	+ 5.6	8	9	+12.5	10	13	+30.0	6	6	.0
31	39	+25.8	14	16	+14.3	8	10	+25.0	10	14	+40.0	4	3	−25.0
15	19	+26.7	11	12	+ 9.1	4	4	.0	7	12	+71.0	12	12	.0
123	159	+29.3	51	58	+13.7	37	49	+32.4	38	76	+100.0	13	14	+ 7.7
54	63	+16.7	20	23	+15.0	12	15	+25.0	17	25	+47.0	17	16	− 5.9
85	120	+41.2	28	35	+25.0	27	41	+51.9	32	63	+76.9	18	25	+38.9
54	70	+29.6	23	25	+ 8.7	14	18	+28.6	14	22	+57.1	12	13	+ 8.3
75	95	+26.7	13	16	+23.1	7	10	+42.9	11	18	+53.6	4	4	.0
68	195	+16.1	74	72	− 2.7	58	63	+ 8.6	59	55	− 6.8	2	2	.0
08	135	+25.0	49	52	+ 6.1	35	43	+22.9	32	38	+18.8	2	2	.0
17	1,467	+20.7	462	470	+ 1.7	365	443	+21.4	296	333	+12.5	32	30	− 6.3
11	16	+45.5	8	10	+25.0	3	4	+33.3	7	8	+14.3	3	2	−33.3
53	68	+30.2	23	25	+ 8.7	17	20	+17.6	23	25	+ 8.7	(1)	(1)	NA

Federal Employment includes Va. & Md. residents of the Washington, D.C. SMSA

These broad shifts, important for the industrial developer, clearly identify geographical areas where future markets for planned employment centers might occur. As ULI Industrial Council member Raymond Sydansk points out, "The rate of increase in U.S. manufacturing capacity will diminish as our population stabilizes. At the same time, the volume of industrial production around the rest of the developing areas of the world will increase. Thus, future industrial development in the United States, with regard to geographic distribution, will be replacement or relocation capacity rather than strong new capacity such as we experienced over the last 30 years."

ULI Industrial Council member Raymond Sayer further suggests, "This will present great opportunities for industrial developments responding to this replacement demand particularly in industries evidencing a high degree of plant obsolescence." At the same time Council member Donald E. Johnson points out, "Areas likely to lose an industrial base may seek mechanisms to control or prevent relocations which would result in loss of employment in the community. Industries may be charged with defending plant relocation proposals."

Financing in the Future

Perhaps the greatest potential for ambiguity in the perception of future trends in industrial development lies in the area of financing. For most of the period following World War II, the general availability of capital funds at reasonable cost and the specific methods of financing projects, although complex and varying, were well understood by the industrial developer. Inflation was constantly upward, but relatively steady and small in percentage. Capital funds were almost always available—with some periodic tightening—and interest rates were steady or rising slightly. However, present and future prospects are unclear. Recent rapid inflation, plus dramatic increases in the cost of money, have produced great uncertainty. Projections for the future vary widely. For the industrial developer, there will be two prime issues—the availability of development capital at what cost, and the types of financing mechanisms which might evolve.

Capital Shortages

The majority of the members of the Industrial Council of the Urban Land Institute perceives a continuing problem of shortage of capital for development. In periods of capital shortages, the higher risks inherent in development activity make it difficult to attract needed capital for the long-term financing required in development. As ULI Industrial Council member Andrew R. Evans suggests, "Lenders will be much more discriminating in providing funds." Developers with good track records will continue to find financing while those with less experience may find it more difficult. ULI Industrial Council member Edmund J. McRickard suggests the shortage of capital "will mean the avoidance of special purpose construction within projects and the design of projects to attract the widest possible market." Such a posture may be difficult to achieve and may constrain flexibility, due to the propensity of communities to want uses and users very specifically identified in advance of public approval.

One of the recent expectations for sources of new capital has been the attraction of offshore funds into the United States. The large surplus funds available to the oil producing countries would appear an obvious source of new capital. The availability of these funds for development in the U.S. may not prove a reality. As ULI Industrial Council member Edmund J. McRickard points out, "These funds may be used to a large extent for economic development within these oil producing countries. To the extent there might be a surplusage for investment abroad, it seems likely, unlike the past when offshore funds were used to receiving a lower rate of return, these sources will demand the same rate of return as domestic sources when investing in the U.S."

ULI Industrial Council member Lowell Eastman affirms, "Industrial developers are likely to suffer from capital shortages and to find well-conceived projects may be difficult to pursue for lack of capital or even may be restricted from access to capital as a result of public policy to reduce demand pressures on the limited available capital." This leads to the point made by R. John Griefen that "long-term, fixed-rate money will be more and more difficult to obtain which means various financing methods will make loan terms more attractive or smaller amounts of capital from various sources will be combined to finance a project."

Financing Methods

Oakleigh Thorne, primary author of the Development Strategy section, offered the Industrial Council the following statement for consideration in discussing future trends in financing development:

The mania for indexing will eventually hit the mortgage industry. Already application of the CPI (Consumer Price Index) in updating leases annually or at renewal is common practice. Two other indices, namely the Wholesale Price Index (WPI) and the GNP deflator are used as indications of growth and inflation in the United States economy. The indexing philosophy, which has been applied for a long time in Brazil, Bolivia, Colombia and to some degree in Israel, undoubtedly will be considered as a tool to be used in the future in this country's mortgage industry.

Currently, all signs in the economy point to a continuing inflation at rates greater than the past despite the federal government's efforts to control it. There would appear to be a substantial demand for escalators or the philosophy of indexing to counter balance the value deflating effect of inflation. For example, the indexing procedure could easily be applied to remaining mortgage loan principal which is unpaid at the end of the year. Some method of indexing would be applied to the remaining unpaid principals left on mortgages to rectify the lower dollar value still to be collected.

In the future the traditional 20, 25, or 30-year amortization mortgages on industrial building may be no longer available, but 40-year amortization periods, with balloon payment 10 to 12 years from the date of origination, possibly will become common. Interest rates may be much higher than in 1974 and 1975. However, the constants will remain fairly stable because of the lengthened amortization period. The use of the balloon mortgage has been implemented much more frequently within the last two to three years. In fact, the time scale of the balloon payment has diminished from somewhere in the vicinity of 20 to 22 years to as few as 12 to 15 years on some recent mortgages. Naturally, as financing becomes more expensive, other players will enter the stage to provide project funds. For example, the pension trust funds and other public development companies will begin to participate in joint ventures, equities, or outright mortgages.

These views generated varied reactions. ULI Industrial Council members Andrew R. Evans, Lowell Eastman, and John O'Donnell noted that variable interest rate mortgages were being offered in California as of 1975, but were quick to add that while this might work in the residential mortgage market it probably would not be applied to industrial and commercial development where

6-30 A branch office of the National Bank of Commerce in Nonconnah Corporate Center, Memphis, Tennessee. Freestanding ancillary uses of this kind are a common feature of most recently developed industrial parks.

other methods might be more attractive.

ULI Industrial Council Chairman Marshall Bennett suggests, "Rents indexed to current dollars with shorter term, fixed-rate leases will become the rule." Council member O. G. Linde suggested with conventional financing more difficult to obtain, alternate financing techniques for some elements of development, might offer some relief. "While assessment districts and bond issues have long been a source of long-term financing of utility and street improvements, use of these techniques will likely need to expand to other elements of the development package. For example, the assessment district techniquc has been used in California for the financing of rail service trackage which has in the past, been financed by the developer and/or the railroads under some rebate arrangement." Council member Robert Walker suggests financing difficulties will lead to wider use of techniques to spread the risk of development and to draw in increments of capital by joint venturing and/or laying-off portions of a development to other developers with special expertise and therefore access to needed capital.

The above suggestions and comments only serve to make clear how uncertain the future is in the area of financing for industrial development. What is clear, is the methods are likely to change, the astute developer will need to be innovative in seeking advantageous financing, lenders will be more cautious and more demanding, and this situation will likely be a constraint against pioneering and innovative (and therefore higher risk) developments.

Summary

A variety of future trends in industrial development have been described. The direction they will take in the future is extremely difficult to project. Many trends are intertwined. All are related to the issues of environmental protection, energy conservation, and national economics. Frequently, a minor change in legislative direction or national concern will have a dramatic impact on a well-defined trend.

Regardless of the direction events take, environmental concerns and energy conservation measures will increasingly influence patterns of industrial development. While widely divergent opinions exist as to the degree of industrial integration with other land uses which should be encouraged, there is no disagreement that such integration is now occurring on a wide scale. Disregarding any positive encouragement, it is likely that more specialized industrial clustering will be built and that concentration of new industry in the suburbs will be offset somewhat by attempts to re-attract industrial development to more urbanized settings. The shift toward industrial locations in smaller metropolitan areas and predominantly rural locations should continue.

Patterns of transportation will focus on the need for flexibility at a minimum cost and maximum efficiency, particularly in relation to fuel consumption. Industrial development has reached a level of sophistication at which marketing techniques are becoming more creative. The social concerns of our society are recognized as being important to economic success. Technological change has been the cornerstone of industrial expansion. Awareness of and responsiveness to the direction of change, coupled with the adoption of new technologies in creative and innovative ways may be more likely to occur in the industrial land use and development field than in any other segment of the land development industry.

The discussion of future trends has not marked a clear route for the industrial developer to follow. Hopefully, the dominant forces for change and the emerging or continuing development techniques have been identified.

APPENDICES
AND
INDEX

APPENDIX A

Selected References & Information Sources

Books

Ahn, Michael. Industrial Bibliography, Washington, ULI-the Urban Land Institute, 1974 (Its Research Report no. 22).

American Bar Association. Special Committee on Environmental Law. *Development and the Environment. Legal Reforms to Facilitate Industrial Site Selection* (Chicago, The American Bar Association, 1974).

Bessire, Howard D. *The Practice of Industrial Development.* El Paso, Texas, Hill Printing Company, 1970.

Boley, Robert E. *Industrial Districts. Principles in Practice.* Washington, ULI-the Urban Land Institute, 1962 (Its Technical Bulletin no. 44).

Boley, Robert E. *Industrial Districts Restudied. An Analysis of Characteristics.* Washington, ULI-the Urban Land Institute, 1961 (Its Technical Bulletin no. 41).

Boykin, James H. *Industrial Potential of the Central City.* Washington, ULI-the Urban Land Institute, 1973 (Its Research Report no. 21).

Friedmann, John and William Alonso (eds). *Regional Policy. Readings in Theory and Applications.* Cambridge, Mass., The MIT Press, 1975.

Howard, Dick (ed). *Guide to Industrial Development.* Englewood Cliffs, N.J., Prentice-Hall, 1972.

Hunker, Henry L. *Industrial Development. Concepts and Principles.* Lexington, Mass., Lexington Books, 1974.

Karaska, Gerald J. and David F. Bramhall (eds). *Locational Analysis for Manufacturing. A Selection of Readings.* Cambridge, Mass., The MIT Press, 1969.

Kinnard, William N. Jr. and Stephen D. Messner. *Industrial Real Estate*, 2nd edition. Washington, Society of Industrial Realtors, 1971.

Pasma, Theodore K. *Characteristics of 63 Modern Industrial Plants.* Washington, U.S. Department of Commerce, Economic Development Administration, 1966.

Peters, Paulhans (ed). *Design & Planning: Factories.* Cincinnati, Van Nostrand Reinhold Company, 1975.

Preston, Richard. *The Universe of Industrial Development and North American Concepts and Practices*, revised edition. Wenham, Mass., American Industrial Development Council Educational Foundation, 1974.

Salzenstein, Marvin A. *Industrial Performance Standards.* Chicago, American Society of Planning Officials, 1971 (Its Planning Advisory Service Report no. 272).

Smith, David M. *Industrial Location. An Economic Geographical Analysis.* New York, John Wiley & Sons, 1971.

U.S. Bureau of the Budget. *Standard Industrial Classification Manual.* Washington, U.S. Government Printing Office, 1972.

Yaseen, Leonard C. *Plant Location*, revised edition. New York, American Research Council, 1960.

Periodicals

AIDC Journal (quarterly. American Industrial Development Council, Inc., 215 West Pershing Road, Suite 707, Kansas City, Missouri 64108).

Area Development (monthly. Halcyon Business Publications, Inc., 114 East 32nd Street, New York, New York 10016).

Growth and Change (quarterly. College of Business and Economics, University of Kentucky, 644 Maxwelton Court, Lexington, Kentucky 40506).

Industrial Development (bimonthly. Conway Publications, Peachtree Air Terminal, 1954 Airport Road, Atlanta, Georgia 30341).

Plants, Sites & Parks (bimonthly. Communications Channels, Inc., 461 Eighth Avenue, New York, New York 10001).

Site Selection Handbook (4 annual volumes). Vol. 1: Real Estate Investment and Development; Vol. 2: Industry's Guide to Geo-Economic Planning; Vol. 3: The Environmental Planning Guide; Vol. 4: Guide to Office and Industrial Parks (Conway Publications, Peachtree Air Terminal, 1954 Airport Road, Atlanta, Georgia 30341).

U.S. Bureau of the Census

Annual Survey of Manufactures (for the years not covered by the Census of Manufactures). Statistics on key measures of manufacturing activity for industries and for geographic areas.

Census of Manufactures (every five years; latest 1972). Includes data on number of establishments; quantity and value of products; quantity and cost of materials; capital expenditures; value of fixed assets; rental payments; inventories, employment, payrolls, manhours; value added by manufacturing; water used in manufacturing.

County and City Data Book (irregular 4-5 year intervals; latest 1972). Social and economic data by regions, divisions, states, SMSAs, counties, cities of 25,000+ and urbanized areas.

Current Industrial Reports. This series includes monthly, quarterly, and annual reports that present data for approximately 5,000 products, by SIC codes.

Statistical Abstract (annual). Current and historical social, economic, and governmental statistics.

Organizations

American Industrial Development Council, 215 West Pershing Road, Kansas City, Missouri 64108.

National Association of Industrial Parks, 1601 North Kent Street, Arlington, Virginia 22209.

National Association of State Development Agencies, 2000 K Street NW, Washington, D.C. 20006.

Society of Industrial Realtors, 925 15th Street NW, Washington, D.C. 20005.

U.S. Department of Commerce, Economic Development Administration, Washington, D.C. 20230.

APPENDIX B

U.S. Economic Adjustment Program

Source: Office of Economic Adjustment—Department of Defense

Completed Major Economic Adjustment Projects
1961-1973

Community/Government Installation	Year of Impact	Civilian Jobs Lost	Civilian Replacement Jobs On-Base	Civilian Replacement Jobs Off-Base	College or Vo-Tech Students on Base
Coden, Alabama—Dauphin Island AFS	1971	26	30		250
Mobile, Alabama—Brookley AFS	1969	12,300	3,200	9,700	1,500
Thomasville, Alabama—Thomasville AFS	1970	18	180		
Kenai, Alaska—Wildwood AFS	1973	65	185	15	240
Fort Smith, Arkansas—Fort Chaffee *	1965	1,003	*	10,900	
Barstow, California—Fort Irwin **	1971	242	**	300	
Benecia, California—Benecia Arsenal	1964	2,321	2,800		
Green Cove Springs, Florida—Atlantic Fleet site	1962	324	500		
Sanford, Florida—Sanford NAS	1968	230	580		1,560
Macon, Georgia—NOP-Macon	1965	958	300		
Decatur, Illinois—Decatur Army Signal Depot	1962	1,310	2,700		
Forest Park, Illinois—Forest Park NOS	1971	1,600	2,000		
Terre Haute, Indiana—DIPEC Facility	1966	253	625	5,400	
Columbus, Indiana—Bakalar AFB	1970	318	282	38	100
Salina, Kansas—Schilling AFB	1965	326	2,413	1,270	838
Wichita, Kansas—Procurement Cutback**	1971	12,300	**	27,300	
Lake Charles, Louisiana—Chennault AFB	1963	252	300		500
New Iberia, Louisiana—New Iberia NAS	1965	85	760		700
Bangor, Maine—Dow AFB	1968	328	1,780	300	900
Presque Isle, Maine—Presque Isle AFB	1961	268	1,900	1,100	380
Springfield, Massachusetts—Springfield Armory	1968	2,400	600		6,300
Watertown, Massachusetts—Watertown Arsenal	1967	2,306	12		
Wadena, Minnesota—Wadena AFS	1971	15	13		
Greenville, Mississippi—Greenville AFS	1965	242	477		244
Neosho, Missouri—Camp Crowder	1970	1,200	2,034	822	1,200
Conrad & Shelby, Montana—Anti-Ballistic Missile Site	1972	138	10	58	
Glasgow, Montana—Glasgow AFB	1968	309	886		200
Lewistown Montana—Lewistown AFS	1971	27	15		175
Hastings, Nebraska—Hastins NAD	1966	240	501	15	4,800
Lincoln, Nebraska—Lincoln AFB	1966	396	1,284	5,780	80
Sidney, Nebraska—Sioux Army Depot	1967	585	275		
Reno, Nevada—Stead AFB	1966	519	630		200
Manchester, New Hampshire—Grenier AFB	1966	138	3,655		
Metuchen, New Jersey—Raritan Arsenal	1964	2,610	6,000		7,500
New Brunswick, New Jersey—Camp Kilmer	1963	139	890		3,180
Gallup, New Mexico—Fort Wingate *	1971	233	59*	283	106
Roswell, New Mexico—Walker AFB	1967	379	2,500		2,200
Newburgh, New York—Stewart AFB	1969	1,011	245		
New York City, New York—Army Pictorial Center	1970	388			
Rome, New York—Rome Air Materiel Area **	1967	3,400	**	2,300	
Schenectady and Voorheesville, New York—Army and General Depots	1966	1,700	2,700		
Port Clinton, Ohio—Erie Army Depot	1966	1,885	800		
Toledo, Ohio—Rossford Arsenal	1964	1,654	4,100	2,000	5,000
Wilmington, Ohio—Clinton County AFB	1971	250	594		1,275
Burns Flat, Oklahoma—Clinton-Sherman Oklahoma	1970	391	698	920	500
Harrisburg, Pennsylvania—Olmstead AFB and Middletown Air Materiel Area	1967	10,050	4,500	7,000	2,000
Lancaster, Pennsylvania—Marietta AFS	1967	750	2,000		
York, Pennsylvania—York NOP	1963	1,092	1,800		
Charleston, South Carolina—Procurement Cutbacks and Reduced Military Activities**	1971	6,195	**	5,700	
Greenville, South Carolina—Donaldson AFB	1963	672	3,200	2,400	600
Edgemont, South Dakota—Black Hills Army Depot	1967	512	100	92	
Smyrna, Tennessee—Sewart AFB	1969	470	2,100		
Amarillo, Texas—Amarillo AFB	1968	1,571	1,458	6,000	2,187
Harlingen, Texas—Harlingen AFB	1962	720	1,400	1,600	1,300
Mexia, Texas—Procurement Cutback**	1971	500	**	633	
San Marcos, Texas—Camp Gary	1964	30	800		2,000
Sherman/Denison, Texas—Perrin AFB	1971	600	714	700	1,325
Sweetwater, Texas—Sweetwater AFS	1971	125	65		400
Waco, Texas—James Connally AFB	1966	833	501		2,800
Moses Lake, Washington—Larson AFB	1966	388	484		330
Madison, Wisconsin—Truax Field	1968	378	432		

* Limited property made available
** Property not available

APPENDIX C

Sample Industrial Park Zone with Performance Standard's, Montgomery County, Maryland

Source: Zoning Administrator, Montgomery County, Rockville, Maryland

Section 111-24. I-3 ZONE INDUSTRIAL PARK. (Ord. 4-32, adopted 11/17/59.)

The purpose of this Industrial Park (I-3) Zone is to provide a protective Zone for a park-like development of industry that is based on the performance of an industry as well as on the type of industry. In order to secure this type of development, the various regulations herein described must be met. These regulations have been established so as to provide a healthful operating environment for industry, for the protection of industry from the encroachment of commercial and residential uses adverse to the operation and expansion of such industry, and to protect industries within the district from the adverse effect of other incompatible industries, and at the same time to reduce to a minimum the impact of industries on surrounding non-industrial land uses; to lessen traffic congestion; to protect the health and safety of the residents or workers in the area; to prevent detrimental effects to the use or development of adjacent properties or the general neighborhood; and to promote the health, safety, morals, comfort and welfare of the present and future inhabitants of the district.

The following regulations shall apply in all I-3 Zones:

A. Use Regulations

1. *Permitted Uses*

No building, structure or land shall be used and no building or structure shall hereafter be erected, structurally altered, enlarged or maintained, except for one or more of the following uses:

Dwellings only in connection with bona fide agricultural operations, or as living quarters for bona fide caretakers and/or watchmen and their families.

Farming or other agricultural uses.

Governmentally owned or operated buildings and uses.

Helistop. (Ord. 5-167, adopted 5/24/66.)

Office buildings for general office purposes, as that term is defined in Sec. 111-18a of the Zoning Ordinance.

Off-street parking of motor vehicles in connection with uses within the zone, subject to the provisions of Subsection J of this section and sections cited therein.

Research, experimental and testing laboratories.

Signs in accordance with Subsection I of this section.

Temporary helistop. (Ord. 5-167, adopted 5/24/66.)

Underground pipe lines, underground electric power and energy transmission and distribution lines, underground or overhead telephone or telegraph lines, overhead electric power and energy transmission and distribution lines not suspended from multi-legged structures, towers and accessory structures.

Warehouses and storage buildings, not including storage of materials and goods specifically prohibited in Section A-3 below.

Uses of a light industrial nature not specifically prohibited in Section A-3 below, including those listed in section 111-22a.

Accessory buildings and uses, swimming pools, recreational facilities and dining facilities for use in connection with the operation of an establishment and primarily for employees.

2. *Special Exceptions*

a. The following uses may be permitted as Special Exceptions in accordance with the provisions applying to such uses within R-A districts, and subject to Sections 111-30 through 111-37.

Airport, airpark, airfield and heliport.

Art or cultural center, non-commercial. (Ord. 4-165, adopted 10/23/62.)

Public utility buildings and utility structures not otherwise permitted, including overhead electric power and energy transmission and distribution lines suspended from multi-legged structures, above-ground pipelines, radio and television broadcasting stations and towers, and accessory structures, subject to Sections 111-30 through 111-37.

Retail Sales and Consumer Service Establishments (not including warehouse sales), accessory to any permitted use, other than agricultural or residential uses, and dealing primarily with employees of establishments permitted as principal uses, provided that such commercial uses shall not occupy more than 5 per cent of the total floor area of all buildings on any lot or group of contiguous lots in common ownership or control.

3. *Prohibited Uses*

The following uses are expressly prohibited in the I-3 Zone:

Recreational or entertainment establishments of a commercial nature.

Fuel storage yard.

Contractors storage yard.

Grist mills.

Lumber yard.

Saw mill.

Stonework.

Abattoir.

Stockyard.

Uses of a heavy industrial nature, including those set forth in Section 111-23a(2) of the I-2 Zone.

Uses expressly prohibited in I-2 Zones as set forth in Section 111-23c.

B. Performance Standards

It is the intent of these regulations to prevent land or buildings, including those permitted by right or special exception, from being used or occupied in any manner so as to create any dangerous, injurious, noxious or otherwise objectionable fire, explosive, radioactive or other hazardous condition; noise or vibration; smoke, dust, odor or other form of air pollution; electrical or other disturbance; glare or heat; liquid or solid refuse or wastes; condition conducive to the breeding of rodents or insects; or other substance, condition or elements (all referred to herein as "Dangeous or Objectionable Elements") in a manner or amount as to adversely affect the surrounding area. Any use permitted under Sections A-1 and 2 may be undertaken and maintained if it conforms to all district regulations including the regulations of this section referred to herein as "Performance Standards." Specifically, all uses shall operate in conformance with the limitations set forth in each subsection below:

1. *Vibration*

No vibration shall be produced which is transmitted through the ground and is discernable without the aid of instruments at or at any point beyond the lot line; nor shall any vibration produced exceed 0.002g peak measured at or beyond the lot line using either seismic or electronic vibration measuring equipment.

2. *Noise*

All noise shall be muffled so as not to be objectionable due to intermittance, beat frequency or shrillness. In no event shall the sound-pressure level of noise radiated continuously from a facility at night-time exceed at the lot line the values given in Table I (set out hereafter) in any octave band of frequency. However, where the lot line adjoins or lies within twenty-five feet of the boundary of a residence district the sound-pressure levels of noise radiated at night-time shall not exceed at the lot line the values given in Table II (set out hereafter) in any octave band of frequency. The sound-pressure level shall be measured with a Sound Level Meter and an Octave Band Analyzer that conform to specifications published by the American Standards Association. (American Standards Sound Level Meters for Measurement of Noise and Other Sounds, Z24.3-1944, American Standards Association, Inc., New York, N.Y., and American Standard Specification for an Octave Band Filter Set for the Analysis of Noise and Other Sounds, Z24.10-1953, American Standards, Inc., New York, N.Y. shall be used).

TABLE I

Maximum permissible sound-pressure levels at the lot line for noise radiated continuously from a facility between the hours of 9 P.M. and 7 A.M.

Frequency Band Cycles Per Second	Sound Pressure Level Decibels re 0.0002 dyne/cm2
20—75	69
75—150	60
150—300	56
300—600	51
600—1,200	42
1,200—2,400	40
2,400—4,800	38
4,800—10,000	35

If the noise is not smooth and continuous and is not radiated between the hours of 9 P.M. and 7 A.M., one or more of the corrections in Table III below shall be added to or subtracted from each of the decibel levels given above in Table I.

TABLE II

Maximum permissible sound-pressure levels at a lot line for noise radiated continuously from a facility between the hours of 9 P.M. and 7 A.M., where the lot line adjoins or lies within twenty-five feet of the boundary of a residence district.

Frequency Band Cycles Per Second	Sound Pressure Level Decibels re 0.0002 dyne/cm2
20—75	65
75—150	50
150—300	43
300—600	38
600—1,200	33
1,200—2,400	30
2,400—4,800	28
4,800—10,000	26

If the noise is not smooth and continuous and is not radiated between the hours of 9 P.M. and 7 A.M., one or more of the corrections in Table III below shall be added to or subtracted from each of the decibel levels given above in Table II.

TABLE III

Type of Operation in Character of Noise	Correction in Decibel
Daytime operation only	plus 5
Noise source operates less than 20% of any one-hour period	plus 5*
Noise source operates less than 5% of any one-hour period	plus 10*
Noise source operates less than 1% of any one-hour period	plus 15*
Noise of impulsive character (hammering, etc.)	minus 5
Noise of periodic character (hum, speech, etc.)	minus 5

*Apply one of these corrections only.

3. *Air Pollution*

a. Definitions. The meaning of terms used in this subsection shall be the same as defined in Chapter 74 of this Code.

b. Visible emissions. There shall not be discharged into the atmosphere from any source any air pollutant in excess of the darkness limitations listed below. This shall include emissions of air pollutant of such opacity as to obscure an observer's view to a degree equal to or greater than do the visible emissions described below. Visible emissions of any kind at ground level past the lot line of the property on which the source of the emissions is located are prohibited:

No. 1 on the Ringelmann Chart at all times, except upon demonstration that the emission contains less than one tenth pound of particulate matter per thousand pounds of dry gases, adjusted to twelve per cent carbon dioxide or fifty per cent excess air.

c. Materials handling. No person shall cause or permit any materials to be handled, transported, or stored in a manner which allows or may allow particulate matter to become airborne.

d. Particulate matter. There shall not be discharged into the atmosphere any particulate matter in excess of the quantities shown below:

(1) From refuse-burning equipment, per one thousand pounds of dry gases, adjusted to twelve per cent CO_2 or fifty per cent excess air, sixty-five hundredths pound for capacities of two hundred pounds per hour or less and thirty hundredths pound for capacities of over two hundred pounds per hour. In any one hour period this shall not exceed two hundred fifty pounds. Refuse shall not be burned in fuel-burning equipment.

(2) From fuel-burning equipment, six tenths pound per million BTU input for installation using less than ten million BTU per hour total input. For installations using more than ten million BTU per hour total input the allowable particulate emission, in pounds per million BTU, is determined by multiplying 8.58 times the input, in BTU per hour, raised to the minus 0.165 power ($E=8.58\,1^{-165}$).

e. Threshold values. There shall not be discharged into the atmosphere any contaminant for which threshold limit values are listed for working atmosphere by the American Conference of Governmental Industrial Hygienists in such quantity that the concentration of the contaminant at ground level at any point beyond the boundary of the property shall at any time exceed the threshold limit.

f. Gaseous emissions. Gaseous emissions shall be controlled by the provisions of Chapter 74, Section 5(d)(1) and (2) of this Code.

4. *Odors*

Any condition or operation which results in the creation of odors of such intensity and character as to be detrimental to the health and welfare of the public or which interferes unreasonably with the comfort of the public shall be removed, stopped or so modified as to remove the odor. (Ord. 6-108, Case F-91, adopted 8/20/68.)

5. *Electromagnetic Radiation*

The following standards shall apply:

a. **General.** It shall be unlawful to operate, or cause to be operated, any planned or intentional source of electromagnetic radiation for such purposes as communication, experimentation, entertainment, broadcasting, heating, navigation, therapy, vehicle velocity measurement, weather survey, aircraft detection, topographical survey, personal pleasure, or any other use directly or indirectly associated with these purposes which does not comply with the then current regulations of the Federal Communications Commission regarding such sources of electromagnetic radiation, except that for all governmental communications facilities, governmental agencies and government owned plants, the regulations of the interdepartment Radio Advisory Committee shall take precedence over the regulations of the Federal Communications Commission, regarding such sources of electromagnetic radiation. Further, said operation in compliance with the Federal Communications Commission or the Interdepartment Radio Advisory Committee regulations shall be unlawful if such radiation causes an abnormal degradation in performance of other electromagnetic radiators or electromagnetic receptors of quality and proper design because of proximity, primary field, blanketing, spurious reradiation, harmonic content, modulation or energy conducted by power or telephone lines. The determination of "abnormal degradation in performance" and "of quality and proper design" shall be made in accordance with good engineering practices as defined in the latest principles and standards of the American Institute of Electrical Engineers, the Institute of Radio Engineers, and the Electronic Industries Association. In case of any conflict between the latest standards and principles of the above groups, the following precedence in the interpretation of the standards and principles shall apply: (1) American Institute of Electrical Engineers, (2) Institute of Radio Engineers, and (3) Electronic Industries Association.

Recognizing the special nature of many of the operations which will be conducted because of the research and educational activities, it shall be unlawful for any person, firm or corporation to operate or cause to be operated, to maintain or cause to be maintained any planned or intentional sources of electromagnetic energy, the radiated power from which exceeds 1000 watts, without the express approval of the Director, Department of Inspection and Licenses. Further, it is required that any person, firm or corporation intending to operate or cause to be operated, to maintain or cause to be maintained any planned or intentional source of electromagnetic energy, the radiated power from which exceeds 10 watts, shall file, at least 30 days prior to such operation, a description of the radiating device and the operating characteristics thereof with the Director, Department of Inspection and Licenses.

b. Electromagnetic Interference. For the purpose of these regulations, electromagnetic interference shall be defined as electromagnetic disturbances which are generated by the use of electrical equipment other than planned and intentional sources of electromagnetic energy which interfere with the proper operation of electromagnetic receptors of quality and proper design. It shall be unlawful to operate or to cause to be operated any source of electromagnetic interference, the radiation or transmission from which exceeds the maximum values tabulated in tables shown below.

c. Method of Measurement: For the purpose of determining the level of radiated electromagnetic interference, standard field strength measuring techniques shall be employed. The maximum value of the tabulation shall be considered as having been exceeded if at any frequency in the section of the spectrum being measured, the measured field strength exceeds the maximum value tabulated for this spectrum section.

For purposes of determining the level of electromagnetic interference transmitted or conducted by power or telephone lines, a suitable, tunable, peak reading, radio frequency voltmeter shall be used. This instrument shall by means of appropriate isolation coupling, be alternately connected from line to line and from line to ground during the measurement. The maximum value of the tabulation shall be considered as having been exceeded if at any frequency in the section of the spectrum being measured, the measured peak voltage exceeds the maximum value tabulated for this spectrum section.

RADIATED

Section of Electromagnetic Spectrum (from-to)	Primary Intended Service	Maximum Field Strength at Edge of Property Containing Interference Source
10 Kilocycles—100 Kc.	Communications Service	500 microvolts/meter
100 Kc.— 535 Kc.	Navigational Aids	300 microvolts/meter
535 Kc.—1605 Kc.	AM Broadcasting	200 microvolts/meter
1605 Kc.— 44 Megacycles	Various Communications Service	200 microvolts/meter
44 Mc.— 88 Mc.	VHF Television—Airport Control	150 microvolts/meter
88 Mc.— 174 Mc.	FM Broadcasting	200 microvolts/meter
174 Mc.— 216 Mc.	VHF Television	150 microvolts/meter
216 Mc.— 580 Mc.	Navigational Aids—Citizens Radio	250 microvolts/meter
580 Mc.— 920 Mc.	UHF Television	300 microvolts/meter
920 Mc.—30,000 Mc.	Various	500 microvolts/meter

BY TRANSMISSION OR CONDUCTION

Section of Electromagnetic Spectrum (from-to)	Primary Intended Service	Maximum Voltage Measured Line to Line or Line to Ground Where Power or Telephone Lines Cross Edge of Property Containing Interference Source
10 Kilocycles—100 Kc.	Communications Services	2.5 Millivolts
100 Kc.— 535 Kc.	Navigational Aids	1.5 Millivolts
535 Kc.—1605 Kc.	AM Broadcasting	1.0 Millivolts
1605 Kc.— 44 Megacycles	Various Communications Services	0.5 Millivolts
44 Mc.— 88 Mc.	VHF Television	0.25 Millivolts
88 Mc.— 174 Mc.	FM Broadcasting—Airport Control	1.5 Millivolts
174 Mc.— 216 Mc.	VHF Television	0.15 Millivolts
216 Mc.— 580 Mc.	Navigational Aids—Citizens Radio	5.0 Millivolts
580 Mc.— 920 Mc.	UHF Television	20.0 Millivolts
920 Mc.—30,000 Mc.	Various	150.0 Millivolts

6. *Fire and Explosion*

All activities and all storage of flammable and explosive materials at any point shall be provided with adequate safety and fire fighting devices in accordance with the Fire Prevention Code of Montgomery County.

7. *Radioactive Materials*

The handling of radioactive materials, the discharge of such materials into air and water, and the disposal of radioactive wastes, shall be in conformance with the regulations of the Atomic Energy Commission as set forth in the Title 10, Chapter one, Part 20—Standards for Protection Against Radiation, as amended; and all applicable regulations of the State of Maryland.

8. *Glare and Heat*

No direct or sky-reflected glare, whether from floodlights or from high temperature processes such as combustion or welding or otherwise, so as to be visible at the lot line shall be permitted. These regulations shall not apply to signs or floodlighting of parking areas otherwise permitted by this ordinance. There shall be no emission or transmission of heat or heated air so as to be discernible at the lot line.

9. *Non-Radioactive Liquid or Solid Wastes*

There shall be no discharge at any point into any public or private sewage disposal system or stream, or into the ground, of any liquid or solid materials except in accordance with the regulations of the Montgomery County Health Department, the Washington Suburban Sanitary Commission, and the Water Pollution Control Commission of the State of Maryland, as applicable.

C. Area Requirements—Net Lot Area—Minimum Floor Area

1. Any tract of land zoned I-3 shall have a minimum area of fifty acres; except, that a tract with an area of not less than two acres may be permitted where such tract adjoins and has a common boundary with an I-1, I-2, or I-3 Zone, or with any industrial zone outside of and adjoining, the county portion of the regional district, or where such tract constitutes part of an area of not less than fifty acres recommended for I-3 zoning on a Master Plan adopted by the planning commission. The above-mentioned fifty acre tract shall include any such tract which is or may be separated into two or more parts by existing or proposed streets of arterial or lesser classification or by utility line properties or easements. (Ord. No. 6-67, adopted 12/5/67, No. F-84.)

2. Within any I-3 zoned area, each main building hereafter erected, together with its accessory buildings, shall be located on a lot having an area of at least two acres, with the exception of public utility buildings and structures necessary for public convenience and service, which may be located on a lot having an area of one acre.

3. The principal building on any lot used for offices, laboratories, warehouses and light industrial uses, except public utility buildings and structures necessary for public convenience and service, shall cover a minimum ground area of ten thousand square feet.

D. Required Yards and Landscaped Areas

1. No parking area, loading or maneuvering area shall be located:

a. less than 200 feet from the boundary of any Residential Zone, or

b. less than 10 feet from any lot line which adjoins an Industrial Zone or Commercial Zone.

2. No principal or accessory building above grade shall be located less than 200 feet from the boundary of any Residential Zone.

3. No principal or accessory building above grade, parking area, loading or maneuvering area shall be located less than the following minimum distances from the street right-of-way line or proposed street right-of-way of the following types of streets or highways as designated on the Master Plan of Highways or by the Montgomery County Planning Board (hereinafter referred to as "Planning Board") where no such plan exists:

(1) Controlled Access Routes	
From a freeway	100 feet
From an express parkway	100 feet
Major highway (with or without service roads)	100 feet
(2) Local Access Routes	
Arterial street separating the Industrial Park from a Residential Zone	100 feet
Arterial street separating the Industrial Park from a Commercial Zone	25 feet
From a local street, arterial street, or a privateway through the Industrial Park Zone	25 feet

4. Whenever property that is zoned Residential or Commercial is owned and/or used by a public utility as a transmission line right-of-way, or is owned and/or used as a railroad right-of-way, the setbacks required in paragraphs 1 and 2 above shall be determined as if said property were zoned Industrial.

5. The required yards set forth above shall be landscaped in accordance with a plan approved by the Planning Board as set forth in Section M-5. In addition to the yards required on the perimeter of the site, not less than five percent of the net lot area shall be landscaped, provided that where landscaping is required in accordance with Section 111-27d, such landscaping shall be included in the five percent required herein. Landscaping shall mean decorative plazas, pools, or the planting of grass, shrubs, or trees, or other comparable surface cover. To the maximum extent possible, driveways within such landscaped areas shall cross said areas by the most direct line, and all planting screens or walls required in Section J. shall be located in accordance with the requirements of Section 111-27. Such landscaped areas shall be maintained at all times. (Ord. No. 6-203, Case No. F-526, adopted 7/28/70.)

E. Street Access and Frontage

1. Each lot shall have a minimum frontage of 150 feet on a street or privateway; provided, however, that the Planning Board may approve a lesser frontage to a minimum of 50 feet for lots located on cul-de-sacs or on street curves or having other extraordinary characteristics. Vehicular access shall be permitted only to one of the following types of streets:

a. Controlled access routes:
Major highways

b. Local access routes:
(1) Major highways
(2) Connecting or secondary (arterial) highway
(3) A local street or privateway connecting only with any of the above highways and not directly connected with any residential street.

2. The designation of any street or highway as to type shall be in conformance with that shown on the officially adopted Master Plan of Highways. In the case of Industrial Park District applications in areas not covered by such plan, the Planning Board may at the time of reviewing any site plan as set forth in Section M, below, designate by type all streets and highways adjacent to and within the proposed district.

3. For the purposes of the above section, streets and highways shall be defined as follows:

a. Freeway—A divided highway for through traffic with full control of access and with grade separations at intersections. (Min. R/W—200′)

b. Express parkways—Same as Freeway except that use is limited to non-commercial traffic and is usually located within a park or a ribbon of parklike development. (Min. R/W—200′)

c. Major Highway, Controlled—A divided highway with parallel service roads which may or may not be continuous and with access connections limited to selected public roads at grade. (Min. R/W—With Service roads—180′. Without service roads—120′)

d. Major Highway—A divided highway with intersections at grade and direct access to abutting property, and on which geometric design and traffic control measures are used to expedite the safe movement of through traffic. (Min. R/W—120′)

e. Connecting or Secondary (Arterial) highway—Same as Major Highway but not necessarily divided. Function is to provide connection between Major Highways and Controlled Access Routes. (Min. R/W—80′)

f. Local Street or Privateway—A way used primarily to connect a lot or lots in the Industrial Park Zone with a public street or road outside the zone. (Min. R/W—70′)

F. Building Height Limit

Within the lot lines, the height of buildings may be equal to the horizontal distance from the nearest lot line.

In no event shall the height of any building exceed one hundred feet. Structures excluded from such height control pursuant to Section 111-4b may exceed the one hundred foot limit, but shall conform with height regulations related to distance from lot lines, above, provided that such horizontal distances shall be measured from the nearest boundary of the I-3 Zone. (Ord. No. 6-203, Case No. F-526, adopted 7/28/70)

G. Lot Coverage

Not more than twenty-five percent of the area of the lot may be covered by buildings, including accessory buildings.

H. Spacing Between Buildings

On the same parcel, no two buildings shall be located closer to one another than a distance equal to the height of the lower building measured perpendicular from the face of each building, provided that when the adjacent walls of such buildings are not fenestrated, the equal distance separation requirement shall not apply. (Ord. No. 6-203, Case No. F-526, adopted 7/28/70)

I. Signs

1. Signs displayed on any lot shall be limited to:

a. Those necessary for directional or informational purposes, but not exceeding four (4) square feet in size per sign.

b. Those necessary to identify the industrial park.

c. Those necessary to identify the use or establishment, and designed as a part of the architectural design of the building or as a part of the site plan for any lot.

2. Signs falling within category b shall not exceed one (1) square foot in size for each five linear feet of frontage of the street on which the sign faces. Signs falling within category c shall not exceed one (1) square foot for each horizontal linear foot of building wall facing on the street on which the sign faces.

3. The design, size of lettering, lighting, etc. of all signs in categories b and c shall accompany the site plan submission and shall meet the approval of the Planning Board as a part of the site plan as set forth in Section M; provided that additional signs may be added and existing signs changed after submission or approval of the site plan if notification thereof and the necessary description as above provided is given to the Department of Inspection and Licenses and the Montgomery County Planning Board and if the additions or changes are approved by the Planning Board. Exterior spot lighting of signs is permissible, but only if shielded so as to direct the light to the sign only.

J. Off-Street Parking Requirements

1. Off-street parking shall be provided in accordance with the general provisions set forth in Section 111-27a, and all permitted uses shall

conform to the schedule in Section 111-27c, except that for an industrial or manufacturing establishment or warehouse or similar use, the minimum requirement shall be one parking space for each one and one-half employees, or one for each two employees on combined major and second shifts, and in addition one visitor parking space for every 10 employees, except that the Planning Board may authorize fewer visitor parking spaces if it finds that a fewer number will be sufficient for the operation anticipated. In addition to the foregoing, one parking space shall be provided for each company-owned or -leased truck, passenger car or other vehicle located or principally based on the premises. No parking spaces may be located within required yards, except that an area equivalent to not more than 5% of the total area of all required parking spaces may be located within a required yard for use as parking space for visitors, selected personnel and minor deliveries. Off-street parking spaces may be grouped in facilities serving more than one lot or establishment.

2. When the lot on which parking spaces are located abuts the rear or side lot line of, or is across the street from, any land in a residential zone other than publicly owned land, a wall, fence or evergreen planting shall be maintained so as to screen substantially the parking lot from view from the nearest property in the residential zone. The screening shall be maintained in good condition at all times. In parking lots of one acre or more, at least 5% of the area of the parking lot shall be devoted to landscaping within the interior of the parking area. No luminaries on parking lots shall be more than ten feet above ground level.

K. Loading Berth Requirements

1. Off-street loading berths shall be provided for all buildings in accordance with the following schedule: for buildings with floor area of ten thousand to twenty-five thousand square feet: one berth. For each additional twenty-five thousand square feet or fraction thereof up to one hundred thousand square feet: one berth. For each additional fifty thousand square feet or fraction thereof above one hundred thousand square feet: one berth.

2. Such loading berths shall be at least fourteen feet wide, forty-eight feet long and fourteen feet in height (unless the Planning Board shall find that only smaller trucks requiring less space will be used for a period of ten years), and may be located either within a building or in the open, but not within required yards. If such berths are not enclosed they shall be located not less than three hundred feet from any residence zone boundary and effectively screened therefrom as in the case of parking areas, above. All access roads to loading berths shall be at least 14 feet wide, except that if tractor trailers would be accommodated, then the roads shall be 14 feet wide for one-way traffic and 22 feet wide for two-way traffic.

L. Enclosed Buildings and Outdoor Storage

1. All permitted uses and accessory activities shall be confined within completely enclosed buildings with the exception of off-street parking spaces, off-street loading berths, accessory fuel storage and employee recreational facilities. In addition, the temporary outdoor storage of materials, equipment or vehicles in an orderly manner is permitted in any areas other than required front, rear, or side yards, provided such outdoor storage does not exceed 15 feet in height or occupy more than ten per cent of the area of the lot, and is effectively screened from any adjacent residence zone boundary, as in the case of parking areas, above.

2. Fuel storage tanks utilized as part of the heating equipment of an establishment shall be located underground or in a building. Bulk storage of gasoline or petroleum products shall not be permitted except as incidental to a laboratory, a production operation or the servicing of company-owned or-leased vehicles and then shall be subject to the Fire Prevention Code for Montgomery County.

M. Application, Review, and Site Plan Approval

1. *Application*

In addition to submitting the plans and other data required in Section 111-52 an applicant for a building permit or certificate of occupancy in an I-3 Zone shall submit a plan of development to the Department of Inspection and Licenses as set forth in paragraph 5 below; and a description of the proposed operation including all machinery, processes, and products; and specifications for the mechanisms and techniques to be used in restricting the emission of dangerous and objectionable elements.

2. *Review by Department of Inspection and Licenses*

Within 10 days of receiving an application, the Department of Inspection and Licenses shall submit the site plan for the proposed development to the Planning Board for its review. In reviewing the application, the Department may secure the advice or assistance of one or more expert consultants selected by the County Council as qualified to advise as to whether a proposed use will conform to the applicable performance standards as set out in Section B above. The assistance of a consultant, if sought, must be obtained within 10 days of the receipt of the application. Such consultant or consultants shall report within 30 days after receipt of such request whether or not the use applied for will be operated in conformance with the performance standard, and if not, what modifications in design or operation would be necessary for conformance. A copy of the report of such consultants shall be furnished to the applicant.

3. *Action by Department*

Following receipt of the consultants' report, if any, and upon approval of the site plan by the Planning Board as set forth in paragraph 5, the Department of Inspection and Licenses shall, within 10 days, either approve, disapprove or require modifications in the plan as submitted. Appeals from the decision of the Department shall be in conformity with the procedure set forth in Section 111-30.

4. *Change in Use*

A new certificate of occupancy shall be required if there are any major structural alterations or substantial variations from the operations referred to in the building permit or previous certificate of occupancy permitting such use. However, any normal replacement or addition of equipment and machinery not affecting the foregoing operations or not changing the degree or nature of dangerous and objectionable elements emitted shall not be considered a change in use.

5. *Site Plan Approval*

No building permit or certificate of occupancy shall be issued for any use except in accordance with a plan of development approved by the Planning Board for the lot or tract on which such use is to be located. Such plan of development shall show the location and height of all buildings and structures; the area devoted to parking facilities and loading berths; all access roads, landscaping and screening plans; areas designated for outdoor storage; proposed signs, and other features required by this ordinance. The Planning Board shall consider the standards and objectives of the Industrial Park regulations and, specifically, the location of buildings, parking and loading areas and other features with respect to the topography of the lot and existing natural features such as water courses and large trees, with the objective of achieving a maximum of coordination between the proposed use and the surrounding uses. The Planning Board shall approve a plan of development if it meets the standards of these regulations, unless it shall find that the plan does not meet the objectives of Section 111-24 or of achieving a maximum of coordination between the proposed use and the surrounding uses. The Board shall submit its finding in writing, together with the reasons therefore, to the applicant and the Director of Inspection and Licenses.

The Planning Board, after receiving the site plan from the Department of Inspection and Licenses, shall be given not more than 45 days for consideration and action on such plan of development or amended plan of development; if no action is taken within such time, then such plan or amended plan shall be deemed to have been approved; provided, however, that the applicant for the Board's approval may waive this time limit and consent to the extension of such period, in which event he shall give notice of the waiver to the Director of the Department of Inspection and Licenses.

N. Enforcement

After occupancy, if there occurs continuous or frequent, even though intermittent, violations of the performance standards and provisions of these regulations for a period of 5 days, without bona fide and immediate corrective work, the Director of Inspection and Licenses shall suspend or revoke the occupancy permit of the use and the operation shall immediately cease until it is able to operate in accordance with these regulations, at which time the occupancy permit shall be reinstated.

O. Saving Clause

Should any subsection, sentence, clause, or phrase of this Section 111-24 be declared invalid by a Court of competent jurisdiction, such decision shall not affect the validity of this section in its entirety or of any part thereof, other than that so declared to be invalid. The County Council of Montgomery County, Maryland, sitting as the District Council for that portion of Montgomery County within the Maryland-Washington Regional District, hereby declares that it would have adopted this section and each subsection, sentence, clause, or phrase thereof irrespective of the fact that any one or more subsections, sentences, clauses, or phrases be declared invalid.

APPENDIX D

Sample Signage Manual, Inverness, Arapahoe County, Colorado

Source: Bennett & Kahnweiler Associates, Chicago, Illinois

Identification

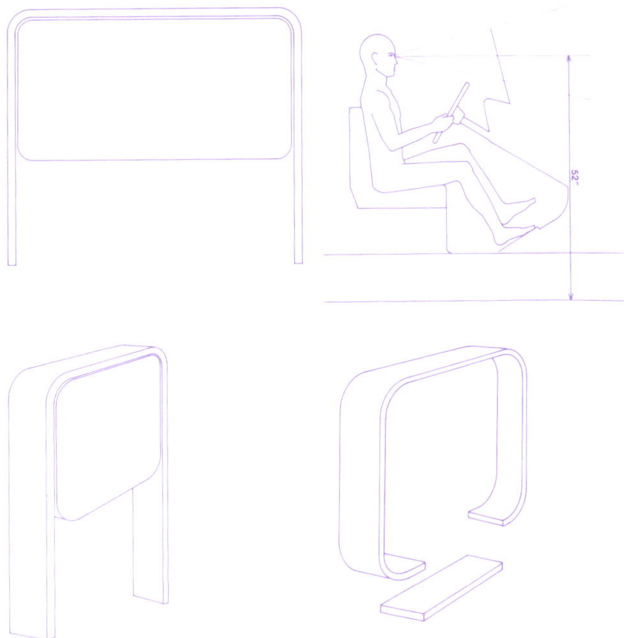

Inverness Signage Manual

A Complete System of Exterior Signage Specifically Designed to Enhance the Efficiency and Appearance of Inverness

Inverness Signage

Effective signage is necessary in any business environment. Both residents and visitors must be informed, directed and controlled. But all too often the needs for signage are distorted into a confusing and ugly assortment of random signs. They not only fail to communicate, they become a visual blight on the surroundings.

Because Inverness is a unique departure in business communities, it was decided early in the planning process that a total system of signage was to be designed. A leading design firm was commissioned to develop a signage program which blends with the character of the land and the concept of the community.

Particular attention was given to a real, but often forgotten quality of signage. While signage may serve strangers most, it is a constant and very visible environmental element to the resident.

So it is vital that the signage at Inverness meet the same high standard of aesthetic quality as the community itself.

Identification

The most direct way to control the impact of signage is to standardize its design. Yet each corporate resident at Inverness needs to identify itself. Some, in fact, have corporate identity programs in which they have great equity. So the need for a variety of graphics is obvious.

The solution to a proper balance between standardized control and individual expression was found in the growing body of data on the visual dynamics of signage communication.

This research has proved that people react most favorably to signage that combines variety and order. In that fact lies the basis of the Inverness signage system.

There is a standard structural system to be used for all Inverness identification signage. But each resident has the freedom to apply his own graphics to his signage. Thus order is achieved through structure, and variety through graphics.

The structure for the identification signage was designed to present a fluid and organic form within the community. Its elegant neutrality and simplicity is a function of the need to blend with a variety of architectural styles within Inverness.

The identification signage is fabricated from high quality rectangular steel channels formed into an elongated tubular frame. Because the frame is hollow, all electrics for illuminated signage can be buried underground and concealed within the frame.

The illuminated main sign box is mounted within the frame. It can be single or double faced. The Inverness identification signage is fitted with long-life neon tubes, serviceable by a specially designed bottom access panel.

A variety of materials and construction methods are available for the execution of graphics on the sign face.

1. A 20 gauge steel face can have routed letters backed with plexiglass, or it can be painted with baked enamel.
2. A poly-carbonate face can be color impregnated or left clear and

Identification

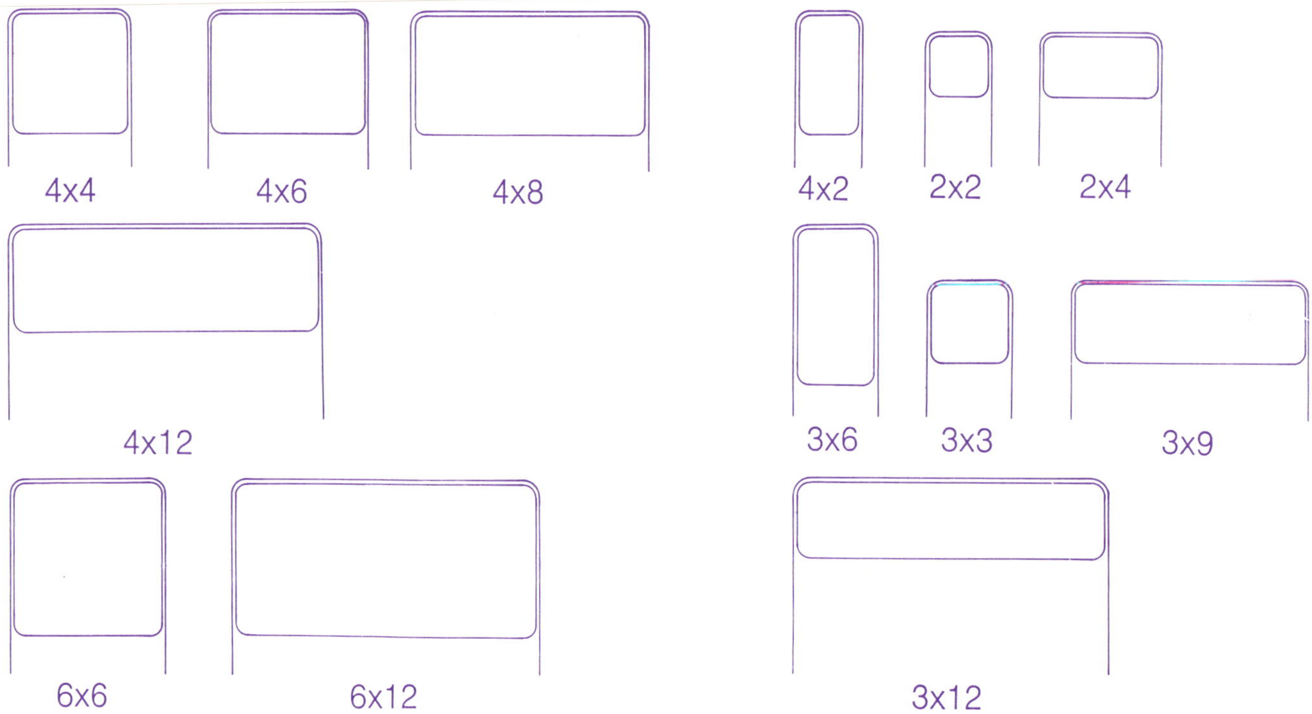

painted with poly-carbonate paint.

The identification signage system was designed to allow for a variety of sizes and shapes to meet differing corporate identity needs, and to be compatible with varying size sites and buildings.

There are no pre-determined rules concerning size selection. However, it is the intention of the program that the size of signage should be in aesthetic balance with the size of the site and its buildings.

The size and placement of all signage should be considered as an integral part of the architectural plans and submitted for the approval of the Architectural Review Committee.

Identification signs should be placed perpendicular to approaching vehicular traffic. There are two basic rules for placement.

1. Identification signs should be placed within the first 20% of the distance between the vehicular entrance and the building, or within the 20% of that distance nearest the building.

2. The sign must be placed so that it does not obscure any other identification, information, or vehicular control signs. In most cases, one identification sign is sufficient. The exceptions are those cases where a site has more than one vehicular entrance on different sides of the building, or where the nature of the site and adjacent roadways is such that more than one sign is required for proper identification.

The height of the identification sign is pre-determined so that the center line of the main panel is always at the optimum viewing height for a person seated in an automobile.

Signs or individual cut-out letters may not be placed on buildings within the complex. The free standing signage offered within this system is the only identification signage acceptable at Inverness.

In some cases, realtors may require site identification. This real estate signage must follow the structural standards for all Inverness site identification. The structure must be either bent pipe or the standard tubular steel channel.

Nomenclature on real estate signage is limited to the nature of the offering (for sale, for lease, etc.), the name of the offering real estate company and the phone number. Company logotypes, symbols, and the names of individual agents are not permitted.

The above sizes are available for identification signage:

Information

The second category in the Inverness signage system is information signage. This includes all information and directional signage other than identification which occurs within a site.

There is a standard design for all information signage in Inverness. This standardization reduces the visual impact of signage on the environment. It also makes it easier for residents and visitors alike to use and understand the signage within the complex.

There are two structures available for information signage. One is a pipe supported, non-illuminated version available in single or double faces. The second is an illuminated light-box version supported by the same structure as the primary identification signage. The main signage panels are constructed of 20 gauge anodized aluminum painted with either flat or reflective paint.

In all cases, the background of the

Information

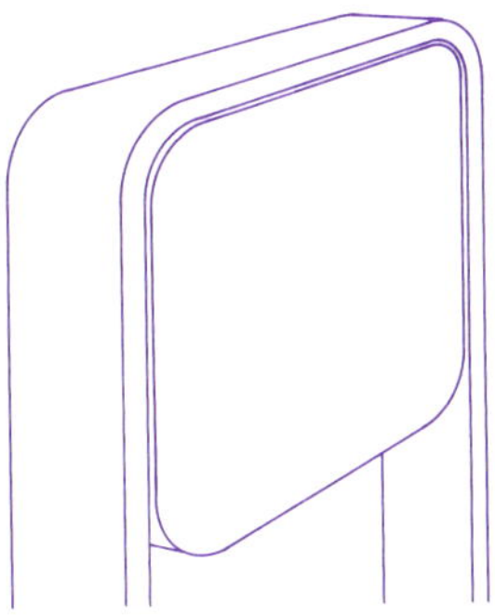

ABCDEFGHIJKLMNOPQRSTUVWXYZ
abcdefghijklmnopqrstuvwxyz
1234567890
&?!().,-:;'

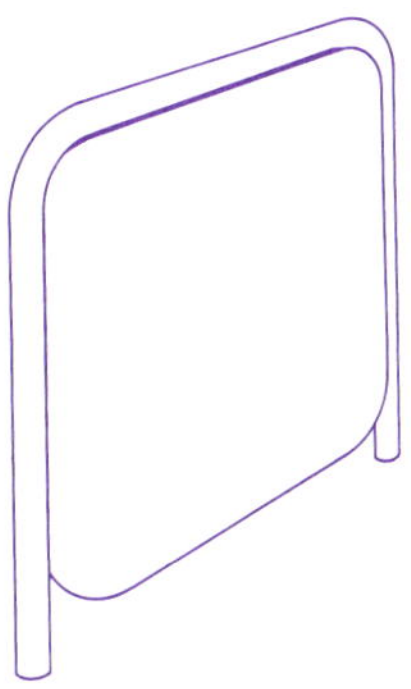

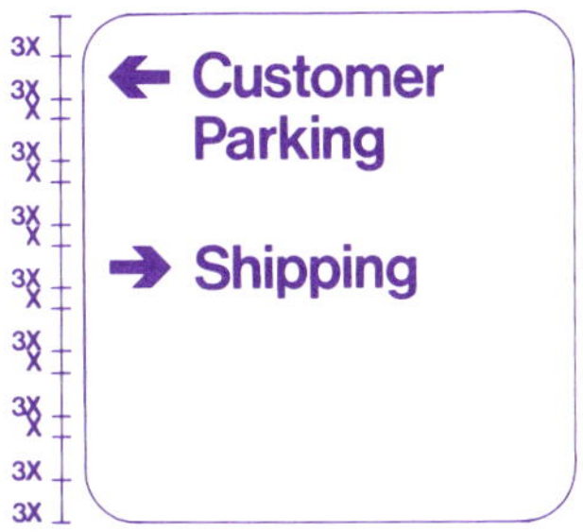

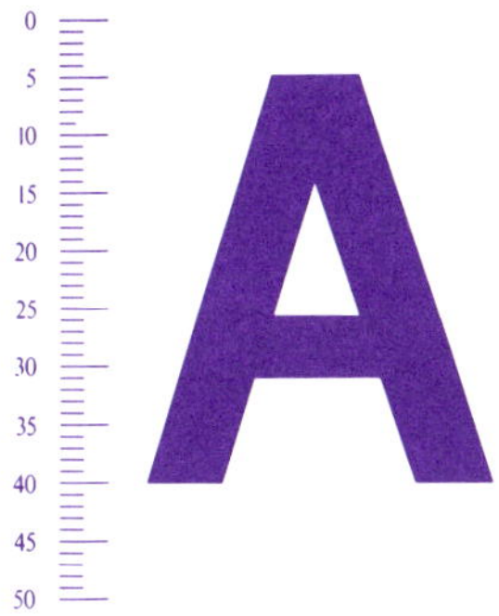

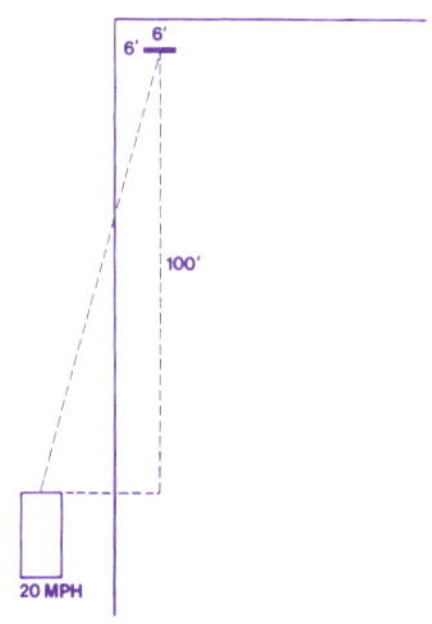

information signage is black, with white lettering.

The typeface chosen for the information signage is the most legible in the world today. This face, called Airport, is a modification of a sans serif typeface known as Standard. The modifications were made to improve legibility in different sizes and at long distances.

There are two critical elements to legibility in addition to the alphabet—letter spacing and copy positioning. When Airport was designed, a unit spacing system was created to define the optimum spacing between all the letterforms and numerals. The signage suppliers involved in the Inverness program have been trained in the use of this spacing system and in the correct rules for copy positioning.

There are some basic type style rules for information signage.

1. Word spacing should be even. Excessive variation in the length of lines should be avoided.

2. All copy is flushed to the left without indentation.

3. Only the first word in a sentence is capitalized unless there are proper names. Numbers under ten are to be spelled out in the body of sign messages.

4. Do not use a period at the end of a heading, sub-heading, title, date or any copy occupying a line by itself.

5. Do not insert the comma between numbers and street name. Insert commas in numbers over four figures.

6. In addition to normal grammatical use, place the colon instead of the dash before listed matter.

The positioning of information signage is critical to its effectiveness. Each site requires careful analysis of vehicular and pedestrian traffic. Decision points must be identified and proper information and directional signage provided. Further detail about site analysis is contained in the Procedures section of this document.

If signage must communicate to vehicular traffic, it must be placed so that it is visible and legible to approaching vehicles according to speed at the following distances:

Speed	Distance
20	100 feet
25	175
30	250

Signage should not be placed nearer to a road than six feet.

All information signage must be perpendicular to approaching traffic. It must be positioned so that there is clear line-of-sight well before the point at which direction must be changed or action taken. Information signage should be positioned to avoid confusing backgrounds, particularly when the signage is directed at vehicular traffic.

Vehicular Control

The third element in the Inverness signage system provides for vehicular control throughout the complex.

Vehicular control signage must meet five basic needs to be effective: it must fulfill a need, command attention, convey a clear and unmistakable meaning, command respect of road users and give adequate time for proper response.

This control system combines the best of European and American design for maximum legibility, recognition and aesthetic quality. Through proper use of color, typography, pictographs and positioning, it provides for safe and convenient access and movement through the community.

Vehicular Control

Procedures

The following are general guidelines for the use of the Inverness Signage System. In every case, a site analysis must be conducted when the building has been designed. This analysis is normally conducted in conjunction with an architect or a design consultant.

The correct size identification sign must be selected according to the size of the site and the building. This is essentially an aesthetic judgment. The placement of the sign, or signs if more than one is indicated under the terms of this covenant, must relate to two factors—the best position for viewing from the roadway and the best visual relationship to the architecture of the building.

Material must be selected and the decision made whether the sign will be illuminated. Illuminated signage is recommended. The graphics of the sign must be designed.

Every resident must be identified with an exterior sign from this system. In the event of multiple tenants on a single site, they should all be identified on the same sign or signs. The exception on this rule is when there are separate entrances for each tenant. In that case there can be separate signs for the tenant using each entrance.

The first step is to analyze exterior pedestrian and vehicular traffic flow to define those points at which people must make decisions. Once the decision points are defined, the signage can be placed on the site plan relative to those decision points according to the distance and placement rules of the covenant. The text for each information sign must be determined, and the decision made as to whether to illuminate some or all of the signs. Illuminated signs are strongly recommended wherever a significant volume of after-dark traffic is anticipated.

In some cases, particularly larger sites, vehicular control signage may be required within a site. That signage is the responsibility of the owner or tenant. Vehicular control, as distinct from information signage, will normally be necessary where there is more than one entrance, one-way internal roadways, cross-walks or complicated loading or parking areas. (The signs will be the same as those used throughout the complex's public roadways.)

The final signage plan should be incorporated into the architectural plans and submitted to the Review Committee. All signage sizes, placement, nomenclature and materials should be defined.

Several signage manufacturing companies have been trained by Inverness management in the details of copy placement, manufacturing and the nature of the covenants. However, residents are free to use a vendor of their choice. Engineering drawings are available for all the signs in the system, along with directions for copy placement on the identification signage.

Requirements for additional signage must be submitted to the Review Committee as they arise.

APPENDIX E

Checklist Outline for Covenants

Source: Harman O'Donnell and Henninger Assoc., Inc., Denver, Colorado

INDUSTRIAL

1. General Property
   a. Owners association
   b. Architectural control committee
   c. Extension of covenants to adjoining lands
   d. Safe, clean conditions and compliance with all government statutes, ordinances, regulations, health/police/fire requirements
2. Development Standards
   a. Building related
      (1) Minimum site size
      (2) Maximum building height
      (3) Lot coverage percent
      (4) Setbacks
         (a) Building
            —To street right-of-ways
            —To adjacent building sites
         (b) Loading docks
      (5) Architectural control
      (6) Loading
      (7) Building mechanical systems maximum distances from established buildings
   b. Landscape related
      (1) Landscaping standards
      (2) Visual quality
      (3) Signs
      (4) Maintenance
      (5) Refuse
         (a) Storage
         (b) Collection
      (6) Visual screen of outdoor storage
      (7) Rights-of-way, easements and utilities
         (a) Water
         (b) Sewer
         (c) Other utilities
         (d) Railroads
      (8) Off-street parking
      (9) Street design
      (10) Special zone lots for planned building groups
      (11) Permanent reference monuments
3. General Restrictions
   a. Nuisances
      (1) Noise
      (2) Smoke
      (3) Odor
      (4) Dust and dirt
      (5) Noxious gases and fumes
      (6) Glare
      (7) Heat
      (8) Vibration
      (9) Radiation
      (10) Fire hazards
      (11) Industrial wastes
   b. Allowed uses and restrictions

APPENDIX F

Sample Covenants—Irvine Industrial Complex, Irvine, California

Source: Irvine Industrial Complex, Irvine, California

DECLARATION OF RESTRICTIONS AND COVENANTS

THIS DECLARATION, made this 10th day of June, 1971, by IRVINE INDUSTRIAL COMPLEX, a California corporation (hereinafter "Declarant");

ARTICLE I RECITALS

1.01 Declarant is the owner of certain real property in the County of Orange, State of California, described in Exhibit "A" which is attached hereto and by reference made a part hereof (hereinafter the "Property").

1.02 In order to establish a general plan for the improvement and development of the Property, Declarant desires to impose on it mutual, beneficial restrictions for the benefit of all the lands in the Property and for the benefit of Declarant and the future owners of those lands.

1.03 The Property is part of a larger area of real property owned by Declarant as of the date of this Declaration in the County of Orange, State of California, described in Exhibit "B" which is attached hereto and by reference made a part hereof (hereinafter the "Retained Complex").

1.04 In order to insure the proper development and use of the Property in relation to the development and use of the Retained Complex, to protect the owner of each parcel within the Property and within the Retained Complex against such improper development and use of parcels within the Property and of the Property as a whole as will depreciate the value of his parcel, to prevent haphazard and inharmonious improvements, and in general to provide adequately for a high type and quality of improvement and use of the Property in accordance with a general plan for the development of the Irvine Industrial Complex as a whole, Declarant desires to subject the Property and each parcel of the Property to certain covenants for the benefit of all property within the Retained Complex. All of the property and each parcel within the Property shall be held, improved and conveyed subject to those covenants which shall be enforceable in accordance with this Declaration by Declarant and by each owner of a parcel or parcels of real property within the Retained Complex.

1.05 The property is presently subject to a Declaration of Restrictions made the 20th day of May, 1965, by Irvine Industrial Complex, as recorded on May 21, 1965, as document number 16662, Official Records, County of Orange, State of California (hereinafter the "General Declaration"). Nothing in this Declaration is intended to alter any of the provisions, rights, or obligations under the General Declaration, but the provisions, covenants and restrictions contained herein shall be independent of and cumulative with the provisions of the General Declaration.

ARTICLE II GENERAL PROVISIONS

2.01 *Establishment of Restrictions and Covenants*

Declarant, owner of the Property and of the Retained Complex, hereby declares that the Property is now held, and shall be transferred, sold, leased, conveyed and occupied subject to the restrictions and covenants herein set forth, each and all of which is and are for and shall inure to the benefit of and pass with each and every parcel of the Property and each and every parcel of the Retained Complex and apply to and bind the heirs, assignees and successors in interest of each and every owner of a parcel or parcels of the Property.

2.02 *Restrictions Operate as Covenants*

Each purchaser of any parcel of the Property covenants and agrees with Declarant, its successors and assigns to use the Property only in accordance with the restrictions herein set forth and to refrain from using the Property in any way inconsistent with or prohibited by the provisions of this Declaration.

2.03 *Purpose of Restrictions and Covenants*

It is the intent and purpose of these covenants and restrictions to allow the location on the Property of general manufacturing activities, provided that such activities are confined within a building or buildings and do not contribute excessive noise, dust, smoke or vibration to the surrounding environment nor contain a high hazard potential due to the nature of the products, material or processes involved. It is the further intent and purpose of these covenants and restrictions to control the user-occupant density on the Property, to expressly prohibit certain uses of the Property and to protect the character of Red Hill Avenue as a Special Landscaped Street.

2.04 *Definitions*

(a) *Area of Elevation.* Total height and length of a building as projected to a vertical plane.

(b) *Building Line.* An imaginary line parallel to the street right-of-way line specifying the closest point from this street right-of-way line that a building structure may be located (except for overhangs, stairs, and sunscreens).

(c) *Lot.* The fractional part of blocks as divided and subdivided on subdivision maps of the Official Records of Orange County, California, as they from time to time become current.

(d) *Right-of-way Line.* When reference is made to right-of-way line it shall mean the line which is then established on either the adopted County Master Plan of Arterial Roads or the filed Tract Map for Minor Roads as the ultimate right-of-way line for roads or streets.

(e) *Side and Front of Lots and Sites.* The Front of a Lot or Site, except a corner Lot or Site, is the portion thereof facing on any street. (Thus a Lot or Site may have two Fronts where, for instance, it faces onto two parallel streets. As to corner Lots or Sites, the narrowest frontage of a Lot or Site facing the street is the Front, and the longest side facing the intersecting street is the Side, irrespective of the direction in which the structures face).

(f) *Sign.* Any structure, device or contrivance, electric or non-electric, and all parts thereof which are erected or used for advertising purposes upon or within which any poster, bill, bulletin, printing, lettering, painting, device or other advertising of any kind whatsoever is used, placed, posted, tacked, nailed, pasted, or otherwise fastened or affixed.

(g) *Site.* All contiguous land under one ownership and tenancy; provided, however, that multiple occupancy of a building in accordance with Paragraph 4.04 of this Declaration does not destroy a parcel's character as a Site.

(h) *Streets.* Reference to all streets or rights-of-way within this ordinance shall mean dedicated vehicular rights-of-way. In the case of private or non-dedicated streets, a minimum setback from the right-of-way line of said streets of ten (10) feet shall be required for all structures. Except for sidewalks or access drives, this area shall be landscaped according to the setback area standards from dedicated streets contained herein.

ARTICLE III
PERMITTED USES

3.01 Each and every parcel of the Property is restricted to the following permitted uses:

(a) Uses primarily engaged in research activities, including but not limited to research laboratories and facilities, developmental laboratories and facilities, and compatible light manufacturing relating to the following examples:

Bio-Chemical
Chemical
Development Facilities for National Welfare on Land, Sea and Air
Film and Photography
Medical or Dental
Metallurgy
Pharmaceutical
X-Ray

(b) Manufacture, research assembly, testing and repair of components, devices, equipment and systems and parts and components such as but not limited to the following examples:

Coils, Tubes, Semi-Conductors
Communication, Navigation Control, Transmission and Reception Equipment, Control Equipment and Systems Guidance Equipment and Systems
Data Processing Equipment and Systems
Glass Edging, Beveling and Silvering
Graphics, Art Equipment
Metering Instruments
Optical Devices, Equipment and Systems
Phonographs, Audio Units, Radio Equipment and Television Equipment
Photographic Equipment
Radar, Infra-Red and Ultra-Violet Equipment and Systems
Scientific and Mechanical Instruments
Testing Equipment

(c) Other manufacturing to include but not limited to the following examples:

(i) Manufacture and/or assembly of the following or similar products:

Aircraft and Related Components
Automobiles and Parts
Boats
Clocks and Watches
Coffins
Ceramic Products
Concrete Products
Electrical Appliances
Farm Equipment
Heating & Ventilating Equipment
Linoleum
Machinery & Machine Tools
Musical Instruments
Neon Signs
Novelties
Oil Well Valves & Repairs
Optical Goods
Refrigeration
Screw Machine Products
Sheet Metal Products
Shoes
Silk Screens
Sporting Goods
Springs
Stencils
Toys
Trailers
Trucks

(ii) The manufacture of products or products made from the following or similar materials:

Aluminum
Bags, Except Burlap Bags or Sacks
Batteries
Boxes, Paper
Brass
Cans
Copper
Glass
Grinding Wheels
Iron
Linoleum
Matches
Mattresses
Paper
Steel
Tin
Tools
Wool
Yarn

(iii) The manufacturing, compounding, processing or treatment of the following or similar items:

Acids, Non-Corrosive
Candles
Cigarettes and Cigars
Detergents
Disinfectants
Dye
Food Products
Lubricating Oil
Pharmaceutical Products
Plastics
Toiletries
Vitamin Products
Waxes and Polishes

(iv) Woodworking Shops, such as: (Provided that, if a planer, router, sticker or moulder is maintained, all doors and windows in the outside walls of the room in which said machinery is located shall be kept closed while said machine is in use.)

Box
Furniture
Wood Products

(v) Distribution and Warehousing Plants

(d) Administrative, professional and business offices associated with and accessory to a permitted use.

(e) Regional or home offices of industries which are limited to a single use and accessory to any of the above industrial developments.

(f) Blue printing, photostating, photo engraving, printing, publishing and bookbinding, provided that no on-site commercial service is associated with said uses.

(g) Cafeteria, cafe, restaurant or auditorium accessory with and incidental to any of the foregoing uses.

(h) Agriculture, as a continuation of the existing land use, and all necessary structures and appurtenances shall be permitted.

(i) General contractor and construction industries relating to the building industry, such as general contractors, electrical contractors, plumbing contractors, etc.

(j) Any other use permitted by applicable zoning ordinances and regulations of government authorities having jurisdiction over the Property, provided Declarant specifically consents to such use in writing.

3.02 All uses listed above are subject to the Development Standards listed in Article IV of this Declaration.

ARTICLE IV
DEVELOPMENT STANDARDS

4.01 *Setbacks*

No building shall be located on any one or more Lots nearer to the front Lot line or nearer to the side Lot line than the minimum setback set forth below:

(a) *Front Yard Setback.* Thirty (30) feet, except that unsupported roofs or sun screens may project six (6) feet into the setback area.

(b) *Side Yard Setback.* Ten (10) feet, except that unsupported roofs and sun screens may project three (3) feet into the setback area; provided that if a single building is constructed on two or more Lots, (or if a Site on which a single building was originally constructed is further subdivided into two or more Lots in accordance with the provisions of Paragraph 5.02 of this Declaration) no side yard setback is required from interior Lot lines; provided further that, in any event, there shall be at least twenty (20) feet of open space between all buildings on the Property which area shall be a Side Yard Setback Area for the purpose of this Declaration. In the case of a corner lot, the street side setback shall be thirty (30) feet, except that unsupported roofs and sun screens may project six (6) feet into the setback area. Interior Lot lines for a corner Lot shall be considered side Lot lines.

(c) *Rear Yard Setback*. No rear yard setback is required.

4.02 *Site Coverage*

Maximum building coverage of fifty percent (50%) of a Site is allowed. Parking structures shall not be calculated as building area; however, said structures shall be used only for the parking of company vehicles, employees' vehicles, or vehicles belonging to persons visiting the subject firm.

4.03 *Number of Buildings per Lot*

For a period of five (5) years from the date of this Declaration, only one building, other than a parking structure to be used only as set forth in Paragraph 4.02 of this Declaration, shall be erected on any one Lot, unless the erection and use of more or less than one building on any one Lot is specifically approved and consented to by Declarant in writing.

4.04 *Number of Tenants per Building*

For a period of five (5) years from the date of this Declaration, no building shall be occupied or used by more than two tenants or users simultaneously without the specific written consent of Declarant, nor shall more than two businesses be conducted simultaneously in any one building without the specific written consent of Declarant. The multiple occupancy and/or use of a building under this Paragraph does not change the character of the parcel, on which the building was constructed and the boundaries of which were used to determine compliance with this Declaration, as a Site for the purposes of this Declaration.

4.05 *Signs*

No Sign shall be erected or maintained on the Property except in conformity with the following:

(a) Signs visible from the exterior of any building may be lighted, but no Signs or any other contrivance shall be devised or constructed so as to rotate, gyrate, blink or move in any animated fashion.

(b) Signs shall be restricted to advertising only the person, firm, company or corporation operating the use conducted on the Site or the products produced or sold thereon.

(c) All Signs attached to the building shall be flush mounted.

(d) Only one (1) single faced or double faced Sign shall be permitted per Street frontage. No Sign or combination of Signs shall exceed one (1) square foot in area for each six hundred (600) square feet of total Site area. However, no Sign shall exceed two hundred (200) square feet in area per face. An additional twenty (20) square feet shall be allowed for each additional business conducted on the site.

(e) A Sign advertising the sale, lease, or hire of the Site shall be permitted in addition to the other Signs listed in this section. Said Sign shall not exceed a maximum area of thirty-two (32) square feet.

(f) No ground Signs shall exceed four (4) feet above grade in vertical height. Also, ground Signs in excess of one hundred (100) square feet in area (single face) shall not be erected in the first twenty (20) feet, as measured from the property line, of any street side setback area. However, the above standards shall not apply to the Community Directional Sign, Special Purpose Sign, Construction Sign, or Future Tenant Identification Sign.

(g) Wall Signs shall be fixture Signs; Signs painted directly on the surface of the wall shall not be permitted.

(h) A wall Sign with the individual letters applied directly shall be measured by a rectangle around the outside of the lettering and/or the pictorial symbol and calculating the area enclosed by such line.

(i) One (1) Construction Sign denoting the architects, engineers, contractor, and other related subjects, shall be permitted upon the commencement of construction. Said Sign shall conform to applicable zoning ordinances and regulations.

(j) A Future Tenant Identification Sign listing the name of future tenants, responsible agent or realtor, and identification of the Irvine Industrial Complex shall be permitted. Said Sign shall conform to applicable zoning ordinances and regulations.

(k) Special Purpose Signs, used to give directions to traffic or pedestrians or give instructions as to special conditions, and Community Directional and/or Identification Signs, used to give directions to and identify areas within the Irvine Industrial Complex, shall be in conformity with applicable zoning ordinances and regulations.

4.06 *Parking*

Each owner of a Site shall provide adequate off-Street parking to accommodate all parking needs for the Site. The intent is to eliminate the need for any on-street parking.

Required off-Street parking shall be provided on the Site of the use served, or on a contiguous Site or within six hundred (600) feet of the subject Site. Where parking is provided on other than the Site concerned, a recorded document shall be filed with the Declarant and signed by the owners of the alternate Site stipulating to the permanent reservation of the use of the Site for said parking.

The following guide shall be used to determine parking requirements:

Office

Three (3) spaces for each one thousand (1,000) square feet of total office space (excluding such areas as pedestrian corridors, rest rooms, elevator shafts, equipment areas).

Manufacture, Research and Assembly

Two (2) parking spaces for each three (3) employees, but in no event less than two (2) spaces for each one thousand (1,000) square feet of gross floor area.

Warehouse

Two (2) parking spaces for each three (3) employees, but in no event less than one (1) space for each one thousand (1,000) square feet of gross floor area for the first twenty thousand (20,000) square feet; one space for each two thousand (2,000) square feet of gross floor area for the second twenty thousand (20,000) square feet; one (1) space for each four thousand (4,000) square feet of gross floor area for areas in excess of the initial forty thousand (40,000) square feet of floor area of the building. If there is more than one shift, the number of employees on the largest shift shall be used in determining parking requirements.

4.07 *Landscaping*

(a) The Front Yard Setback Area of each Site shall be landscaped with an effective combination of street trees, trees, ground cover and shrubbery. All unpaved areas not utilized for parking shall be landscaped in a similar manner. The entire area between the curb and a point ten (10) feet in back of the front property line shall be landscaped, except for any access driveway in said area. Notwithstanding the above, the entire area between the curb and the building setback line of any property fronting on Red Hill Avenue shall be landscaped.

(b) Side and Rear Yard Setback Areas not used for parking or storage shall be landscaped utilizing ground cover and/or shrub and tree materials.

(c) Undeveloped areas proposed for future expansion shall be maintained in a weed-free condition and shall be landscaped if required by Declarant.

(d) Areas used for parking shall be landscaped and/or fenced in such a manner as to interrupt or screen said areas from view from access Streets, freeways, and adjacent properties. Plant materials used for this purpose shall consist of lineal or grouped masses of shrubs and/or trees.

4.08 *Loading Areas*

No loading shall be allowed which is visible from adjacent Streets. On other than Red Hill Avenue, Street side loading shall be allowed provided the loading dock is set back a minimum of seventy (70) feet from the Street right-of-way line, or one hundred ten (110) feet from the street centerline, whichever is greater. Said loading area must be screened from view from adjacent Streets.

4.09 *Storage Areas*

All outdoor storage shall be visually screened from access Streets, freeways, and adjacent property. Said screening shall form a complete opaque screen up to a point eight (8) feet in vertical height but need not be opaque above that point. Outdoor storage shall be meant to include parking of all company owned and operated motor vehicles, with the exception of passenger vehicles. No storage shall be permitted between a frontage street and the Building Line.

4.10 *Refuse Collection Areas*

All outdoor refuse collection areas shall be visually screened from access Streets, freeways, and adjacent property by a complete opaque screen. No refuse collection areas shall be permitted between a frontage Street and the Building Line.

4.11 *Telephone and Electrical Service*

All "on site" electrical lines (excluding lines in excess of 12 KV) and telephone lines shall be placed underground. Transformer or terminal equipment shall be visually screened from view from Streets and adjacent properties.

4.12 *Nuisances*

No portion of the Property shall be used in such a manner as to create a nuisance to adjacent Sites, such as but not limited to vibration, sound, electro-mechanical disturbance and radiation, electro-magnetic disturbance, radiation, air or water pollution, dust emission of odorous, toxic or noxious matter.

4.13 *Service Stations*

No service station shall be constructed or operated on the Property.

ARTICLE V
ADDITIONAL RESTRICTIONS

5.01 *No Access from Red Hill Avenue*

No Lot or Site shall have direct access onto Red Hill Avenue and no driveway or other provision for access other than dedicated streets shall be constructed onto Red Hill Avenue.

5.02 *No Lot Splits*

Each and every Site shall consist of at least one whole and entire Lot. No owner shall initiate action to reduce the size of any Lot or further subdivide any Lot within five (5) years from the date of this Declaration.

ARTICLE VI
ENFORCEMENT

6.01 *Abatement and Suit*

Violation or breach of any restriction and covenant herein contained shall give to Declarant, and every other owner of property for whose benefit these restrictions and covenants are expressly made, the right to enter upon the property upon or as to which said violation or breach exists and to summarily abate and remove, at the expense of the owner or lessee thereof, any structure, thing or condition that may be or exist thereon contrary to the intent and meaning of the provisions hereof, or to prosecute a proceeding at law or in equity against the person or persons who have violated or are attempting to violate any of these restrictions and covenants to enjoin or prevent them from doing so, to cause said violation to be remedied or to recover damages for said violation.

6.02 *Deemed to Constitute a Nuisance*

The result of every action or omission whereby any restiction or covenant herein contained is violated in whole or in part is hereby declared to be and to constitute a nuisance, and every remedy allowed by law or equity against an owner, either public or private, shall be applicable against every such result and may be exercised by Declarant or by any owner of property for whose benefit these restrictions and covenants are made.

6.03 *Attorney's Fees*

In any legal or equitable proceeding for the enforcement or to restrain the violation of this Declaration or any provision hereof, the losing party or parties shall pay the attorney's fees of the prevailing party or parties, in such amount as may be fixed by the Court in such proceedings. All remedies provided herein or at law or in equity shall be cumulative and not exclusive.

6.04 *Inspection*

Declarant may from time to time at any reasonable hour or hours, enter and inspect any property subject to these restrictions to ascertain compliance therewith.

6.05 *Failure to Enforce Not a Waiver of Rights*

The failure of Declarant or any other property owner to enforce any restriction herein contained shall in no event be deemed to be a waiver of the right to do so thereafter nor of the right to enforce any other restriction or covenant.

ARTICLE VII
TERM, TERMINATION AND MODIFICATION

7.01 *Term*

This Declaration, every provision hereof and every covenant and restriction contained herein shall continue in full force and effect for a period of twenty-five (25) years from the date hereof, unless otherwise specifically provided.

7.02 *Termination and Modification*

This Declaration, or any provision hereof, or any covenant or restriction contained herein, may be terminated, extended, modified or amended, as to the whole of said property or any portion thereof, with the written consent of the owners of sixty-five percent (65%) of the property specified in Exhibit "B" hereto based on the number of square feet owned as compared to the total number of square feet so specified, provided, however, that so long as Declarant owns at least twenty percent (20%) of the property specified in Exhibit "B" hereto, no such termination, extension, modification or amendment shall be effective without the written approval of Declarant thereto. No such termination, extension, modification or amendment shall be effective until a proper instrument in writing has been executed and acknowledged and recorded in the office of the Recorder of Orange County, California.

ARTICLE VIII
MISCELLANEOUS PROVISIONS

8.01 *Constructive Notice and Acceptance*

Every person who now or hereafter owns or acquires any right, title or interest in or to any portion of said property is and shall be conclusively deemed to have consented and agreed to every covenant, condition and restriction contained herein, whether or not any reference to this Declaration is contained in the instrument by which such person acquired an interest in said property.

8.02 *Rights of Mortgagees*

All restrictions and other provisions herein contained shall be deemed subject and subordinate to all mortgages and deeds of trust now or hereafter executed upon land subject to these restrictions, and none of said restrictions shall supersede or in any way reduce the security or affect the validity of any such mortgage or deed of trust; provided, however, that if any portion of said property is sold under a foreclosure of any mortgage or under the provisions of any deed of trust, any purchaser of such sale, and his successors and assigns, shall hold any and all property so purchased subject to all of the restrictions, covenants and other provisions of this Declaration.

8.03 *Mutuality, Reciprocity; Runs with Land*

All restrictions, covenants and agreements contained herein are made for the direct, mutual and reciprocal benefit of each and every part and parcel of said property; shall create mutual, equitable servitudes upon each parcel in favor of every other parcel; and shall create reciprocal rights and obligations between the respective owners of all parcels of the Property and privity of contract and estate between all grantees of said parcels, their heirs, successors and assigns.

In addition, all restrictions contained herein shall operate as covenants running with the land for the benefit of the land described in Exhibit "B" hereto and shall inure to the benefit of all grantees of said land, their heirs, successors and assigns, and shall apply to and bind the grantees of any and all parcels of the Property, their heirs, successors, and assigns.

8.04 *Paragraph Headings*

Paragraph headings, where used herein, are inserted for convenience only and are not intended to be a part of this Declaration or in any way to define, limit or describe the scope and intent of the particular paragraphs to which they refer.

8.05 *Effect of Invalidation*

If any provision of this Declaration is held to be invalid by any court, the invalidity of such provision shall not affect the validity of the remaining provisions hereof.

8.06 *Addition of Territory*

Declarant may at any time or from time to time during the pendency of these restrictions add all or a portion of the land described in Exhibit "B," which is attached hereto and by reference made a part hereof, to the Property which is covered by this Declaration, and upon the recording of a notice of addition to territory containing the provisions set forth in Section 8.07 of this Article VIII, the covenants contained in this Declaration shall apply to the added land in the same manner as if it were originally covered by this Declaration; and thereafter, the rights, powers and responsibilities of the parties to this Declaration with respect to the added land shall be the same as with respect to the original land, and the rights, privileges, duties and liabilities of the owners, lessees and occupants of parcels within the added land shall be the same as in the case of the original land.

8.07 The notice of addition to territory referred to in Section 8.06 of this Article VIII shall contain the following provisions:

(a) A reference to this Declaration, which reference shall state the date of recording hereof and the book or books of the records of Orange County, California, and page numbers, where this Declaration is recorded;

(b) A statement that the provisions of this Declaration shall apply to the added territory in the manner set forth in Section 8.06 of this Article VIII; and

(c) An exact description of the added territory.

IN WITNESS WHEREOF, the undersigned has executed this Declaration on the date first hereinabove written.

IRVINE INDUSTRIAL COMPLEX
(Signed)
John E. Murphy, *President*
(Signed)
Lansing E. Eberling, *Secretary*

STATE OF CALIFORNIA
COUNTY OF ORANGE ss.

On June 10, 1971, before me, the undersigned, a Notary Public in and for said State, personally appeared JOHN E. MURPHY, known to me to be the President, and LANSING E. EBERLING, known to me to be the Secretary, of the corporation that executed the within instrument, known to me to be the persons who executed the within instrument on behalf of the corporation therein named, and acknowledged to me that such corporation executed the within instrument pursuant to its by-laws or a resolution of its board of directors.

WITNESS my hand and official seal.

(Signed)
GERALDINE L. WAGNER
Notary Public, State of California
Principal Office in Orange County.

APPENDIX G

Sample Covenants—Hunt Valley Business Community

Source: Maryland Properties Inc., Hunt Valley, Maryland

ARTICLE I
Land Use

A. No land shall be used for any purpose other than as permitted in an "M. L. Zone—Manufacturing, Light" under Section 253 entitled "Use Regulations" of the Baltimore County Zoning Regulations in force and effect on the date of this Deed as the same may be hereafter from time to time amended, expressly including within the uses permitted motels, motor hotels and motor courts, but expressly excluding any use for which a special exception is required under Section 502 of the Baltimore County Zoning Regulations as the same may be hereafter from time to time amended, unless such special exception be finally granted and such use be approved in writing by Developer. The Developer reserves the right, however, further to limit or restrict the use of a particular lot or lots under the provisions of Article III A. hereof.

B. No use will be made of any lot or any portion thereof or any building or structure thereon at any time, nor shall any materials or products be manufactured, processed or stored thereon or therein, which shall, in the opinion of the Developer, cause an undue fire hazard to adjoining properties, or which shall constitute a nuisance or cause the emission of noxious odors or gases or smoke, or cause noises or other conditions which might injure the reputation of the lot in question or neighboring properties or which shall constitute a violation of any law of the United States, the State of Maryland, or Baltimore County, or any regulation or ordinance promulgated thereunder.

ARTICLE II
Setbacks

No Building, or any part thereof or projection therefrom, shall be erected nearer than fifty (50) feet from any primary road in the Park (a primary road being a public right-of-way sixty (60) feet or more in width granted, or to be granted, to Baltimore County), nor nearer than thirty (30) feet from any secondary road in the Park (a secondary road being a public right-of-way less than sixty (60) feet in width granted, or to be granted, to Baltimore County), nor nearer than thirty (30) feet from any side or rear boundary line of the parcel on which the building is erected.

ARTICLE III
Plans and Specifications

A. No building, fence, wall, sign, advertising device, roadway, loading facility, outside storage facility, parking area, site grading, planting, landscaping, facility for industrial waste or sewage disposal, nor any other improvement shall be commenced, erected or constructed, nor shall any addition thereto or change or alteration therein be made (except to the interior of a building), nor shall any change in the use of any premises be made, until the plans and specifications therefor, showing the nature, kind, shape, heights, materials, color scheme, lighting and location on the lot of the proposed improvements, grading, landscaping or alterations and the proposed use or change in the use of the premises, shall have been submitted to and approved in writing by the Developer and a copy of such plans and specifications as finally approved lodged permanently with the Developer. The Developer shall have the right to refuse to approve any such plans or specifications or proposed use of the premises for any reason which the Developer, in its sole discretion, may deem in the best interests of the Park and the owners or lessees or prospective owners or lessees of other properties therein.

B. No parking will be permitted on the streets in the Park and each lot owner shall provide on his property necessary and adequate parking facilities and private driveways as approved by the Developer under paragraph A of this Article III.

C. Construction and alteration of all improvements in the Park shall be in accordance with the requirements of all applicable Building, Zoning and other Codes and Regulations.

ARTICLE IV
Maintenance

A. Each lot owner shall at all times keep his premises, buildings, improvements and appurtenances in a safe, clean, neat and sanitary condition and shall comply with all laws, ordinances and regulations pertaining to health and safety. Each lot owner shall provide for the removal of trash and rubbish from his premises.

B. During construction it shall be the responsibility of each lot owner to insure that construction sites are kept free of unsightly accumulations of rubbish and scrap materials, and that construction materials, trailers, shacks and the like are kept in a neat and orderly manner.

C. The Developer agrees to maintain all undeveloped land owned by it within the Park in a manner compatible with the provisions of this Article IV.

ARTICLE V
Amendment, Terminations, Extension

Each condition, restriction and covenant as herein provided shall terminate and be of no further effect on January 1, in the year 2000, provided that at any time prior to that date, the owners (excluding mortgagees and the holders of other security devices who are not in possession, lessees and tenants) of a majority of the acreage in the Park (excluding highways) may by written declaration signed and acknowledged by them and recorded among the Land Records of Baltimore County alter, amend, revoke or extend indefinitely, or for a limited duration, any or all of said conditions, restrictions and covenants and such declaration may provide for further amendment, alteration, revocation or extension as herein provided or in any other manner; provided, however, that no such alteration or amendment shall affect any plans, specifications or use theretofore approved by Developer under Article III hereof or any improvements theretofore or thereafter made pursuant to such approval.

ARTICLE VI
Waiver of Invalidation

Any waiver of failure to enforce any provision of these covenants and restrictions in a particular situation shall not be deemed a waiver or abandonment of such provision as it may apply in any other situation or to the same or a similar situation at any other location in the Park or of any other provision of these covenants and restrictions. Invalidation by Court adjudication of any provision of these covenants and restrictions shall not affect the validity of any other provision, and all other provisions thereof shall remain in full force and effect.

ARTICLE VII
Covenants Run with Land; Enforceability

A. The foregoing covenants and restrictions shall run with, bind and inure to the benefit of the parcel hereby conveyed and the land in the Park retained by the Developer, and the Developer covenants and agrees to hold such land retained by it subject to all the terms, provisions and conditions of these covenants and restrictions and that any and all sales, leases, mortgages or other dispositions of such remaining land or any part thereof shall be subject to these covenants and restrictions. The Developer reserves the right however from time to time hereafter to delineate, plat, grant or reserve within the remainder of the Park not hereby conveyed such public streets, roads, sidewalks, ways and appurtenances thereto, and such easements for drainage and public utilities, as it may deem necessary or desirable for the development of the Park (and from time to time to change the location of the same) free and clear of these restrictions and covenants, and to dedicate the same to public use or to grant the same to Baltimore County and/or to appropriate public utility corporations.

B. Said covenants and restrictions shall be jointly and severally enforceable by the Developer and its successors and assigns and by the Grantee, and its successors and assigns, provided however that only the Developer or its assignees, under Article VIII hereof, shall have the right to exercise the discretionary powers herein reserved to the Developer.

C. Violation of any of said restrictions or conditions, or breach of any covenant or agreement herein contained shall give the Developer or its assignees under Article VIII, in addition to all other remedies, the rights (but not the obligation) to enter upon the land as to which such violation or breach exists and summarily to abate and remove any erection or thing or correct any condition that may constitute such violation or breach at the expense of the then owner of such land, which expense shall be a lien on such land enforceable in Equity; provided however that no such entry shall be made unless the violation or breach has not been remedied and corrected within thirty days after delivery of notice of such violation or breach from the Developer or its said assignees to the occupant of the premises on which the violation or breach has occurred or in the alternative within thirty days after mailing such notice, by first class mail, postage prepaid, to the record owner of such premises at his or its last known address.

ARTICLE VIII

Nominees and Successors of Developer

The Developer may from time to time delegate any or all of its rights, powers, discretion and duties hereunder to such agent or agents as it may nominate. It may also permanently assign any or all of its powers and duties (including discretionary powers and duties), obligations, rights, title, easements and estates reserved to it by this deed to any one or more corporations, associations or persons that will accept the same. Any such assignment shall be in writing recorded among the Land Records of Baltimore County and the assignee shall join therein for the purpose of evidencing its acceptance of the same, and such assignee shall thereupon have the same rights, title, powers, obligations, discretion and duties as are herein reserved to the Developer, and the Developer shall thereupon be released therefrom.

GENERAL PLANNING STANDARDS AND REGULATIONS FOR HUNT VALLEY BUSINESS COMMUNITY

1. Generally, the use of concrete block or cinder block for outside facing of exterior walls will not be permitted nor will any frame structures be permitted.

2. A scale drawing in color of any sign, bill-board, trademark or advertising device to be used on any lot or the exterior of any building or structure will be submitted to Maryland Properties, Inc., in triplicate for the written approval by Maryland Properties, Inc. Normally the occupant's trade mark and/or trade name may be displayed on the building in the manner in which they are generally used by the occupant.

3. Not more than fifty per cent (50%) of any lot area shall be covered by buildings.

4. All present and future vehicle parking, including trucks, trailers, employee and visitor parking, shall be provided on the premises and shall comply with all provisions of the applicable Baltimore County Zoning Regulations. All parking areas are to be paved to provide dustfree all-weather surfaces with macadam, concrete or any approved material other than gravel. No parking area will be permitted within building set back lines (fifty (50) feet on primary roads and thirty (30) feet on secondary roads) except that lots bounded by more than one road may have parking areas within the set-back lines along roads other than the one on which the building fronts if, in the judgment of the Developer, the parking area is set back a reasonable distance and is properly screened from both front and side roads. Off-street parking spaces will be provided in accordance with the following:

one space—size 10′ x 20′ for automobiles per 1,000 sq. ft. of warehouse space
one space—size 10′ x 20′ for automobiles per 600 sq. ft. of manufacturing space
one space—size 10′ x 20′ for automobiles per 400 sq. ft. of office space.

5. No loading docks shall be permitted on the front of any building and, except where a lot is bounded by three or more roads, no loading docks shall be permitted on the side of any building facing a road.

6. No material, supplies, or products shall be stored or permitted to remain on the premises outside a permanent structure without the prior written consent of Maryland Properties, Inc. Approval of outside storage will be granted only where storage is screened from view by a masonry wall, or other appropriate screen, six (6) feet in height or rising two (2) feet above the stored material, whichever is higher.

7. No building shall be painted, repainted, stuccoed or be surfaced with any material unless and until Maryland Properties, Inc., approves the color and/or material in writing.

8. All set-back areas facing roads between the front building line and the curb, with the exception of driveways, sidewalks, and other walk ways shall be used exclusively for the planting and growing of trees, shrubs, lawns and other ground covering or material as approved by Maryland Properties, Inc. If developed lots are not properly maintained, Maryland Properties, Inc., may undertake such maintenance as may be necessary, at the expense of the occupant.

9. Occupants of the Park shall not cause or make any excessive noise, odors, harmful sewage or vibration that could be deemed objectionable to other occupants and that would conflict with the purposes or restrictions of the Park, and shall not create or maintain a nuisance. Each occupant must provide for trash disposal from his building.

10. All unused land area that is planned for future building expansion or other purposes shall be maintained and kept free of unsightly plant growth, stored material, rubbish and debris.

11. These general planning standards and regulations are subject to amendment, modification or termination by Maryland Properties, Inc. at any time.

INDEX